16 Springer Series in Chemical Physics

Edited by Vitalii I. Goldanskii

W0232213

Springer Series in Chemical Physics
Editors: V. I. Goldanskii   R. Gomer   F. P. Schäfer   J. P. Toennies

V. L. Broude   E. I. Rashba
E. F. Sheka

Spectroscopy of Molecular Excitons

With 135 Figures

Springer-Verlag
Berlin Heidelberg GmbH

Professor Dr. Vladimir L. Broude †
Professor Dr. Emmanuel I. Rashba

L. D. Landau Institute of Theoretical Physics, Academy of Sciences of the USSR
SU-117334 Moscow, USSR

Professor Dr. Elena F. Sheka

Institute of Solid State Physics, Academy of Sciences of the USSR
SU-142432 Chernogolovka, USSR

Series Editors

Professor Vitalii I. Goldanskii

Institute of Chemical Physics
Academy of Sciences
Vorobyevskoye Chaussee 2-b
Moscow V-334, USSR

Professor Robert Gomer

The James Franck Institute
The University of Chicago
5640 Ellis Avenue
Chicago, IL 60637, USA

Professor Dr. Fritz Peter Schäfer

Max-Planck-Institut für
Biophysikalische Chemie
D-3400 Göttingen-Nikolausberg
Fed. Rep. of Germany

Professor Dr. J. Peter Toennies

Max-Planck-Institut für Strömungsforschung
Böttingerstraße 6-8
D-3400 Göttingen
Fed. Rep. of Germany

ISBN 978-3-642-88222-7
DOI 10.1007/978-3-642-88220-3

ISBN 978-3-642-88220-3 (eBook)

Library of Congress Cataloging in Publication Data. Broude, V. L. (Vladimir L'vovich), 1924–1978. Spectroscopy of molecular excitons. (Springer series in chemical physics ; 16). Bibliography: p. Includes index. 1. Exciton theory. 2. Crystals — Spectra. 3. Molecular spectroscopy. I. Rashba, E. I. (Emmanuil Iosifovich), 1927–. II. Sheka, E. F. (Elena Fedorovna) III. Title. IV. Series: Springer series in chemical physics ; v. 16. QC176.8.E9B7613 1985 539'.6 85–4785

© Springer-Verlag Berlin Heidelberg 1985
Softcover reprint of the hardcover 1st edition 1985

2153/3130-543210

Preface

Low-temperature spectroscopy of organic molecular crystals came into being in the late 20s, just when quantum physics of solids as a whole began to develop vigorously. Already in the early works, two experimental facts of prime importance were discovered: the presence of a multitude of narrow bands in the low-temperature spectrum of a crystal, and a close relationship between the spectrum of the crystal and that of the constituent molecules. These findings immediately preceded the celebrated paper of Frenkel in which he went beyond the framework of Bloch's scheme and advanced the exciton concept. Subsequent investigations showed that the most interesting features of the spectra of molecular crystals are associated with excitons, and then the spectroscopy of molecular excitons began to form gradually on the basis of the spectroscopy of organic crystals. The molecular exciton became synonymous to the Frenkel exciton in a molecular crystal.

In view of the difficulties involved in the analysis of rich spectra containing many tens of bands, the spectroscopy of molecular crystals had long been connected most closely with the spectroscopy of molecules. It had developed independently, to a large extent, from the other branches of solid-state physics. This was also emphasized by the difference in experimental techniques, the specific properties of the objects, etc. As a result, there was some lag in ideas and concepts. Suffice it to say that twenty-five years ago, when a considerable body of data on the electron dispersion laws in semiconductors and Fermi surfaces of metals had already been accumulated, no information on the band structure of molecular excitons was available, and possible ways of their experimental determination were not even discussed in the literature. Interpretation of the material was also exclusively qualitative, and experimental methods for exerting an influence on spectra were practically non-existent. The following fact will aid in understanding the situation: among the experimental data then available there was not a single curve, i.e. the dependence of some parameter on another, with the exception, of course, of absorption curves which were invariably assumed to have either a Lorentzian or Gaussian shape.

The present book is, to some extent, a report on the development of the spectroscopy of molecular excitons during the subsequent years. It is practically impossible at the moment to embrace all its aspects comprehensively enough in a book of limited size. Broad reviews inevitably duplicate each other. Therefore, we have decided to confine ourselves from the very outset to phenomena which are most intimately connected with the exciton band structure, i.e. which either provide information on it or can be described quantitatively on the basis of the band-structure data obtained by other methods. This criterion was applied both in selecting the experimental material and in presenting the theory. In our exposé we tried to follow a pattern where theory and experiment would be most closely intertwined and would complement one another, and where the practical methods for the reconstruction of the band structure from experimental data thus far elaborated would be shown "in action". Finally, we proceeded from the fact that the spectroscopy of molecular excitons is an inseparable part of solid-state physics and tried to demonstrate the unique opportunities offered by it in investigating some general problems, such as the properties of disordered systems and electron-phonon interaction.

The three of us began the work on this book together, but after the untimely death in 1978 of V.L. Broude, who has contributed so heavily to the development of the spectroscopy of excitons, we had to finish our work without him. Accordingly, we alone are responsible for any faults that may be found in the final version of the book.

June 1983   *E.I. Rashba · E.F. Sheka*

Contents

1. Experimental Background

Molecular crystals have extensively been studied by experimental methods. This introductory chapter reviews the results relating to the energy and the optical spectra of molecular crystals and their constituent molecules. It also deals with the basic classification of spectra, formulates the range of problems under investigation and introduces the terminology used throughout the monograph.

1.1 Molecular Crystals

Organic molecular crystals are characterized by two types of interatomic interactions. Within the molecule, the atoms are bound by valence forces, while the intramolecular interactions are of the van der Waals type. Accordingly, the equilibrium interatomic distances in a molecule usually range from 1 to 1.5 Å, while the distances between the nearest atoms of the neighbouring molecules range from about 2.2 to 3.3 Å. The difference in interaction energies appreciably exceeds the difference in distances. The intramolecular interaction energy equals several electron volts (eV) per bond, while the intermolecular interaction energy equals several tenths of one electron volt per pair of molecules.

The difference between the intra- and the intermolecular interaction energies is so large that the molecules of the crystal retain their individuality to a great extent. Thus, it is possible to treat *molecular crystals* as a special type of a solid [1.1]. The retained individuality of the molecules manifests itself in the fact that the interatomic distances remain practically unchanged during crystallization, and that entire molecules are evaporated during sublimation.

For the same reason there is a close relationship between the spectra of the component molecules and those of the molecular crystals. This experimental fact, upon which the spectroscopy of molecular crystals is based, was originally discovered by OBREIMOV and DE HAAS [1.2], and PRINGSHEIM and KRONENBERGER [1.3,4]. Our presentation of crystal spectroscopy will be preceded by a brief description of molecular spectra (for more details, see [1.5-8]).

1.2 Electronic Spectra of Molecules

Of great interest for the electronic spectroscopy of molecules and crystals are aromatic organic molecules, such as benzene, naphthalene, and anthracene, which contain π-electrons. These electrons are rather weakly bound to the nuclei compared to the σ-electrons. The electronic transitions in the visible and near ultraviolet regions of the electromagnetic spectrum (up to 5-6 eV) involve the excitation of these electrons. The energies of the first three singlet electronic transitions for the above-mentioned molecules are listed in Table 1.1. The symmetry operations and the characters of the irreducible representations of the symmetry groups of these molecules (D_{6h} and D_{2h}) are listed in Tables 1.2,3. Standard symbols for the point groups are used [1.13].

The energy spectrum of the excited states of a molecule includes *vibrational* and *electron-vibrational (vibronic) excitations*. The frequency of a vibration, ν, is usually much lower than the distance, measured in units of frequency, between successive electronic levels. From Tables 1.1,4 it can be seen that this condition is fairly well fulfilled by the molecules of the aromatic series. Thus, the entire spectrum can be divided into individual *electronic transitions*, each of which contains a *pure electronic* component and *vibronic* components. The theory of vibronic spectra can be developed using the *adiabatic approximation* [1.14,15]. In the *harmonic approximation*, the potential energy of a vibration is proportional to the squares of the displacements of the nuclei, and the individual normal vibrations can be considered independently.

The vibrational Hamiltonian of the ground electronic state of a molecule is

$$H = (\nu/2)(q^2 - \partial^2/\partial q^2) \quad , \tag{1.1}$$

where q is a dimensionless coordinate[1]. The energy of the electronic ground state is chosen to be zero. The Hamiltonian of the excited electronic state is

$$H = E_e + (\nu/2)(q^2 - \partial^2/\partial q^2) + \sqrt{2}\gamma\nu q + \delta q^2 \quad , \tag{1.2}$$

where E_e is the electronic excitation energy at $q = 0$. The last two terms de-

---

[1] Henceforth, we shall let $\hbar = 1$ and measure frequencies in units of energy.

Table 1.1. Parameters of singlet electronic transitions in aromatic molecules

Molecule	Symmetry of electronic transition[a]	Energy cm^{-1} E_e(calc)[b]	E_{mol}(exp)[c]	Oscillator strength calc[b]	exp[c]
Benzene	$^1A_{1g} \to {}^1B_{2u}$	39200	38089	0	0
	$^1A_{1g} \to {}^1B_{1u}$	47200	44200	0	0
	$^1A_{1g} \to {}^1E_{1u}$	53550	53640	1.08	-
Naphthalene	$^1A_g \to {}^1B_{3u}$	32400	32020	0	10^{-4}
	$^1A_g \to {}^1B_{2u}$	36200	35905	0.21	0.11
		46200	45500	1.99	1.70
Anthracene	$^1A_g \to {}^1B_{2u}$	26000	27560	0.27	0.30
	$^1A_g \to {}^1B_{3u}$	29323	-	0	-
		37020	38700	2.57	1.60

a) The classification of a transition was made in accordance with Tables 1.2,3
b) The calculated electronic transition energies E_e - see (1.5) - and oscil-
    lator strengths of a rigid molecule were taken from [1.9,10]
c) The experimental values for the electronic terms E_{mol} - see (1.5) - and for
    the oscillator strengths were taken from [1.9-12]

scribe the *vibronic coupling (linear and quadratic[2])* with respect to q. The
linear interaction shifts the equilibrium position of the oscillator to the
point $q_0 = -\sqrt{2}\gamma$ and reduces the potential energy by the *Franck-Condon energy*

$$E_{FC} \approx \gamma^2 \nu \quad ; \tag{1.3}$$

assuming that $\delta \ll \nu$. The quantity γ is the dimensionless *linear coupling constant*. The quadratic interaction changes the vibrational frequency of the
electronically excited state by

$$\Delta_\nu \equiv \nu^* - \nu \approx \delta \quad , \tag{1.4}$$

(for more details see Appendix A). This quantity Δ_ν will be called the vibrational frequency shift (henceforth, the *frequency shift*). Symmetry considera-

---

2 The electronic excitation is also accompanied by off-diagonal quadratic
   terms in the potential energy, which leads to a mixing of normal vibrational
   modes (*Dushinsky effect*). Since the mixing is usually small, it shall be neglected.

Table 1.2. Symmetry operations and characters of the irreducible representations of the point group D_{6h}

D_{6h}	E	$2C_6(z)$	$2C_3$	C_2	$3C_2'$	$3C_2''$	i	σ_h	$3\sigma_v$	$3\sigma_d$	$2S_6$	$2S_3$	p
A_{1g}	1	1	1	1	1	1	1	1	1	1	1	1	
A_{1u}	1	1	1	1	1	1	-1	-1	-1	-1	-1	-1	
A_{2g}	1	1	1	1	-1	-1	1	1	-1	-1	1	1	
A_{2u}	1	1	1	1	-1	-1	-1	-1	1	1	-1	-1	P_z
B_{1g}	1	-1	1	-1	1	1	1	-1	-1	1	1	-1	
B_{1u}	1	-1	1	-1	1	1	-1	1	1	-1	-1	1	
B_{2g}	1	-1	1	-1	-1	-1	1	-1	1	-1	1	-1	
B_{2u}	1	-1	1	-1	-1	-1	-1	1	-1	1	-1	1	
E_{1g}	2	1	-1	-2	0	0	2	-2	0	0	-1	1	
E_{1u}	2	1	-1	-2	0	0	-2	2	0	0	1	-1	P_x,P_y
E_{2g}	2	-1	-1	2	0	0	2	2	0	0	-1	-1	
E_{2u}	2	-1	-1	2	0	0	-2	-2	0	0	1	1	

Table 1.3. Symmetry operations and characters of the irreducible representations of the point group D_{2h}

D_{2h}	E	$\sigma(xy)$	$\sigma(xz)$	$\sigma(yz)$	i	$C_2(z)$	$C_2(y)$	$C_2(x)$	p
A_g	1	1	1	1	1	1	1	1	
A_u	1	-1	-1	-1	-1	1	1	1	
B_{1g}	1	1	-1	-1	1	1	-1	-1	
B_{1u}	1	-1	1	1	-1	1	-1	-1	P_z
B_{2g}	1	-1	1	-1	1	-1	1	-1	
B_{2u}	1	1	-1	1	-1	-1	1	-1	P_y
B_{3g}	1	-1	-1	1	1	-1	-1	1	
B_{3u}	1	1	1	-1	-1	-1	-1	1	P_x

tions for non-degenerate molecular levels show that linear interactions are only possible for *totally symmetric* (TS) *vibrations*; for *non-totally symmetric* (NTS) *vibrations* $\gamma = 0$. Quadratic coupling takes place in every vibration ($\Delta_\nu \neq 0$).

The frequency of a pure electronic transition (*00-transition*) is equal to

$$E_{mol} = E_e - E_{FC} + \Delta_\nu/2 \quad .$$ (1.5)

This quantity will be called the *electronic term of the molecule*.

4

Electron-vibrational *(vibronic)* transitions are allowed because of the vibronic interactions. If the matrix element of the dipole moment of the transition at $q = 0$ is not too small, its dependence on q can be neglected *(Condon approximation)*. Hence, the total intensity of the electronic transition I_{tot} is distributed between the pure electronic component ($n = 0$) and the vibronic components involving n TS vibrations:

$$I(n)/I_{tot} = \left[\gamma^{2n}\exp(-\gamma^2)\right]/n! \quad .$$ (1.6)

When $\gamma \sim 1$, a series of vibronic transition occurs. If, however, an electronic transition at $q = 0$ is forbidden or its probability is low, NTS vibrations dominate and allow the vibronic transitions due to the intramolecular admixture of higher-frequency electronic transitions *(non-Condon approximation)* to take place. Mixing is either a result of the dependence of the matrix element of the transition on q *(Herzberg-Teller effect)* [1.16] or a result of nonadiabaticity [1.17] (i.e., the breakdown of the Born-Oppenheimer approximation). The vibronic transition with an NTS vibration may itself produce a series built up on the TS vibration.

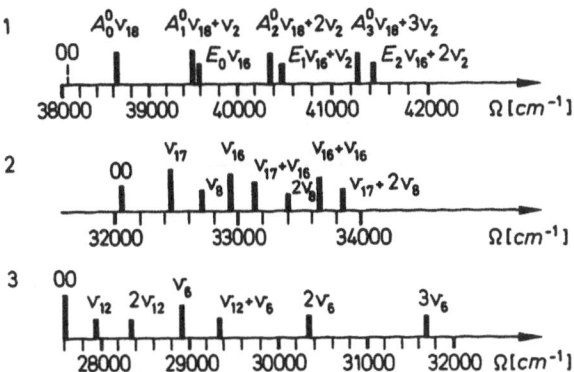

Fig.1.1. Lowest electronic transitions of the aromatic molecules (Table 1.4): 1) benzene, 2) naphthalene, and 3) anthracene

Figure 1.1 illustrates the principal features of the vibronic spectra of several aromatic molecules. The spectra have been arranged so that the intensity of the 00-transition increases from top to bottom. The transition parameters are given in Table 1.4.

The 00-transition of the highly symmetrical benzene molecule ($^1A_{1g} \rightarrow {}^1B_{2u}$; D_{6h}) is not observed because it is symmetry forbidden. All of the associated vibronic transitions with TS vibrations are also forbidden. The vibronic transitions involving NTS vibrations, ν_{18} and ν_{16} of e_{2g} symmetry, are the strongest

5

Table 1.4. Parameters of the vibronic transitions in aromatic molecules

Molecule	Electronic transition[a]	Vibronic transition[a]	Oscillator strength	Vibration characteristics					γ^2	
				Symmetry type	ν [cm^{-1}]	ν^* [cm^{-1}]	Number of repetitions in the series	Δ_ν [cm^{-1}]	experimental	calculated
Benzene [1.22]	$^1A_{1g} \to {}^1B_{2u}$	$00 + \nu_2$	0	a_{1g}	992	923	–	-69	1.41 absorpt. / 1.31 emiss.	1.30[c] / 1.20
		$00 + \nu_{18}$	0.001	e_{2g}	606	520	1	-86	0	0
		$00 + \nu_{16}$	0.001	e_{2g}	1586/1606	1480	1	ν-116	0	0
		$00 + \nu_{18} + n\nu_2$	0.005[b]	–	–	–	4	–	–	–
Naphthalene [1.60]	$^1A_g \to {}^1B_{3u}$	$00 + n\nu_8$	0.00015[b]	a_g	758	702	2	-56	0.2	–
		$00 + \nu_{17}$	0.001	b_{1g}	509	438	1	-71	0	0
		$00 + \nu_{16}$	0.001	b_{1g}	936	911	1	-25	0	0
		$00 + \nu_{17} + n\nu_8$	0.0015[b]	–	–	–	3	–	–	–
Anthracene [1.12]	$^1A_g \to {}^1B_{2u}$	$00 + n\nu_{12}$	0.07[b]	a_g	396	394	2	-2	0.55[d]	–
		$00 + n\nu_6$	0.08[b]	a_g	1403	1403	5	0	1.00	1.07[e]

<u>Table 1.4.</u> (continued)

a) The vibrations are numbered according to their serial number for each type of symmetry from high to low frequencies. The sequence of symmetry types in the complete vibrational representation Γ_v is as follows [1.18]:

Benzene: $\Gamma_v = 2a_{1g} + a_{2g} + a_{2u} + 2b_{1u} + 2b_{2g} + 2b_{2u} + e_{1g} + 3e_{1u} + 4e_{2g} + 2e_{2u}$ (D_{6h})

Naphthalene: $\Gamma_v = 9a_g + 4a_u + 8b_{1g} + 4b_{1u} + 3b_{2g} + 8b_{2u} + 4b_{3g} + 8b_{3u}$ (D_{2h})

Anthracene: $\Gamma_v = 12a_g + 5a_u + 11b_{1g} + 6b_{1u} + 4b_{2g} + 11b_{2u} + 6b_{3g} + 11b_{3u}$ (D_{2h})

b) Oscillator strength of a set of n vibronic transitions

c) [1.19]

d) [1.20]

e) [1.21]

in the spectrum (Table 1.4). Their intensities determined by the mixing of the $^1B_{2u}$ state with a higher $^1E_{1u}$ state nevertheless are low (the oscillator strength is $\sim 10^{-3}$, see Table 1.4), due to the Herzberg-Teller effect [1.16] or due to the deviation from adiabaticity [1.17]. Each of these vibronic transitions leads to a series of composite transitions involving the TS vibration ν_2; they are usually denoted by A_i^0 and E_i, respectively [1.22]. The intensity distribution in these series reveals a large shift in the equilibrium position of nuclei in the electronically excited state for the vibration ν_2. The corresponding coupling constant is $\gamma^2 \approx 1.3$.

The 00-transition of the naphthalene molecule (Fig.1.1) ($^1A_{1g} \to {}^1B_{3u}$; D_{2h}) is very weak (Table 1.4). The vibronic transitions with TS vibrations are just as weak. In this series, the transitions involving the vibrations ν_{17} and ν_{16} (of b_{1g} symmetry) are the most intensive ones due to the mixing of the $^1B_{3u}$ state with the $^1B_{2u}$ state. The TS vibration ν_8 breaks up these transitions into comparatively short series, which implies a small coupling constant.

The 00-transition of the anthracene molecule (Fig.1.1) ($^1A_{1g} \to {}^1B_{2u}$; D_{2h}) is allowed and has a high oscillator strength. Along with it, one also observes intense repetitive transitions, mainly with the TS vibrations ν_{12} and ν_6. The vibration ν_6 dominates the vibronic spectrum and has a large coupling constant $\gamma^2 \approx 1$ (Table 1.4). Against the background of such transitions it is naturally difficult to observe the weak transitions with NTS vibrations.

1.3 Comparing the Electronic Spectrum of a Crystal and That of a Molecule

Polyatomic molecules, which contain planar aromatic carbon rings, usually form anisotropic crystals of relatively low crystal systems: triclinic, monoclinic,

7

and less often, orthorhombic. The optical properties of such a crystal are described by the refractive indices that correspond to a triaxial ellipsoid of
dielectric permeability. Spectral investigations in which the incident light
is polarized along one of the principal crystallographic axes are the most
informative ones, since their results can easily be related to the values of
the dielectric permeability tensor. Consequently, the experimental data for
crystals will only be given for special orientations of the plane of polarization of the incident light with respect to the crystallographic axes.

Two different classes of phenomena are expected due to the formation of
a crystal from free molecules. The first class includes phenomena associated
with the arrangement of the molecules and their deformation in the crystal
field. These effects are not only observed in the formation of a perfect one-
component crystal, but also when individual guest molecules are incorporated
into the host lattice. In the condensed state the molecules lose their high
symmetry. The site symmetry in crystal lattices typical of organic compounds is
low (C_1, C_s, and C_i; rarely C_{2h}). It is determined by the local symmetry group.
Actual shifts of nuclei due to the lowering of the molecular symmetry are
small (0.01 - 0.03 Å). However, the lowering of the molecular symmetry in the
perfect crystal or in the matrix allows previously forbidden transitions to
take place, changes the intensity of weak transitions and lifts degeneracies.

The second class of phenomena is observed in the formation of a perfect
crystal and is associated with the presence of translational symmetry. The
band states of solids belong to this class. Their description requires the
introduction of various quasi-particles, such as excitons or phonons.

1.3.1 Model of an Oriented Gas

The similarity between a spectrum of a crystal and that of a free molecule
naturally leads to a model for the interpretation of the crystal spectrum -
the *oriented gas model* (OGM). In this model the crystal is represented by a
group of oriented, rigidly fixed molecules. In its simplest version the effect of the crystal field on the molecular spectrum is ignored. An improved
version takes into account the lowering of the molecular symmetry and the
change in the intensities of the transitions[3]. Henceforth, the abbreviation
OGM will always imply this more general model.

Dilute solutions of molecules in a crystalline matrix and the widely known
Shpolsky matrices are examples of real systems described by the OGM.

---

[3] Sometimes the term *oriented sites model* is used for this version of OGM.

A comparison of the spectra of a gas of free molecules with that of a crystal yields information on the polarization of the electronic transitions in the molecule. It also reveals the changes in the spectrum due to the lowering of the symmetry of the molecule in the local crystal field, such as the removal of degeneracy and the allowance of previously forbidden transitions. In the Heitler-London approximation these phenomena and the change in intensity of weak transitions are usually described by the phrase: the *configurational mixing of the electronic states* of free molecules *due to intermolecular interaction*. With reference to molecular crystals, this effect was studied by CRAIG [1.23] and is termed the *Craig effect*.

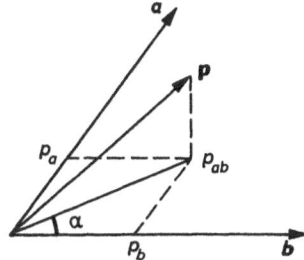

Fig.1.2. Projection of the dipole moment **p** of a molecule on the **a**- and **b**-directions of the crystal lattice (a and b are assumed to be mutually orthogonal)

One can introduce the notion of a *polarization ratio*. It is an important quantity in the spectroscopy of molecular crystals, since it gives the ratio of the intensities of absorbed light of different polarizations. Using the notations of Fig.1.2, one can easily relate the polarization ratio to the orientation of the dipole moment of the transition, **p**. This ratio is equal to $P_{a/b} = p_a^2/p_b^2 = \tan^2\alpha$, where the angle α determines the orientation of the projection of the vector **p** onto the plane **ab**.

For strong transitions, the orientation of the dipole moment **p** relative to the molecular axes is slightly different in the condensed state. The change in the orientation of the dipole moment **p** may be quite large for weak transitions. As an example let us consider the naphthalene crystal. The dipole moment of its lowest pure electronic transition $^1A_{1g} \rightarrow {}^1B_{3u}$ is oriented along the long axis of the molecule. Accordingly, the calculated value of the polarization ratio is $P_{b/a} = 0.24$ when the incident light is polarized perpendicular and parallel to **b**, respectively (b is the monoclinic axis). The experimental value is $P_{b/a} = 160$. This 670-fold change in $P_{b/a}$ is due to the Craig effect. A change in $P_{b/a}$ of the same order of magnitude is also observed when naphthalene is dissolved in crystalline durene [1.24]. Investigations of this matrix system enabled McCLURE to determine, for the first time, the symmetry of the first electronic transition in the naphthalene molecule.

The oscillator strength acquired by a forbidden or a weak transition E_1, as a result of mixing with a strong transition E, is of the order of $\delta f_1 \sim [D/(E-E_1)]^2 f$, where D is the vapour-crystal shift in the spectrum. D serves as an estimate of the crystal field ($D \sim 1000$ cm^{-1}). The role of mixing is significant if $\delta f_1 \gg f_1$. For the first electronic transition in benzene this inequality holds, since $f_1 \approx 0$. In naphthalene $\delta f_1 \sim f_1$.

The OGM assumes that no excitation energy is exchanged between molecules. Therefore it cannot, in principle, describe the specific features of a spectrum associated with translational symmetry and the excitation of cooperative states.

1.3.2 Electronic Spectra of a Crystal

Cooperative effects show up most prominently in the polarization of the spectrum of *intrinsic absorption*. A study of the spectra of molecular crystals in polarized light was initiated by OBREIMOV and PRIKHOTJKO [1.25]. They observed [1.26-29] *sharply polarized bands* which represent the most interesting phenomenon in the intrinsic absorption spectrum of a molecular crystal.

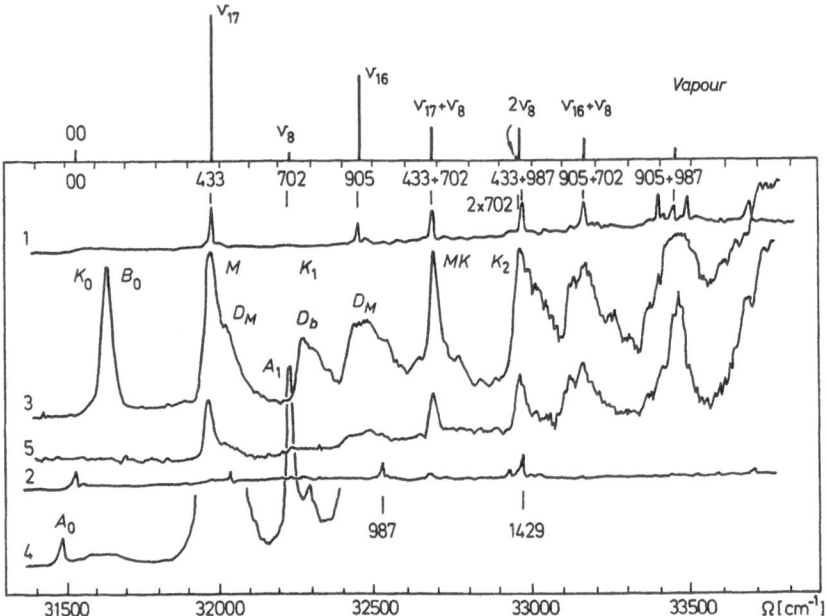

Fig.1.3. Absorption spectra of naphthalene - vapour (top), a crystalline solution in durene (C =0 .1%) [1.24], and a crystal (T = 20 K) [1.30].
1) perpendicular and 2) parallel **c** components of the solution; 3) parallel and 5) perpendicular **b** components of the crystal, crystal thickness d = 0.5 μm; 4) perpendicular **b** component of the crystal spectrum, d = 2 μm

Figure 1.3 compares the absorption spectrum of naphthalene in the vapour phase and in a durene matrix, and the spectrum of a naphthalene crystal (the latter two in polarized light). The vapour and solution spectra practically coincide. The solution spectrum, displaced with respect to that of the vapour, exhibits an enhanced 00-transition due to the Craig effect.

The position of the crystal spectrum is close to that of the solution spectrum, but the bands contained in it fall into two groups. Doublets of *sharply polarized* bands are observed for the 00-transition and for the vibronic transitions involving the TS vibration ν_8. The vibronic bands involving NTS vibrations remain unsplit and *weakly polarized*. The term introduced by PRIKHOTJKO [1.29] for these two types of bands - *crystal and molecular* - will be used in the following text. The *band multiplets* of the vibronic spectrum, which include n TS vibrations, are denoted by the symbol K_n (K_n bands), noting that n = 0 corresponds to the pure electronic transition. In addition, the symbols A_n, B_n, and C_n shall be used for the individual bands of a multiplet, thus indicating their polarization in accordance with the notation of the crystallographic axes. The narrow vibronic absorption bands involving NTS vibrations will be denoted as M bands. The impurity bands will be specified by primes (for example, K_0', K_0''... are pure electronic, and K_n', K_n'', ... M', M" ... are vibronic bands). Vibronic transitions that involve a TS or NTS phonon will be called K and M transitions, respectively. Two letters will denote composite and overtone vibronic transitions (for example, MK and K_2, respectively).

An important feature of a crystal spectrum is the broad-band (D band) absorption, which follows practically all of the vibronic bands on the high-frequency side. As the crystal thickness increases, the spectrum becomes more and more "pillar-like" [1.29,30], with "pillars" of approximately equal widths. This is due to the characteristics of the vibronic absorption of a molecular crystal, to be considered in Chap.6. Additional broadening and the appearance of side-bands can be observed in the region of the 00-transition and in the vibronic spectrum. This is a result of the interaction of the electronic excitation with the lattice vibrations of the crystal. The absorption spectrum of the naphthalene crystal is rich in details due to the crystal itself (from K_n bands to vibronic "pillars" and side-bands), which makes it an excellent object of investigation.

Figure 1.4 illustrates the absorption spectra of a benzene crystal in polarized light. These spectra are quite similar to those of the naphthalene crystal. As in naphthalene, one can make out the K_n bands, associated with a series of vibronic transitions involving the TS vibration ν_2, and the M bands of various transitions involving NTS vibrations.

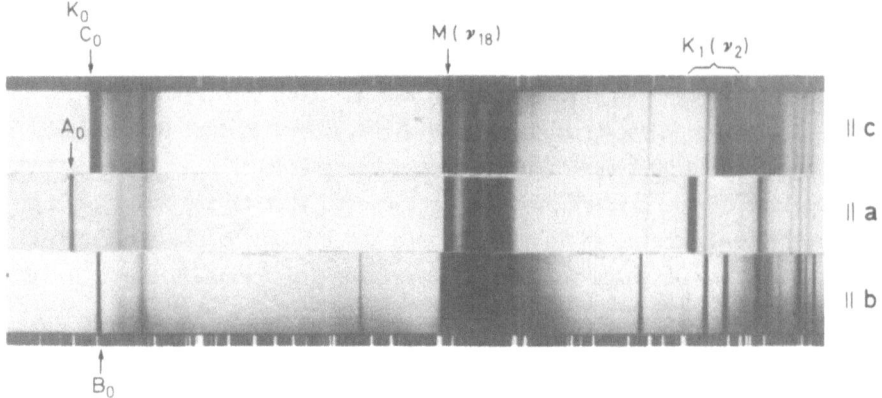

Fig.1.4. Lower portion of the absorption spectrum of the benzene crystal in polarized light at T = 4.2 K, d = 5 μm [1.32]

The absorption spectra of benzene and naphthalene crystals simultaneously exhibit K_n and M bands. This is due to the fact that both have comparable and relatively low intensities and are the result of configurational mixing (either inter- or intramolecular). The role of both of these effects is small in crystals with strong electronic transitions. Therefore, their absorption spectrum consists of practically only K_n transitions. An example of such a spectrum is the absorption spectrum of the anthracene crystal (Fig.1.5).

The polarization of the K bands along the crystallographic axes clearly shows that the absorption cannot be attributed to individual molecules, but that it is of a *collective nature* [1.28,34].

The collective nature of pure electronic excitations is due to the coincidence between the levels of different molecules, i.e., due to the *resonance*

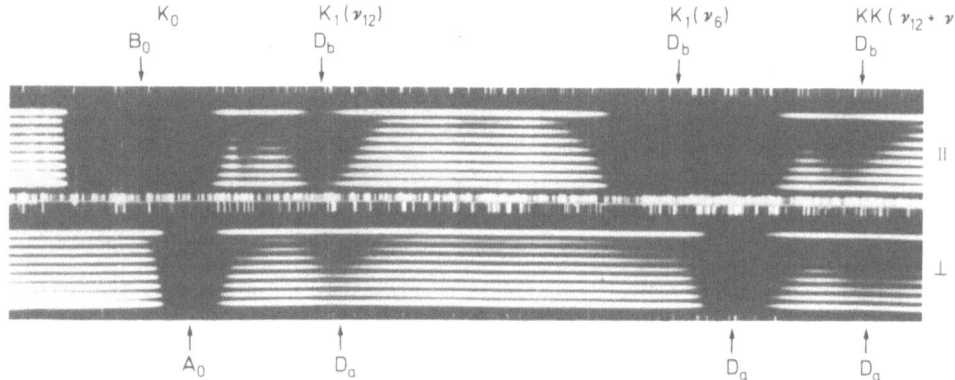

Fig.1.5. Lower portion of the absorption spectrum of the anthracene crystal in polarized light at T = 70 K, d = 2 μm [1.33]

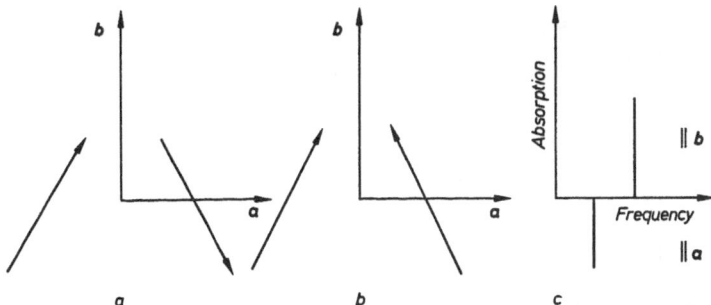

Fig.1.6a-c. Two-oscillator model; **a** and **b** are the symmetry axes. (a) and (b) exhibit two normal modes, (c) shows a sharply polarized absorption spectrum

between them. How this resonance gives rise to sharply polarized bands can clearly be seen in the system of two crossed oscillators (Fig.1.6) The system may be treated as a model for the primitive cell of a crystal. Both oscillators possess two normal vibrational modes that correspond to an absorption that is sharply polarized along the symmetry axes a and b. In passing from a cell to a crystal the normal vibrations are transformed to waves of intramolecular excitation, i.e., to Frenkel excitons that travel through the crystal. In this case the exciton absorption bands of the crystal remain narrow and exhibit the same polarization as the normal modes of the primitive cell due to the conservation of the quasi-momentum (see Fig.1.6). The observation of sharply polarized bands, particularly the K_0 bands of 00-transitions, led to the development of the spectroscopy of molecular excitons. The theory of Frenkel excitons in molecular crystals was first developed by DAVYDOV [1.34-36]. The formation of electronic excitons imposes a collective character on the whole vibronic spectrum of crystals in the region of K_n and M transitions. These phenomena will be discussed in Chap.6. Certain aspects of molecular excitons are treated in several reviews [1.37-44].

1.4 Isotopic Effect in Electronic Spectra

The *isotopic effect* plays an important role in the spectroscopy of molecular crystals. Efficient methods of studying the structure of electronic spectra are based on this effect. This is because vibrational frequencies and amplitudes are sensitive to isotopic substitution. We shall restrict ourselves to the hydrogen-deuterium (H-D) substitution. To avoid an excessive repetition of the bulky names of isotopically substituted molecules, we shall use the notation given in Table 1.5.

Table 1.5. Symbols for the isotopic molecules used in the text

Molecule	Chemical formula	Symbol
Benzene	C_6H_6	bd_0
Monodeuterobenzene	C_6H_5D	bd_1
Dideuterobenzene	$C_6H_4D_2$	bd_2
Trideuterobenzene	$C_6H_3D_3$	bd_3
Tetradeuterobenzene	$C_6H_2D_4$	bd_4
Pentadeuterobenzene	C_6HD_5	bd_5
Hexadeuterobenzene	C_6D_6	bd_6
Hexamethylbenzene	$C_6(CH_3)_6$	$hmb-d_0$
Deuterohexamethylbenzene	$C_6(CD_3)_6$	$hmb-d_{18}$
Naphthalene	$C_{10}H_8$	nd_0
Monodeuteronaphthalene	$-\begin{cases} \alpha-C_{10}H_7D \\ \beta-C_{10}H_7D \end{cases}$	$\begin{cases} \alpha-nd_1 \\ \beta-nd_1 \end{cases}$
Dideuteronaphthalene	$-\begin{cases} C_{10}H_6(\alpha D)_2 \\ C_{10}H_6(\beta D)_2 \end{cases}$	$\begin{cases} \alpha-nd_2 \\ \beta-nd_2 \end{cases}$
Trideuteronaphthalene	$C_{10}H_5(\alpha D)_3$	$\alpha-nd_3$
Tetradeuteronaphthalene	$-\begin{cases} C_{10}H_4(\alpha D)_4 \\ C_{10}H_4(\beta D)_4 \end{cases}$	$\alpha-nd_4$ $\beta-nd_4$
Pentadeuteronaphthalene	$C_{10}H_3(\alpha D)(\beta D)_4$	$\alpha\beta-nd_5$
Octadeuteronaphthalene	$C_{10}D_8$	nd_8
Anthracene	$C_{14}H_{10}$	ad_0
Decadeuteroanthracene	$C_{14}D_{10}$	ad_{10}

The change in vibrational frequencies is reflected in the change in the intervals in the vibronic spectra of isotopic molecules. The change in frequency upon deuteration is considerable (sometimes it reaches 40-50% [1.6]). This leads to substantial changes in the structure of the spectrum, so that the correlation of identical transitions of isotopic molecules becomes less obvious.

An isotopic effect, such as the shift of the 00-transition energy E_{mol}, is also closely associated with a shift in vibrational frequencies [cf. the term $\Delta_v/2$ in (1.5)]. This shift is called the *isotopic shift of the electronic term* Δ_{ex}, or simply the isotopic shift. It is nearly a linear function of the number of substituted hydrogen atoms. On the average the substitution of one atom is accompanied by a shift of 33 cm^{-1} in the benzene molecule, 15 cm^{-1} in naphthalene, and 6 cm^{-1} in anthracene. However, upon substitution in the α and β positions of the naphthalene molecule the shift equals 11 and 19 cm^{-1}, respectively.

Isotopic substitution leads to a change in the intensity of individual bands for two reasons: the change in the integral intensity of the electronic transition and the change in the parameters of the vibronic interaction, in particular γ^2, which leads to a redistribution of intensity between separate vibronic bands.

A qualitative decrease in the integral absorption intensity of the first singlet electronic transition of deuterated naphthalene has been observed in the vapour and solid-solution spectra [1.45,46]. In crystal spectra the intensity decreases by a factor of about 1.5 when the molecule is completely deuterated (i.e., perdeuterated) [1.47]. The effect decreases as the number of substituted atoms is decreased and is practically never observed for molecules with one or two deuterium atoms. The same is true for benzene.

Figure 1.7 shows the absorption spectra of the electronic transition $^1A_{1g} \rightarrow {}^1B_{3u}$ in the crystals for the nd_0 and nd_8 isotopic forms of naphthalene. The relatively strong change in intensity of the different vibronic transitions (three fold for the transition $00 + \nu_8$) should be noted. Such an effect was also observed in the triplet (S→T) absorption spectra [1.48].

A change in molecular symmetry upon partial isotopic substitution is evident in the vapour spectrum of bd_1. The lower C_{2v} symmetry of bd_1 (as compared to D_{6h} for bd_0) is accompanied by a splitting of the doubly degenerate vibrations of benzene and by an increase in the number of TS vibrations. This leads to the appearance of new bands in the vibronic spectrum of bd_1 [1.49]. However, the magnitude of the deformation of the nuclear skeleton is still insufficient to allow the 00-transition, $^1A_{1g} \rightarrow {}^1B_{3u}$, to occur in bd_1.

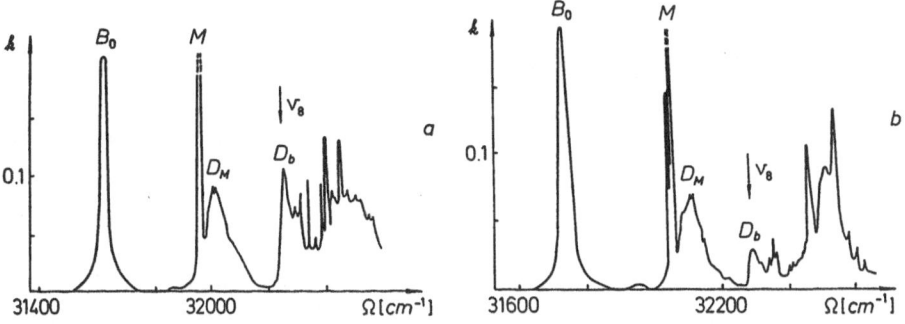

Fig.1.7a,b. Absorption spectra of crystals of nd_0 (a) and nd_8 (b) for the electronic transition $^1A_{1g} \rightarrow {}^1B_{3u}$. The light was polarized parallel to **b** [1.47] (for an explanation of the vibronic spectra, see Sect.6.3)

1.5 Spectra of Doped Crystals

From the very beginnings of molecular crystal spectroscopy it has been known that a relatively small increase in the thickness of a crystal results in an intensive low-energy spectrum due to the presence of impurities [1.50]. The original goal was to identify the impurities in order to obtain pure crystals. This problem has been solved for the most part for benzene, naphthalene and anthracene. But it was established that the spectrum of the impure crystal contains considerable information about the pure crystal and may be used for finding methods of investigating the host crystal. This is how the *spectroscopy of imperfect crystals* evolved as a method for studying the *energy spectrum of the perfect crystal*. This subject will be treated in detail in Chaps.3-6. In the present section we shall only consider the classification of the *impurity absorption* in relation to the structure of the *impurity centre*. We restrict ourselves to pure electronic absorption, whose bands are denoted by the symbols K_0', K_0'' ... (Sect.1.3).

A typical absorption spectrum of an electronic impurity is shown in Fig.1.8. We chose the simplest case of a doped crystal, i.e, the isomorphic replacement of host molecules by isotopic molecules. In this case we do not have to deal with the nonequivalence of the positions of the guest molecules in the crystal matrices, which would complicate the spectra under Shpolsky-effect conditions [1.52]. Nevertheless, this spectrum also has a complicated structure.

The distinguishing feature of the spectrum is the intensive K_0' band. Its intensity varies linearly with the degree of impurities, and thus, one can attribute it to an absorption by single guest molecules. The bands 2 and 3 in Fig.1.8 are associated with the accompanying impurities. In contrast to the K_0' band, the intensity of the K_0'' band increases super-linearly with the concentration of bd_0, which suggests that it is due to an absorption by an aggregate (a pair in this case) of molecules. Thus, the impurity absorption is generally made by distinct centres, some of which consist of single molecules (they will be called *monomer centres*), while others consist of molecular aggregates (*aggregate centres*).

If an impurity is nonisotopic but replaces the host molecules isomorphically the electronic absorption spectrum has a similarly complicated structure (for example, the spectrum of p-dibromobenzene impurity in a crystal of p-dichlorobenzene [1.53]).

Adding a nonisotopic or, as it is usually called, a *chemical impurity* under close-packing conditions causes a deformation of the adjacent host molecules [1.54]. This gives rise to "*defect*" *molecules*, which form a new type of im-

16

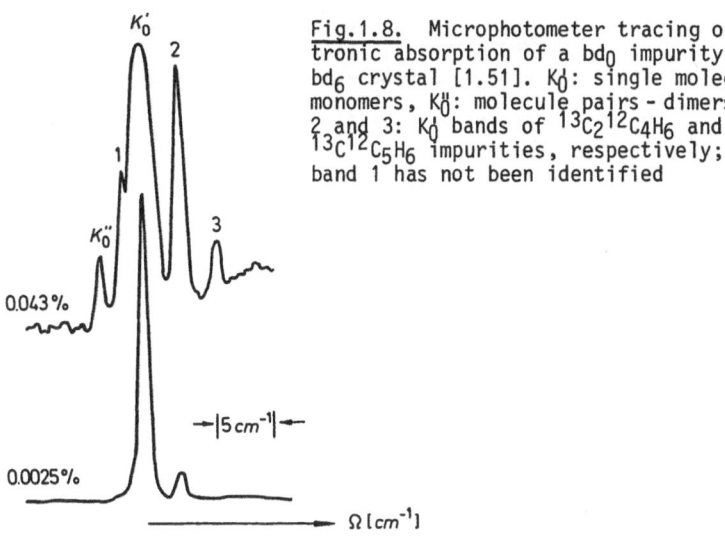

Fig.1.8. Microphotometer tracing of electronic absorption of a bd_0 impurity in a bd_6 crystal [1.51]. K_0': single molecule - monomers; K_0'': molecule pairs - dimers; 2 and 3: K_0' bands of $^{13}C_2{}^{12}C_4H_6$ and $^{13}C^{12}C_5H_6$ impurities, respectively; band 1 has not been identified

purity centre. The pattern of absorption by *defect centres* is quite similar to the one illustrated in Fig.1.8.

Thus, regardless of the chemical nature of the centres, the monomer K_0' band dominates the impurity absorption spectra of dilute crystal solutions. The bands due to aggregate centres are much weaker, although they can be reliably observed down to very low concentrations. Hence, zero-phonon impurity bands are implied [1.55].

At the same time, the quantitative characteristics of impurity absorption depend on the nature of the centres. Let us consider the K_0' bands. The impurity centres can be divided into deep and shallow ones. Impurity absorption by the deep centres is described by the OGM. Its intensity is determined by the oscillator strength of the transition in the impurity molecule, and its polarization ratio is determined by the orientation of the molecule in the lattice. The K_0' band in Fig.1.8 exhibits such properties.

Impurity absorption of quite a different nature, typical of shallow centres, was discovered in the spectra of deuteronaphthalene crystals [1.56]. Figure 1.9 exhibits the spectra of several crystals. In this series of spectra the magnitude of the isotopic shift in the electronic level of the impurity molecule, Δ_{ex}, changes by a factor of 2.6. The distance of the impurity band from the intrinsic absorption spectrum changes by a factor of 6, the polarization ratio (which should have been constant according to the OGM) by a factor of 17, and the absolute intensity of the impurity absorption of light polarized perpendicular to **b** (Fig.1.9a) becomes comparable with the intrinsic absorption intensity. The impurity absorption spectrum differs qualitatively from the pattern

17

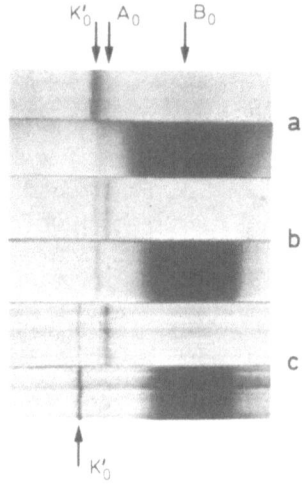

Fig.1.9a-c. Absorption spectra of shallow impurity centres in isotopically doped crystals of deuteronaphthalene [1.57] in polarized light at $T = 20$ K: A_0, B_0 are exciton absorption bands of the host crystal; (a) nd_0 in α-nd_4; (b) nd_0 in β-nd_4; (c) nd_0 in nd_8 (impurity concentration $\approx$5%)

expected according to the OGM. This phenomenon, which was predicted in [1.58], is frequently called the *Rashba effect*. A similar impurity-absorption pattern was observed for monomer defect centres [1.59].

As in the case of a perfect crystal, the properties of impurity absorption also deviate from the OGM due to the collective nature of the excited states of the guest molecule and the neighbouring hosts. In the perfect crystal the propagation of the excitation through the crystal takes place under conditions of strict resonance. For shallow impurity levels, the propagation of the excitation takes place under *quasi-resonance* conditions. The resonance detuning is determined by the quantity Δ_{ex}. The closeness to resonance determines the sensitivity of the impurity absorption with respect to the structure of the energy spectrum of the host crystal.

1.6 Electronic Spectra of Crystals and the Spectroscopy of Molecular Excitons

The spectroscopy of molecular excitons evolved as a result of joint effort in the field of experiment and theory aiming at the quantitative interpretation of the electronic spectra of molecular crystals. The following chapters will deal with its mathematical tools and the underlying basic principles to demonstrate the effectiveness of the quantitative analysis of the spectra. We restrict ourselves to electronic spectra with discrete vibronic structures. They are the most interesting ones from a theoretical point of view and at the same time are typical enough for molecular crystals. In the following discussion we neglect the crystal vibrations that are due to the external degrees of freedom of the molecules (*external phonons*). The distinct structure of the electronic spectrum and the narrowness of individual bands suggest that the interaction of the electrons with the external phonons is relatively weak. Under such conditions this interaction determines both the detailed shape of the individual bands and the appearance of phonon side-bands but does not change the basic structure of the spectrum. Thus, the *rigid lattice approximation* accurately describes the main features of the structure of the spectrum.

Figure 1.10 illustrates an absorption spectrum whose principal details can now be explained by the spectroscopy of molecular excitons. It is seen that this spectrum has a rich structure that practically includes all of the main features of the experimentally observed spectra.

The electronic spectrum of a crystal can be divided into two parts: the *exciton spectrum* that covers the range of pure electronic transition, and the *vibronic spectrum*. Each of these spectra exhibits distinct regions in the frequency scale. Their designations are clear from the notations for the various transitions given in Sect.1.3.

The main concept in molecular-exciton spectroscopy is the exciton which describes the transfer of electronic excitation in a crystal. Only the region of pure electronic excitation is directly associated with the exciton spectrum. As we shall see later, the vibronic spectrum of a crystal can conveniently be regarded to be the result of the interactions of the exciton with the intramolecular vibrational excitations of the crystal (*internal phonon*).

In the following discussion we shall always simultaneously consider the *energy* and *optical* spectrum of a crystal. Our task will be to find out to what

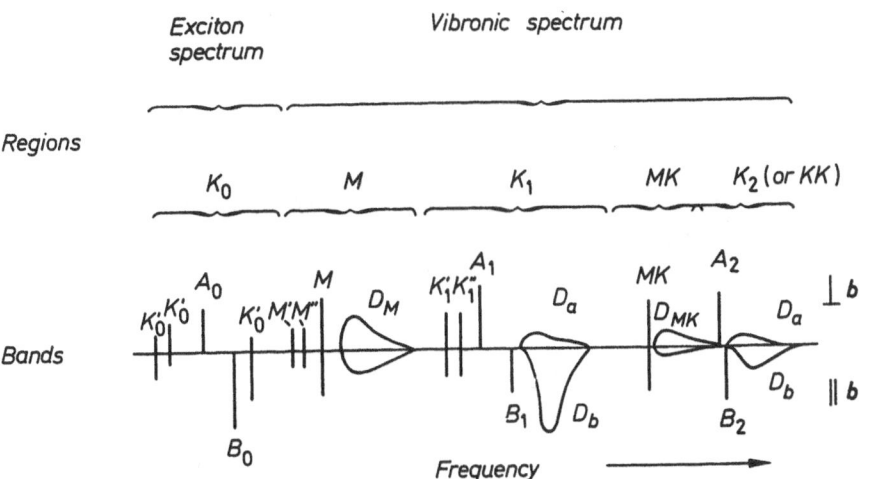

Fig.1.10. Absorption spectrum of a molecular crystal in terms of molecular exciton spectroscopy

extent the structure of the energy spectrum allows one to describe the structure of the absorption spectrum and, vice versa, how experimentally determined optical absorption spectra (and, occasionally, fluorescence spectra) can be used to quantitatively determine the parameters of the energy spectrum.

The energy spectrum of a crystal in the exciton region consists of the band spectrum of excitons in the perfect crystal and the discrete spectrum of local excitons bound to impurities. The absorption spectrum generally consists of a multiplet of sharply polarized intrinsic absorption bands of the A_0 and B_0 type (Fig.1.10) and of the structural spectrum of impurities (K_0', K_0'' ... bands located on both the low- and the high-frequency side of the intrinsic absorption). The description of the spectra in the K_0 region and the properties of excitons in the perfect and an imperfect crystal will be found in Chaps.2-4.

The K_0 region of the crystal spectrum is followed by M, K_1, MK, K_2, and other vibronic absorption regions. As in the K_0 region, the primed bands in these regions denote discrete impurity levels. The consideration of impurity vibronic states is so far restricted to the regions of the M and K_1 transitions. The scheme of Fig.1.10 suggests the existence of two types of vibronic bands: relatively narrow ones (M, A_1-B_1, MK, A_2-B_2), and broader ones (D_M, D_a, D_b, D_{MK}), which lead to the "pillar-like" absorption spectra of thick crystals. The spectroscopy of vibronic transitions is treated in Chaps.5 and 6.

Such a description is naturally incomplete. First of all, consideration has only been given to a restricted region of the spectrum that corresponds to the *Frenkel singlet excitons*. Therefore, the *triplet excitons* due to the lowest electronic excitation in aromatic molecules, remain outside of the scope of our treatment. Nevertheless, many of the ideas and methods described below can also be applied to them. Furthermore, higher-frequency transitions, corresponding to the formation of *charge-transfer excitons* and of *electron-hole pairs*, are entirely excluded from our consideration. Finally, definite preference is given to those problems where a comprehensive understanding of the real physical situations has been achieved so far due to the synthesis of theory and experiment.

2. Exciton Spectra of Perfect Crystals

This chapter, as well as the preceding one, is of an auxiliary nature. It briefly formulates the basic principles of the theory of molecular excitons in a rigid lattice. Both the energy and optical spectra of excitons are considered. The material is presented in terms of Green's functions and introduces the terminology used throughout the subsequent text. The concluding section summarizes the data on the energy spectrum of excitons in molecular crystals. The optical spectra are analyzed and discussed in the following chapters.

2.1 Molecular Excitons

An analysis of the experimental data in Chap.1 shows that the lower energy levels of the molecules and of the crystals formed by them are strongly related. It can also be ascertained that the intermolecular interaction energy, estimated from the shift between the vapour and crystal spectrum or from the splitting of the bands in the K_0 multiplet, is much less than the difference between successive electronic levels. Thus, the classification of the electronic spectrum of a crystal is based on the lower levels of isolated molecules. Each intramolecular electronic level is then correlated to a corresponding exciton band. Such an exciton is called a *Frenkel exciton* [2.1] or a *molecular exciton*. Two distinguishing features of a Frenkel exciton are the lack of internal degrees of freedom and the lack of internal quantum numbers. Its state is completely determined by the serial number of the exciton band to which it belongs, and by the wave vector **k** (quasi momentum).

There exists an extensive literature on the theory of Frenkel excitons, particularly on their electrodynamics (see, for instance, [2.2-4] and the references listed there). This chapter does not claim to give a comprehensive survey of this literature. Its purpose is to give a concise description of the basic principles and to present a number of specific results, so that the reader rarely has to resort to any other sources of literature.

2.2 Wave Functions and the Energy Spectrum of Excitons

In developing the theory of Frenkel excitons, a simple and, at the same time, effective approximation can be made: the wave function of an exciton is expressed in terms of a linear combination of functions that correspond to the excitation of single molecules into states belonging to a definite molecular term. Such an approximation is called the *Frenkel model*.

The most interesting peculiarities of the exciton spectra of molecular crystals are due to the presence of several molecules in the primitive cell. The Frenkel excitons in crystals with such a primitive cell are described by the Davydov theory [2.2,5].

It is convenient to specify the position of a molecule in the crystal using a vector **n** which specifies the position of the primitive cell to which the molecule belongs and an index α ($\alpha = 1, 2, \ldots, Z$), which specifies the position of the molecule in the primitive cell; Z is the number of molecules in the primitive cell[1]. Molecules **n**α and **m**α with $\mathbf{n} \neq \mathbf{m}$ are said to be *translationally equivalent*, while molecules **n**α and **m**β with $\alpha \neq \beta$ are said to be *translationally nonequivalent*. Even if a crystal consists of identical molecules, the individual sublattices that correspond to different values of α may find themselves in different crystal fields. As a result, the levels of the molecules in these sublattices will be shifted differently. If these sublattices can be transformed into each other through operations of the space group of the crystal, the degeneracy of the levels is retained. Such sublattices are said to be *symmetrically dependent*. The sublattices of the crystals discussed in this book (benzene, naphthalene, and anthracene) are all symmetrically dependent. Therefore, the theory will be presented with reference to such crystals. However, it can readily be generalized to include crystals with symmetrically independent sublattices (for example, stilbene and tolane).

For the sake of simplicity, we assume that the electronic levels of the molecule are nondegenerate. The theory is, of course, also applicable to degenerate levels, since the index that labels the states of the degenerate term can be considered to be included in α. The wave function of the crystal in which a molecule located at the site **n**α is excited (or, more briefly, with an excited site **n**α) will be denoted by $|\mathbf{n}\alpha\rangle$.

Further on, it will be convenient to operate in the secondary quantization representation by introducing the creation and annihilation operators of elec-

---

[1] In all the crystals which are considered below the primitive cell coincides with the unit cell. Hence, in what follows we shall use the latter for brevity.

tronic excitations, $a_{n\alpha}^{+}$ and $a_{n\alpha}$. Then,

$$|n\alpha> = a_{n\alpha}^{+}|0> \quad , \tag{2.1}$$

where $|0>$ is the ground state of the crystal ("exciton vacuum"). The Hamiltonian of the crystal in the representation of secondary quantization is

$$H = H_e + H_{ex} \quad , \tag{2.2}$$

where

$$H_e = E_\rho \sum_{n\alpha} a_{n\alpha}^{+} a_{n\alpha} \quad , \quad E_\rho = E_{mol} + D \quad ; \tag{2.3}$$

$$H_{ex} = \sum_{n\alpha \neq m\beta} M_{n\alpha m\beta} a_{n\alpha}^{+} a_{m\beta} \quad . \tag{2.4}$$

E_{mol} is the energy of the electronic term in the isolated molecule [including the shift due to intramolecular vibronic interactions; see (1.5) and Appendix A], and D is the energy of the matrix (vapour-crystal) shift. Assuming that the molecules interact with one another pairwise,

$$D_\alpha = \sum_{\substack{m\beta \\ m\beta \neq n\alpha}} D_{n\alpha m\beta} \quad . \tag{2.5}$$

For a crystal with symmetrically dependent sublattices, all of the $D_\alpha = D$ and (2.3) is written for this case. The quantity E_ρ will be called the *electronic term of the crystal*.

The term H_{ex} describes the transfer of excitation between sites and is specific for excitons. The coefficients $M_{n\alpha m\beta}$ will be called *excitation transfer integrals* or, simply, *transfer integrals*. They are already assumed to be renormalized for the intramolecular vibronic interaction (Sect.6.1). Due to translational symmetry,

$$M_{n\alpha m\beta} = M_{\alpha\beta}(n - m) \quad , \tag{2.6}$$

and the Hamiltonian H is diagonalized by a transformation to the momentum representation, whose basis functions are

$$|\mu k> = \sum_{n\alpha} \psi_{\mu k}(n\alpha)|n\alpha> \quad , \tag{2.7}$$

where

$$\psi_{\mu k}(n\alpha) = (1/\sqrt{N})B_\alpha(\mu k)\exp(ikn_\alpha) \quad ; \tag{2.8}$$

n_α is the position vector of the site $n\alpha$, and N is the number of unit cells in the normalization volume. The wave vector k runs over the entire Brillouin zone. The functions $\psi_{\mu k}$ are the wave functions of the excitons in the site representation. To determine the coefficients $B_\alpha(\mu k)$ it is necessary to insert the functions $|\mu k>$ into the Schroedinger equation

$$H|\mu \mathbf{k}> = E_\mu(\mathbf{k})|\mu \mathbf{k}> \quad , \tag{2.9}$$

which can be conveniently written in the following form:

$$H_{ex}|\mu \mathbf{k}> = \varepsilon_\mu(\mathbf{k})|\mu \mathbf{k}> \quad ,$$

$$E_\mu(\mathbf{k}) = E_\rho + \varepsilon_\mu(\mathbf{k}) \quad . \tag{2.10}$$

As a result the following set of equations is obtained:

$$\sum_\beta L_{\alpha\beta}(\mathbf{k})B_\beta(\mu \mathbf{k}) = \varepsilon_\mu(\mathbf{k})B_\alpha(\mu \mathbf{k}) \quad , \quad \text{where} \tag{2.11}$$

$$L_{\alpha\beta}(\mathbf{k}) = \sum_{\mathbf{n}-\mathbf{m}} M_{\mathbf{n}\alpha\mathbf{m}\beta}\exp[-i\mathbf{k}(\mathbf{n}_\alpha - \mathbf{m}_\beta)] \quad . \tag{2.12}$$

The eigenvectors define the wave functions of the excitons, $\psi_{\mu \mathbf{k}}$, and the eigenvalues $\varepsilon_\mu(\mathbf{k})$ define their energy spectrum or *dispersion law*. It is significant that each molecular level in a crystal with Z molecules per unit cell yields Z branches in the exciton spectrum ($\mu = 1, 2, \ldots, Z$) or Z exciton bands. This phenomenon is called the *Davydov splitting*.

The condition for the orthonormalization of the eigenvectors of (2.11),

$$\sum_\alpha B_\alpha^*(\mu \mathbf{k})B_\alpha(\mu'\mathbf{k}) = \delta_{\mu\mu'} \quad , \tag{2.13}$$

ensures that the functions $\psi_{\mu \mathbf{k}}$ are normalized. The corresponding completeness condition is

$$\sum_\mu B_\alpha^*(\mu \mathbf{k})B_\beta(\mu \mathbf{k}) = \delta_{\alpha\beta} \quad . \tag{2.14}$$

It also follows from (2.11-13) that

$$\varepsilon_\mu(\mathbf{k}) = \sum_{\substack{\mathbf{n}-\mathbf{m} \\ \alpha\beta}} B_\alpha^*(\mu \mathbf{k})M_{\mathbf{n}\alpha\mathbf{m}\beta}B_\beta(\mu \mathbf{k})\exp[-i\mathbf{k}(\mathbf{n}_\alpha - \mathbf{m}_\beta)] \quad . \tag{2.15}$$

The above equations determine the energy spectrum $E_\mu(\mathbf{k})$ of the exciton. Its centre of gravity or first moment is equal to

$$\bar{E} = \frac{1}{NZ} \sum_{\mu \mathbf{k}} E_\mu(\mathbf{k}) = E_\rho + \frac{1}{NZ} \sum_{\mu \mathbf{k}} \varepsilon_\mu(\mathbf{k}) = E_\rho + \frac{1}{NZ} Tr\{H_{ex}\} = E_\rho \quad . \tag{2.16}$$

The trace of H_{ex} is equal to zero since, according to (2.4), the site representation of H_{ex} does not contain any diagonal elements. Thus, the electronic term E_ρ is the *centre of gravity of the energy spectrum* of the exciton.

The detailed shape of $\varepsilon_\mu(\mathbf{k})$ depends on the transfer integrals $M_{\mathbf{n}\alpha\mathbf{m}\beta}$, which, however, cannot be calculated in a general case. For singlet excitons it is essential that the coefficients $M_{\mathbf{n}\alpha\mathbf{m}\beta}$ always include, along with an exponentially decreasing contribution, a contribution associated with multipole interactions, which decreases according to a power law and dominates at large distances. If the molecules interact with one another pairwise and the transition is dipole allowed, the main term for large distances R is

$$M_{n\alpha m\beta} \approx \frac{p_\alpha p_\beta}{R^3} - 3\frac{(Rp_\alpha)(Rp_\beta)}{R^5} \quad , \quad R = n_\alpha - m_\beta \quad , \tag{2.17}$$

which corresponds to a dipole-dipole interaction. Here, p_α is the matrix element of the dipole moment of the electronic transition in the molecules of the αth sublattice. For the sake of simplicity, p_α is assumed to be real. Since the coefficients $M_{n\alpha m\beta}$ decrease slowly, only with the cube of R, the sums $L_{\alpha\beta}$ in (2.12) diverge as $k \to 0$ [2.6]. This is a well-known problem, which also arises when the dynamics of ionic lattices is considered. It can be solved, according to the Ewald method, by separating the longitudinal electric macrofield induced by a long-range Coulomb interaction [2.7]. This field can be conveniently defined in exciton theory as the field corresponding to the average density of the dipole moment of the excitonic transition [2.8]. The field induced by the sublattice β at the point r is equal to

$$E_\beta(r) = -\frac{4\pi}{v}\frac{k(kp_\beta)}{k^2}\exp(ikr) \quad , \tag{2.18}$$

$k = |k|$ and v is the volume per unit cell. Hence, according to (2.12), the contribution of the long-range interactions to $L_{\alpha\beta}$ is

$$L_{\alpha\beta}^{LR}(k) = -p_\alpha E_\beta = (4\pi/v)(kp_\alpha)(kp_\beta)/k^2 \quad . \tag{2.19}$$

This expression exhibits a discontinuity when k is small. It retains the dependence on the direction of k as $|k| \to 0$. Excitons whose Hamiltonian takes into account the longitudinal macrofield are called *Coulomb excitons* [2.2,9].

It is also convenient to introduce *mechanical excitons*. The nonanalytical part of their Hamiltonian can be removed in the following way:

$$\mathcal{L}_{\alpha\beta}(k) = L_{\alpha\beta}(k) - L_{\alpha\beta}^{LR}(k) \quad . \tag{2.20}$$

This procedure is only unambiguous when $k \to 0$ since one can simultaneously subtract the arbitrary continuous function of k. Such a difficulty does not in fact arise, because mechanical excitons may be introduced only when $k \sim 1/\lambda \ll 1/d$, where λ is the wavelength of the light, and d is the lattice constant. The wave functions and the spectrum $E_\mu(k)$ of mechanical excitons are determined from a set of equations similar to the ones in (2.11):

$$\sum_\beta \mathcal{L}_{\alpha\beta}(k)B_\beta(\mu k) = [E_\mu(k) - E_\rho]B_\alpha(\mu k) \quad . \tag{2.21}$$

The quantities $B_\alpha(\mu k)$ satisfy conditions similar to the ones given in (2.13, 14).

For weak transitions the p_α are small, and so is the dipole-dipole interaction. The spectra of the Coulomb and mechanical excitons resemble one another even more so when the contribution of the long-range interactions is
reduced. Under such conditions only the interactions between a molecule and
its nearest neighbours have to be taken into account in the sums $L_{\alpha\beta}$ (2.12).
The model that is based on this approximation is called the *restricted Frenkel
model*. Within the framework of this model, the wave functions $\psi_{\mu k}$ are usually
written for a crystal with a single molecule in the unit cell:

$$\psi_k(n\alpha) = (1/\sqrt{NZ}) \exp(ikn_\alpha) \quad . \tag{2.22}$$

However, **k** runs over a region whose volume is Z times that of the Brillouin
zone.

In the optical absorption spectrum of a crystal, only transitions to the
state **k** = **K** are allowed, where **K** is the wave vector of the photon. Since
$K \ll d^{-1}$, it can be assumed that $K \approx 0$. The Z absorption bands with energies
$E_\mu(K)$ form an *exciton multiplet*. The appearance of an exciton multiplet from
a nondegenerate intramolecular level can be called a *Davydov splitting of the
optical spectrum*. The magnitude of the Davydov splitting Δ_{Dav} is defined as
the difference between the centres of gravity of the absorption bands seen
in certain *spectrum components* (i.e., for different polarizations of the
light). Certain multiplet components may be degenerate, depending on the symmetry of the crystal. Individual components of a multiplet are polarized along
the crystallographic axes in accordance with the symmetry of the crystallographic point group; some transitions may be forbidden. The selection rules
can be determined and related to the symmetry of the intramolecular transitions according to group theory [2.10,11] described in [2.2,12]. It is
obvious that the exciton multiplets correspond to Prikhotjko's K_0 bands
(Sect.1.3).

From (2.19) it can be seen that the exciton dispersion law in the exciton-
energy band corresponding to an isolated exciton-absorption band that is
polarized parallel to the crystallographic direction **a** has the following form
for small values of $|k|$:

$$\varepsilon(k) = \varepsilon_t + \frac{4\pi}{v} p_a^2 \cos^2\theta \quad , \tag{2.23}$$

where θ is the angle between **k** and **a** and p_a is the dipole moment of the transition per unit cell. For a transverse (**k** $\perp$ **a**) exciton, $\varepsilon = \varepsilon_t$, while
$\varepsilon = \varepsilon_\ell = \varepsilon_t + 4\pi p_a^2/v$ for a longitudinal (**k** $\parallel$ **a**) exciton. The quantity

$$\varepsilon_{\ell t} = 4\pi p_a^2/v \tag{2.24}$$

determines the discontinuity in $\varepsilon_\mu(k)$ at $k = 0$ (*longitudinal-transverse splitting*). The resulting singularity is described by (2.23), which can be generalized on the basis of electrodynamical considerations [2.13].

2.3 Exciton Green's Function

Many aspects of exciton theory can best be considered in terms of Green's functions [2.14]. Exciton Green's function is defined by

$$G_{n\alpha m\beta}(t,t') = -i\langle Ta_{n\alpha}(t)a_{m\beta}^+(t')\rangle \quad , \qquad (2.25)$$

where the angular brackets denote an averaging over the "exciton vacuum", i.e., over the crystal state without any excitons, the $a_{n\alpha}(t)$ represent the exciton operators in the Heisenberg representation, and T is the time ordering. By going over to the frequency representation and by expanding the product of the operators into a complete system of eigenfunctions $\psi_\lambda(n\alpha)$ of the exciton Hamiltonian, one obtains (Appendix B)

$$G_{n\alpha m\beta}(\Omega) = \int_0^\infty G_{n\alpha m\beta}(t,t') \exp[i\Omega(t-t')]d(t-t')$$

$$= \sum_\lambda \frac{\psi_\lambda(n\alpha)\psi_\lambda^*(m\beta)}{\Omega - E_\lambda + i0} = (\omega - H + i0)_{n\alpha m\beta}^{-1} \quad , \qquad (2.26)$$

where the infinitesimal imaginary term i0 guarantees the convergence of the integral at the upper limit.

Using (2.8) for the perfect crystal yields

$$G_{n\alpha m\beta}^0(\Omega) = \frac{1}{N}\sum_k G_{\alpha\beta}^0(\Omega k)\exp[ik(n_\alpha - m_\beta)] \quad , \qquad (2.27)$$

where $G_{\alpha\beta}^0(\Omega k)$ is the Green's function in the k representation:

$$G_{\alpha\beta}^0(\Omega k) = \sum_\mu \frac{B_\alpha(\mu k)B_\beta^*(\mu k)}{\Omega - E_\mu(k) + i0} \quad . \qquad (2.28)$$

The expression for $G_{n\alpha m\beta}^0$ can be substantially simplified, owing to the symmetrical dependence of the sublattices, if the sites $n\alpha$ and $m\beta$ coincide. To do this, we must set $n\alpha = m\beta$ in (2.27), take an average over α, and make use of the normalization condition (2.13):

$$G_0(\Omega) \equiv G_{n\alpha n\alpha}^0 = \frac{1}{NZ}\sum_{\mu k}\frac{1}{\Omega - E_\mu(k) + i0} = \frac{v}{Z(2\pi)^3}\sum_\mu \int\frac{d^3k}{\Omega - E_\mu(k) + i0} \quad . \qquad (2.29)$$

If we divide G_0 into the real and the imaginary part,

$$G_0(\Omega) = G_0' + iG_0'' \quad , \qquad (2.30)$$

and make use of the general expression

$$(\Omega - H + i0)^{-1} = (\Omega - H)^{-1} - \pi i \delta(\Omega - H) \quad , \qquad (2.31)$$

we obtain

$$\rho(\Omega) \equiv -\frac{1}{\pi} G_0''(\Omega) \quad , \qquad G_0'(\Omega) = \int \frac{\rho(\Omega')}{\Omega - \Omega'} d\Omega' \quad , \qquad (2.32)$$

where the integration is understood in the sense of the Cauchy main value.
Here, $\rho(\Omega)$ represents the *normalized density of the states* in the energy spectrum. The electronic term E_ρ is the centre of gravity of the density of states:

$$E_\rho = \int \Omega \rho(\Omega) d\Omega \quad , \qquad \int \rho(\Omega) d\Omega = 1 \quad . \qquad (2.33)$$

Outside the spectrum, $G_0 = G_0'$. The energy and frequency calculated from E_ρ will be denoted by $\varepsilon = E - E_\rho$, $\omega = \Omega - E_\rho$.

These equations can also be generalized to include imperfect crystals. From (2.26) it follows that the general equation is valid:

$$\rho(\Omega) = -\frac{1}{\pi NZ} \sum_{n\alpha} \text{Im}\,\{G_{n\alpha n\alpha}(\Omega)\} = -\frac{1}{\pi NZ} \text{Im}\{\text{Tr}[G(\Omega)]\} \quad . \qquad (2.34)$$

The contribution to the density that is made by states with a definite quasi momentum **k** is proportional to the quantity

$$S_{\alpha\beta}(\Omega\mathbf{k}) = -\text{Im}\,\{G_{\alpha\beta}(\Omega\mathbf{k})\} = -\frac{1}{N} \text{Im}\,\{\sum_{nm} G_{n\alpha m\beta}(\Omega) \exp[-i\mathbf{k}(\mathbf{n}_\alpha - \mathbf{m}_\beta)]\} \quad , \qquad (2.35)$$

called the *spectral density* or *spectral function*. It is related to $\rho(\Omega)$ as follows:

$$\rho(\Omega) = \frac{1}{\pi} \frac{1}{NZ} \sum_{\mathbf{k}\alpha} S_{\alpha\alpha}(\Omega\mathbf{k}) \quad . \qquad (2.36)$$

In the perfect crystal G is described by (2.28) and the dependence of S on Ω is that of a δ function.

2.4 Conductivity Tensor of a Crystal

The macroscopic optical properties of a crystal are determined by the complex conductivity tensor σ_c, which is also used to determine the tensor of the complex dielectric permeability

$$\kappa_c(\Omega\mathbf{K}) = 1 + 4\pi i \sigma_c(\Omega\mathbf{K})/\Omega \quad , \qquad (2.37)$$

and the real part of the conductivity tensor $\sigma = \text{Re}\{\sigma_c\}$, which directly describes the absorption. It is assumed that the vector potential is equal to

$$\mathbf{A}(\mathbf{r}t) = \mathbf{A}\exp\left[i(\mathbf{K}\mathbf{r} - \Omega t)\right] \quad . \tag{2.38}$$

Therefore, the problem of the microscopic theory is reduced to the calculation of σ_c.

The calculation of the conductivity can conveniently be started with the general Kubo formula, which expresses the linear response, i.e., the current induced by the vector potential $\mathbf{A}(\mathbf{r}t)$ of the external field through the correlator of currents [2.14]. In the site representation, it has the following form:

$$j_i(\mathbf{n}\alpha,t) = -\frac{e^2\mathcal{N}}{mc}A_i(\mathbf{n}\alpha,t) + \frac{i}{c}\int_{-\infty}^{t}\sum_{\mathbf{m}\beta}<J_i(\mathbf{n}\alpha,t)J_j(\mathbf{m}\beta,t')$$

$$- J_i(\mathbf{m}\beta,t')J_j(\mathbf{n}\alpha,t)>A_j(\mathbf{m}\beta,t')dt' \quad , \tag{2.39}$$

where $\mathcal{N}$ is the number of electrons per molecule, c is the light velocity, and m is the mass of the electron. The contribution to the current operator of the site $\mathbf{n}\alpha$ associated with the exciton transition under consideration is equal to

$$J(\mathbf{n}\alpha,t) = j_\alpha a_{\mathbf{n}\alpha}(t) + j_\alpha^* a_{\mathbf{n}\alpha}^+(t) \quad . \tag{2.40}$$

In the above equation we took advantage of the fact that for a perfect crystal $j_{\mathbf{n}\alpha} = j_\alpha$. The intramolecular matrix elements of the currents j_α are related to the matrix elements of the dipole moment $\mathbf{p}_\alpha$ for the same transition by the usual expression

$$j_\alpha = -iE_{mol}\mathbf{p}_\alpha \quad , \tag{2.41}$$

where E_{mol} is the frequency of the electronic transition in the molecule (Sect.1.2). In general the angular brackets in (2.39) denote an averaging over Gibbs' distribution. In this sense, the correlator in (2.39) is a temperature-time one. However, the electronic excitation energy greatly exceeds the temperature T, i.e., $E_{mol} \gg T$. Therefore, the average in (2.39) must be understood to be an average over the exciton vacuum, but Gibbs' averaging over the low-frequency excitations, such as phonons, is also included in it.

Singling out the resonance contribution to the current in (2.39) and using (2.25,26) yields

$$j(\mathbf{n}\alpha,t) = \tilde{j}(\mathbf{n}\alpha,t) - \frac{1}{c}\sum_{\mathbf{m}\beta}\int_{-\infty}^{+\infty} G_{\mathbf{n}\alpha\mathbf{m}\beta}(t,t')\exp\left[i(\mathbf{K}\mathbf{m}_\beta - \Omega t')\right]\widehat{j_\alpha j_\beta^*}A(\mathbf{m}\beta,t')dt'$$

$$= \tilde{j}(\mathbf{n}\alpha,t) - \frac{1}{c}\sum_{\mathbf{m}\beta} G_{\mathbf{n}\alpha\mathbf{m}\beta}(\Omega)\exp\left[-i\mathbf{K}(\mathbf{n}_\alpha - \mathbf{m}_\beta)\right]\widehat{j_\alpha j_\beta^*}A(\mathbf{m}\beta,t) \quad . \tag{2.42}$$

The symbol $\widehat{j_\alpha j_\beta^*}$ represents the diadic product of the vectors j_α and j_β^*, which is defined as follows: $(\widehat{j_\alpha j_\beta^*})_{ij} \equiv (j_\alpha)_i(j_\beta^*)_j$.

The contribution to the current, $\tilde{\mathbf{j}}$, is a smooth function of the frequency. If the contributions to the absorption corresponding to different transitions are not superimposed, $\tilde{\mathbf{j}}$ describes the polarization current, and the relevant contribution to the complex conductivity is imaginary.

To obtain the complex conductivity tensor σ_c one must use the equation $\mathbf{E} = -\dot{\mathbf{A}}/c$ and a Fourier transformation with respect to the coordinates must be performed. As a result, the conductivity tensor σ of a perfect crystal in which an interaction between excitons and phonons is absent, i.e., when (2.28) can be used for G, is equal to

$$\sigma(\Omega\mathbf{K}) = -(1/\Omega v)\, \mathrm{Im}\left\{ \sum_{\substack{\mathbf{n}-\mathbf{m} \\ \alpha\beta}} G^0_{\mathbf{n}\alpha\mathbf{m}\beta}(\Omega) \exp\left[-i\mathbf{K}(\mathbf{n}_\alpha - \mathbf{m}_\beta)\right] \widehat{\mathbf{j}_\alpha \mathbf{j}_\beta^*} \right\}$$

$$= -\frac{1}{\Omega v}\, \mathrm{Im}\left\{ \sum_\mu \frac{\widehat{\mathbf{j}_\mu \mathbf{j}_\mu^*}}{\Omega - E_\mu(\mathbf{K}) + i0} \right\}, \qquad \text{where} \tag{2.43}$$

$$\mathbf{j}_\mu = \sum_\alpha B_\alpha(\mu, \mathbf{K}=0) \mathbf{j}_\alpha \; . \tag{2.44}$$

Similarly, for the real part of the tensor $\kappa_c(\Omega\mathbf{K})$ we obtain

$$\kappa(\Omega\mathbf{K}) = \kappa_0(\Omega) - \frac{4\pi}{\Omega^2 v} \sum_\mu \frac{\widehat{\mathbf{j}_\mu \mathbf{j}_\mu^*}}{\Omega - E_\mu(\mathbf{K})} \; , \tag{2.45}$$

where the function $\kappa_0(\Omega)$ denotes the smooth part of the dielectric permeability. Although κ resonates on the frequencies of mechanical excitons, the refractive indices of the electromagnetic waves, as determined by Maxwell's equations, resonate on the frequencies of Coulomb excitons.

In writing (2.43,45) we took account of the following. In (2.42) the entire macroscopic field of the electromagnetic wave, including the transverse and the longitudinal components, was considered to be a perturbation. Therefore, the longitudinal macrofield must be excluded from the nonperturbed Hamiltonian. Accordingly, the parameters of the mechanical, not of the Coulomb excitons, must be inserted into the tensors σ_c and κ_c (Sect.2.2). Therefore, (2.43,45) do not contain Green's function G^0 but instead contain the similarly constructed G^0 function in which E_μ is replaced by E_μ and B_α is replaced by B_α. PEKAR [2.13] has shown that (2.45) contains the dispersion law of Coulomb excitons in the long-wavelength range. This law is determined by the roots $\Omega = \Omega(\mathbf{K})$ of the equation

$$\mathbf{K}\kappa(\Omega\mathbf{K})\mathbf{K} = 0 \; . \tag{2.46}$$

The electrodynamics of a crystal in the region of exciton transitions exhibits many interesting features. In a perfect crystal without an exciton-phonon interaction there is no exciton absorption of light, although the dielectric permeability shows a resonance growth according to (2.45). In the long-wave (electrodynamical) region the undamped normal excitations of a crystal are not excitons, but are mixed *"light-mechanical" waves* which obey a complicated dispersion law defined by the tensor $\kappa(\Omega K)$ and Maxwell's equations. Such waves are called *polaritons*, or *light-excitons*. The polariton concept was first introduced during the development of Born's classical dynamical theory of lattices [2.7,15,16] and then generalized to include excitons at the quantum level [2.17,18]. This concept was applied for molecular excitons by AGRANOVICH [2.18], using the Bogolyubov-Tjablikov method. In the polariton concept, the absorption of light is treated as a scattering of polaritons on imperfections in the crystal, on phonons, etc.

According to (2.45), the dielectric permeability depends strongly on **K** (*spatial dispersion*) due to the presence of the function $E_\mu(\mathbf{K})$ in the resonance denominator. Therefore, the electrodynamical equations possess additional solutions called *Pekar waves* [2.19].

The theory of these phenomena has been generalized in the books by AGRANOVICH and GINZBURG [2.9a,b] and PEKAR [2.9c] (see also [2.2,3,20]). Most of the experimental data refer to semiconductors rather than to molecular excitons. These phenomena are very important in crystals with intensive exciton transitions and broad exciton energy bands. For crystals which exhibit weak transitions and relatively narrow exciton energy bands (Sect.2.5), e.g., for crystals of benzene and naphthalene, a simpler theory is sufficient to explain most of the experimental data. Namely, if we replace the term i0 in the denominator of (2.43) by the damping iγ due to the scattering of excitons, we can neglect the dependence of E_μ on **K**, if $K \sim 1/\lambda \ll 1/d$. Hence, the theory can be greatly simplified; and from now on we shall operate within the framework of this approximation. In particular, when the damping γ is weak, (2.43) transforms to

$$\sigma(\Omega) = (\pi/\Omega v) \sum_\mu \widehat{\mathbf{j}_\mu \mathbf{j}_\mu^*} \delta(\Omega - E_\mu) \quad , \quad E_\mu \equiv E_\mu(\mathbf{K}=0) \quad . \qquad (2.47)$$

The conductivity tensor can be expressed somewhat differently if we introduce the Fourier components of the current density operator:

$$\mathbf{J}(\mathbf{k}) = (1/\sqrt{Nv}) \sum_{m\beta} \mathbf{J}(m\beta) \exp(i\mathbf{km}_\beta) \quad . \qquad (2.48)$$

As a result the first equality in (2.43) can be rewritten as follows:

$$\sigma_{ij}(\Omega K) = (\pi/\Omega) \sum_\mu \langle 0|J_i^+(\mathbf{K})|\mu\mathbf{K}\rangle\langle\mu\mathbf{K}|J_j(\mathbf{K})|0\rangle \delta[\Omega - E_\mu(\mathbf{K})] \quad , \qquad (2.49)$$

where the $|\mu\mathbf{K}\rangle$ can be obtained from (2.7,8) by replacing B_α by B_α. From (2.49) it is easy to obtain (2.47).

2.5 Exciton Bands of Molecular Crystals

The entire quantitative analysis of the exciton spectra is based on the dispersion law $\varepsilon_\mu(\mathbf{k})$ and the density of states in the energy spectrum $\rho(\varepsilon)$. Therefore, we will briefly summarize the data on $\varepsilon_\mu(\mathbf{k})$ for the exciton bands of the first excited singlet states of benzene, naphthalene, anthracene and hexamethylbenzene.

There are two ways of determining $\varepsilon_\mu(\mathbf{k})$. One is based on the calculation of $M_{n\alpha m\beta}$ by using the molecular wave functions, i.e., from the first principles. The difficulties that are encountered in calculating $M_{n\alpha m\beta}$ are due to the absence of sufficiently accurate molecular wave functions and the change in the interactions between molecules due to the dielectric screening by the higher excited states. The latter difficulty is of a more fundamental nature because it requires, as a minimum, a knowledge of the spatial dispersion of the dielectric permeability in the frequency range near the exciton transitions. To overcome the first difficulty, the interaction is usually expanded into a series in terms of multipoles. For the nearest neighbours the multipole expansion parameter has the order of magnitude $a/d \sim 1$, where a is the size of the molecule and d is the distance between the molecules. Therefore, multipole expansion is inapplicable in this case. At the same time, the nearest-neighbours contribution is dominant for weak transitions, when short-range interactions prevail. Even for strong transitions, when the contribution of the long-range interaction is large, the interaction with the nearest neighbours makes up about one half of the whole effect. Therefore, one cannot rely heavily on the calculations using the multipole approximation. As a result, the calculations that were made this way for the energy bands of benzene and naphthalene did not yield satisfactory results [2.21,22,40]. They did, however, play a substantial role in the early stages of the investigations by helping to estimate the order of magnitude of $\varepsilon_\mu(\mathbf{k})$. More detailed calculations of the Davydov splitting and oscillator strength have been recently carried out by SCHLOSSER and PHILPOTT [2.41]. They cover a number of dipole allowed transitions in naphthalene, anthracene, tetracene and pentacene crystals.

The second way to determine $\varepsilon_\mu(\mathbf{k})$ is based on the reconstruction of $\varepsilon_\mu(\mathbf{k})$ from the experimental data. The techniques that can be used are described in Chaps.3-6. These methods have produced mutually correlated data for the spectra of imperfect crystals and for vibronic spectra. Most of the data given below were obtained in this way.

2.5.1 Benzene

The benzene crystal belongs to the orthorhombic system, space group D_{2h}^{15}. The unit cell of the crystal contains four molecules, whose arrangement is shown in Fig.2.1a. The symmetry axes are screw axes and the planes of symmetry are the glide planes. The corresponding operations permute the four sublattices, which are symmetrically dependent. The Brillouin zone is shown in Fig.2.1b, where $\mathbf{a}^*$, $\mathbf{b}^*$ and $\mathbf{c}^*$ are the basis vectors of the reciprocal lattice. The letters refer to the symmetry elements (lines and points) of the zone.

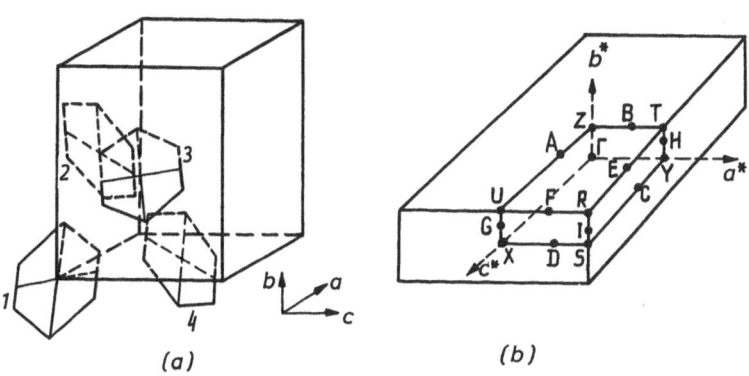

(a)   (b)

Fig.2.1a,b. Arrangement of molecules in the benzene crystal [2.23] (a) and the Brillouin zone of the crystal [2.24] (b)

Benzene is a typical example of a substance that exhibits weak electronic transitions. The lowest electronic transition is symmetry forbidden for the free molecule. It is allowed in a crystal field with a local symmetry C_i due to configurational mixing (the Craig effect). Its intensity is low, and the oscillator strength f is $\sim 10^{-3}$. The contribution of the dipole interactions to the formation of the exciton band, which can readily be estimated from (2.24), is $E_{dd} \sim E_{at}(d_{at}/d)^3 f$, where E_{at} and d_{at} have the following orders of magnitude: $E_{at} \sim 10$ eV and $d_{at} \sim 10^{-8}$ cm. As a result, $E_{dd} \sim 1$ cm^{-1}, which is negligibly small. The energy spectrum is the result of interactions which fall off as slowly as, or still slower than, those between octopoles. Actually, a sufficient accuracy is already attained when the interaction of a few nearest neighbours is taken into account (2.12):

$$\varepsilon_{1,2,3,4}(\mathbf{k}) = 2M_a \cos(\mathbf{k}\cdot\mathbf{a}) + 2M_c \cos(\mathbf{k}\cdot\mathbf{c}) \overset{+}{\underset{-}{}} 4M_{12} \cos(\mathbf{k}\cdot\mathbf{a}/2) \cos(\mathbf{k}\cdot\mathbf{b}/2)$$

$$\overset{+}{\underset{-}{}} 4M_{13} \cos(\mathbf{k}\cdot\mathbf{b}/2) \cos(\mathbf{k}\cdot\mathbf{c}/2) \overset{-}{\underset{+}{}} 4M_{14} \cos(\mathbf{k}\cdot\mathbf{a}/2) \cos(\mathbf{k}\cdot\mathbf{c}/2) \quad , \quad (2.50)$$

where M_a, M_c, M_{12}, M_{13} and M_{14} are the integrals of the transfer of excitation between the molecule 1 and its neighbours which are spaced by distances **a**, **c**, $(\mathbf{a}+\mathbf{b})/2$, $(\mathbf{b}+\mathbf{c})/2$, and $(\mathbf{a}+\mathbf{c})/2$, respectively, from molecule 1 (Fig.2.1a). Equation (2.50) can readily be derived without solving the system in (2.11). This can be done because of the symmetrical dependence of the sublattices. In fact, under these conditions the energy spectrum is the same as the one for the orthorhombic face-centred lattice with a single molecule in a primitive cell that is four times smaller than the initial unit cell.

The symmetry of the exciton states at every point of the Brillouin zone is determined by the irreducible representations of the factor group of the wave vector (or **k** vector) group (WVG) [2.25,26]. The dimensionality of the corresponding irreducible (projective) representations is equal to 2 along the prism edges (points E, F, and I) and the medians of the faces of the Brillouin zone (points A, B, C, D, G, and H) (Fig.2.1b) because of the comparatively high symmetry of the benzene crystal [2.27]. Hence, the exciton states are degenerate. Since all of the faces are perpendicular to the two-fold screw axes, an additional double degeneracy arises due to the time inversion symmetry [2.25]. As a result, the bands are doubly degenerate on the faces and quadruply degenerate on their medians and on the prism edges.

Figure 2.2 illustrates the exciton bands of the benzene crystal arising from the B_{2u} electronic state of the molecule which were calculated for different directions in the Brillouin zone using (2.50). The calculations are based on the parameters $M_a = -0.25$, $M_c = -0.75$, $M_{12} = -1.55$, $M_{13} = 3.93$, and $M_{14} = 3.28$ cm^{-1} [2.28], which provide the best description of the positions of the bands of the exciton multiplet (the A_0, B_0, and C_0 bands in Fig.1.4) and the density of the exciton states (Fig.5.9). The energy $\varepsilon_\mu(\mathbf{k})$ is determined from the electronic term $E_\rho = 37835$ cm^{-1}.

The symmetry of the exciton states of the crystal at the point Γ is given in Fig.2.2 in terms of the irreducible representations of the point group D_{2h} (which is isomorphic to the factor group of WVG). The representations B_{2u}, B_{3u}, and B_{1u} correspond to allowed transitions which are polarized along the a, c and b axes of the crystal (the A_0, C_0, and B_0 bands). The transition to the A_u band is forbidden.

Since the exciton bands on the faces of the Brillouin zone are degenerate, they form a single continuum about 60 cm^{-1} wide. It is seen from comparison of Fig.2.1b and Fig.2.2 that wherever the spectrum is quadruply degenerate, the dispersion of the exciton bands is small (for example, along the E direction). The reason is that under these conditions (2.50) only includes those terms that correspond to an interaction between translationally-equivalent

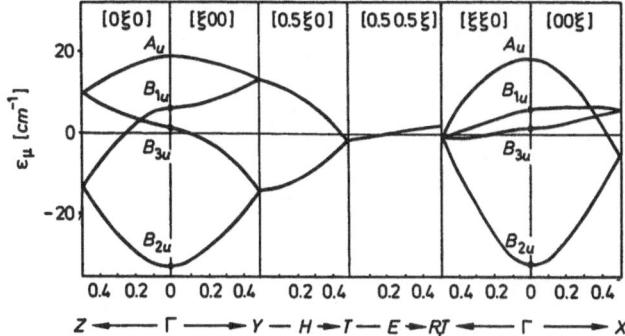

Fig.2.2. Exciton dispersion law in the benzene crystal (solid circles at the zone centre indicate the positions of the centres of gravity of the A_0, B_0 and C_0 bands of the exciton triplet)

molecules; and these are small for the benzene crystal. The characteristic shape of the density-of-states distribution for benzene is a narrow peak. It arises since for much of the k space, formed by the vicinity of the "skeleton" (i.e., of the edges of the Brillouin zone and the medians of its faces), the dispersion of $\varepsilon_\mu(\mathbf{k})$ is small.

2.5.2 Naphthalene

The lattice of naphthalene is monoclinic and its space group is C_{2h}^5. The unit cell contains two molecules whose arrangement is shown in Fig.2.3a. The symmetry operations associated with the screw axis b and the glide plane ac permute the symmetrically dependent sublattices. The Brillouin zone is shown in Fig.2.3b. As in the benzene crystal, the nearest-neighbours interaction prevails for the electronic state B_{3u}. In an isolated naphthalene molecule the transition is weak ($f \sim 10^{-4}$). In the crystal, it is enhanced by configurational mixing; per unit cell $f_b \approx 4 \cdot 10^{-3}$ and $f_a \approx 2 \cdot 10^{-5}$. The contribution of a dipole-dipole interaction to the exciton dispersion law can be estimated from the longitudinal-transverse splitting $\varepsilon_{\ell t}$ ($\varepsilon_{\ell t} \approx 4$ cm^{-1}, Sect.2.2) near the B_0 band using (2.46) and the experimental data on the low-temperature dielectric permeability of naphthalene [2.30,31].

The exciton dispersion law, in the approximation of the interaction of a few nearest neighbours, has the following form:

$$\varepsilon_{1,2}(\mathbf{k}) = 2M_a \cos(\mathbf{k} \cdot \mathbf{a}) + 2M_b \cos(\mathbf{k} \cdot \mathbf{b}) + 2M_{a+c} \cos[\mathbf{k} \cdot (\mathbf{a} + \mathbf{c})]$$

$$\pm 4\{M_{12} \cos(\mathbf{k} \cdot \mathbf{a}/2) + M'_{12} \cos[\mathbf{k} \cdot (\mathbf{a}/2 + \mathbf{c})]\} \quad , \tag{2.51}$$

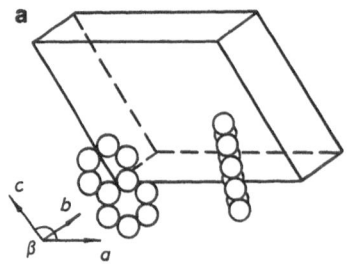

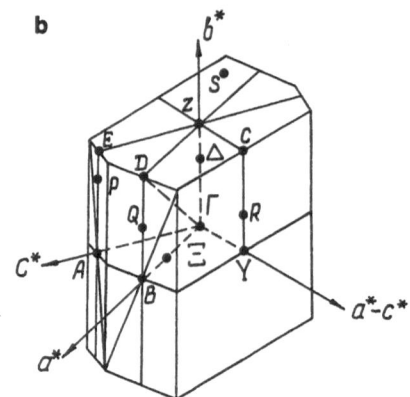

Fig.2.3a,b. Arrangement of molecules
in the unit cell of the naphthalene
crystal [2.29] (a) and the Brillouin
zone of the crystal [2.24] (b)

where the transfer integrals M_{12} and M'_{12} correspond to the distances between
the translationally-nonequivalent molecules, which are $(\mathbf{a}+\mathbf{b})/2$ and
$(\mathbf{a}+\mathbf{b})/2+\mathbf{c}$, respectively.

The integral $M_{12} \approx 20$ cm^{-1} is the largest of the transfer integrals $M_{\alpha\beta}$. The
information on the other integrals is unreliable. The available data are pre-
sented in Sect.3.4 (Table 3.5).

Figure 2.4 shows the exciton bands of a naphthalene crystal which were
calculated using (2.51) and the set of transfer integrals listed in Table 3.5.
The energy ε_μ is given with respect to the electronic term $E_\rho = 31557$ cm^{-1}.
The symmetry of naphthalene crystals is substantially lower than that of
benzene. Nevertheless, band degeneracy also exists. At selected points on the
boundary of the Brillouin zone (for example, at Y and B) the $\mathbf{k}$ vector symmetry
is described by two-fold projective representations. Since the translatory

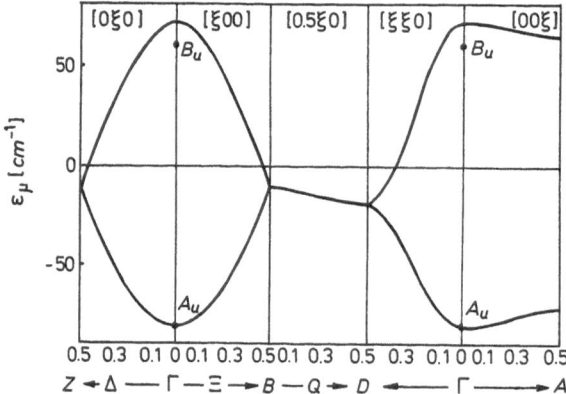

Fig.2.4. Exciton dispersion law in the naphthalene crystal (solid circles
at the zone centre denote the positions of the centres of gravity of the A_0
and B_0 bands of the exciton doublet

motion caused by the glide plane takes place along the **a** axis, the behaviour of the spectrum at the point B and A is quite different. At the first point the bands are degenerate, while at the second they are not. Furthermore, a two-fold degeneracy arises on both the prism bases and the lines R and Q due to the time inversion symmetry. At the point Γ, the factor group of WVG coincides with that of the crystallographic point group, i.e., C_{2h}. A_u and B_u correspond to representations that determine the polarization of the electronic transition perpendicular and parallel to **b** (A_0 and B_0 bands).

The exciton spectrum of the crystal is a single-continuum approximately 160 cm^{-1} wide because the two bands stick together at the boundaries of the Brillouin zone. Similar to the benzene spectrum, the dispersion of the naphthalene energy spectrum is small throughout the region where the bands stick together, since it is determined by the transfer integrals between translationally-equivalent molecules (for example, in the Q direction, Fig.2.4). On the whole, the dispersion in the **c*** direction is small. Therefore, the spectrum is quasi two dimensional. The corresponding density of states is shown in Fig.3.19.

2.5.3 Anthracene

The crystalline structure of anthracene is the same as that of naphthalene. The unit cell contains two molecules whose arrangement is similar to the one shown in Fig.2.3a. The Brillouin zone of the crystal is shown in Fig.2.3b.

An optical transition to the lowest excited electronic state B_{2u} is symmetry allowed, is polarized parallel to the short axis of the molecule, and is intensive ($f \approx 0.3$). The longitudinal-transverse splitting $\varepsilon_{\ell t} \sim 200 - 400$ cm^{-1}.

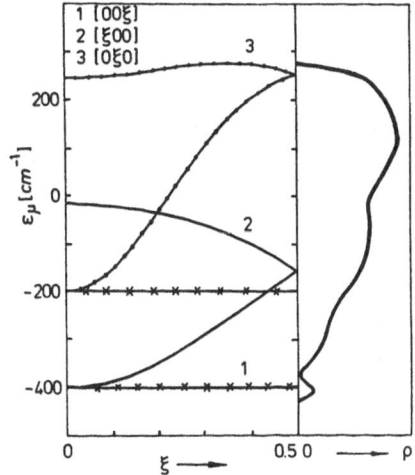

Fig.2.5. Dispersion law and density of states in the exciton bands of the anthracene crystal [2.35,36] (dipole moment of transition p = 0.61 Å·e.)

Thus, the contribution of the long-range interaction to the exciton spectrum
is considerable. There are practically no experimental data on the structure
of the exciton bands. However, they have been repeatedly calculated using the
dipole-dipole approximation [2.2,32-35]. The results of the last two calculations are similar. Figure 2.5 illustrates the dispersion law and the density
of states of the excitons. The energy can be determined from the electronic
term that corresponds to the centre of gravity of the density of states. The
total width of the exciton spectrum is apporximately 700 cm^{-1}. However, it
may actually be one and a half to two times smaller. The exact position of
the electronic term is unknown. The probable value is $E_\rho = 25330$ cm^{-1} (see
Sect.4.5, case C).

2.5.4 Hexamethylbenzene (Low-Temperature Modification)

At present there is no consent of opinion concerning the lattice structure
and the interpretation of the spectrum of the low-temperature modification
of hexamethylbenzene [2.37]. The most recent data [2.38,39] suggest that the
lattice is rhombohedral and contains one molecule per unit cell. The arrangement of the aromatic rings is outlined in Fig.2.6. The symmetry of the first

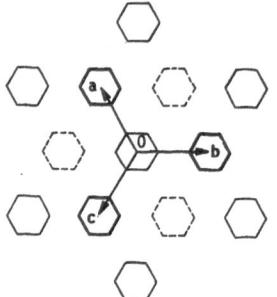

Fig.2.6. Arrangement of molecules in the low-temperature modification of the hexamethylbenzene crystal
[2.38] (molecules belonging to three successive layers
are shown; the basis vectors **a**, **b** and **c** connect molecules located in neighbouring layers)

excited state is B_{2u}, just as in benzene. The transition is forbidden in
the molecule but is allowed in the crystal. The exciton absorption spectrum
consists of a single band $\Omega = 35134$ cm^{-1} (T = 2 K). The transition in the crystal
is weak. Therefore, the exciton band is the result of short-range interactions.
The interplanar transfer integrals M_a, M_b and M_c are all equal to -3.3 cm^{-1}
and the intraplanar ones of the type, M_{a-b} (-0.4 cm^{-1}) are the largest.
The band width is approximately equal to 40 cm^{-1}. The calculated dispersion
law $\varepsilon(\mathbf{k})$ in the Brillouin zone and the calculated density of states $\rho(\varepsilon)$
(assuming that only interactions between pairs of nearest neighbours are
important) are illustrated in Fig.2.7. The spectrum is given with respect
to the electronic term $E_\rho = 35156$ cm^{-1}.

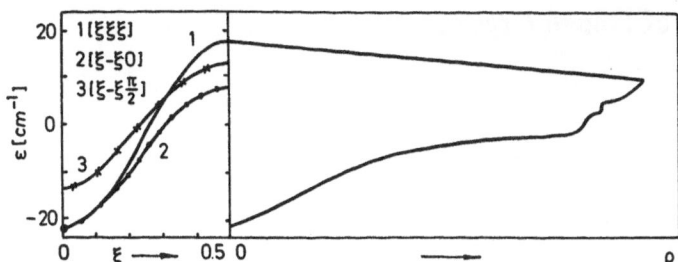

Fig.2.7. Dispersion law and density of states in the exciton band of the hexamethylbenzene crystal [2.38] (the solid circle at the zone centre denotes the position of the centre of gravity of the K_0 absorption band)

3. Exciton Spectra of Doped Crystals

The spectra of the local excitons that appear in crystals which contain isolated impurities or defects are substantially affected by the same collective
interactions that are responsible for the exciton spectra of perfect crystals.
This is particularly true for shallow local excitons, whose levels are close
to the exciton band. We shall develop a theory for the phenomena associated
with the manifestation of the collective interactions and analyze the possibilities of obtaining information on the structure of the exciton bands of
perfect crystals from the spectra of local excitons. The spectra of local
excitons that exist near the various impurity centres (isotopic, aggregate
and defect centres) are considered in detail with reference to the benzene
and naphthalene crystals.

3.1 Imperfect Crystals

This and the following chapters will consider in detail the present state
of the art of studying the exciton spectra of imperfect molecular crystals.

Imperfect molecular crystals can be divided into two groups. The first
group consists of crystals that contain isolated impurity centres. Such centres
may be made up of isolated guest molecules, aggregates of guest molecules
(dimers, trimers, etc.), or of individual lattice defects. The centres may
possess a complicated internal structure, because the deformation of the molecules around a guest molecule or a defect is often substantial. Excitons that
are bound to isolated centres will be called *local excitons*.

The second group consists of *mixed crystals*, i.e., binary or more complicated systems whose components have comparable concentrations. We shall actually deal with components that only differ in their isotopic composition. They
exhibit an isomorphic structure, form a continuous series of crystalline solutions, and therefore can be studied over a wide range of concentrations.

The spectra of imperfect crystals occupy a special, extremely important
niche in the spectroscopy of molecular excitons. The reason for this is that

the optical spectra of imperfect crystals are one of the main sources of information on the energy spectrum of excitons in perfect crystals. Indeed, as will be shown later on in this chapter, the spectra of local excitons enables one to determine the positions of the edges of the exciton bands, the distribution of the density of states in these bands, and the approximate form of the exciton dispersion law. The spectra of mixed crystals in the high concentration range are the most reliable sources of information on the genesis of exciton bands.

The exceptional possibilities offered by the spectroscopy of imperfect molecular crystals for the investigation of the general properties of disordered systems are of interest to solid-state physicists because crystals that are made up of large organic molecules and simultaneously possess structural narrow-band spectra are absolutely unique physical objects from the purely experimental point of view. Indeed, the substitution of individual atoms, primarily their isotopic substitution, makes it possible to slightly shift the energy levels of a molecule, while the other parameters are practically left unaltered. The replacement of a distinct number of nonequivalent atoms allows one to practically continuously shift the energy levels of a molecule and to study the optical spectra of such systems. The availability of such detailed experimental data provides excellent opportunities for comparing theory with experiment and for appraising various models and approximations. This also applies to mixed molecular crystals with comparable concentrations of components. Their absorption spectra also clearly exhibit the general features of spectra which are known now as one- and two-mode (and which were apparently first studied in detail just in the spectra of molecular excitons), the existence of the cluster structure, etc. Various aspects of the spectroscopy of molecular excitons in imperfect crystals have been considered in a number of reviews [3.1-8].

This chapter discusses the theory and the experimental data on the spectra of imperfect crystals in the low-impurity concentration range, i.e., when it is possible to recognize individual impurity centres. Such crystals will be called *doped crystals* (Sect.1.5). We shall start our discussion with a consideration of isotopic impurity centres, for which the theory is particularly simple and straightforward and considerable experimental data exist. Then, we shall consider the spectra of aggregate centres which are constructed from several isotopic impurities. Finally, the spectra of the defect centres caused by the introduction of chemical impurities are also considered. Certain features of the spectra of these systems have already been briefly discussed in Sect.1.5.

Within the framework of sufficiently general models the theory of local excitons that arise near isolated centres is formulated in terms of degenerate (or local) perturbations. This theory was first developed with reference to solid-state theory by LIFSHITZ [3.9] and then by KOSTER and SLATER [3.10]. It has been widely used by various authors [3.11].

3.2 The Theory of the Absorption of Light by Local Excitons: Isotopic Impurities

3.2.1 General

Individual impurity centres are isolated in dilute crystalline solutions and can be considered independently. If the impurity level is located far from the exciton band, the electronic excitation of the impurity is practically entirely concentrated on the guest molecule. Such a system corresponds to an oriented gas. As the impurity level approaches the exciton band, a quasi-resonance takes place between a guest molecule and the hosts. The quasi-resonance becomes important when the distance between the levels of the guest and host molecules is of the order of the width of the exciton band. Under these conditions the excitation is no longer concentrated on the guest molecule, but is extended to the neighbouring host molecules. Thus, a *partial delocalization of the excitation* can be observed. The most distinctive manifestations of the delocalization are the gigantic changes in the intensity and the polarization of the absorption which are observed for a wide range of positions of the impurity levels [3.12,13]. The experimental manifestations of the excitation delocalization indicate the width of the exciton energy band and the position of its edges.

In this section we consider isotopic guest molecules as the simplest objects which reveal the basic regularities most clearly.

3.2.2 Isotopic Substitution Model

Generally the *ideal isotopic substitution model* is used to describe isotopic guest molecules. Isotopic substitution changes the energy of the electronic excitation of a molecule by Δ_{ex} (Sect.1.4). In this and the following chapter this isotopic shift is simply denoted by Δ. All of the matrix elements $M_{n\alpha m\beta}$, which represent the interactions between the molecules, and the matrix elements of the dipole moment $\mathbf{p}_\alpha$ or the current $\mathbf{j}_\alpha$, which represent the interactions with the electromagnetic wave, are not changed upon isotopic substitution. The introduction of an isotopic guest molecule does not deform the neighbouring host molecules. Some of the experimental facts which help to determine the accuracy of this model, in particular, with respect to the smallness of the relative change in $M_{n\alpha m\beta}$ and p_α, were given in Sect.1.4. Other facts will be introduced later on in this chapter. The magnitude of the isotopic shift Δ is small compared to the distance E_e between successive molecular electronic levels. However, in the exciton dynamics the isotopic shift does not compete with E_e, but does compete with the *half-width of the exciton band $\mathcal{M}$*. In molecular crystals this quantity is usually small compared to E_e. Section 3.3 shows that Δ and $\mathcal{M}$ are of a comparable order of magnitude in a number of crystals.

In the adopted model, the introduction of an isotopic guest to replace one of the host molecules (at the site $n_0\alpha_0$) only changes one matrix element in the Hamiltonian of the crystal. The perturbation operator has the following form:

$$(H_{imp})_{n\alpha m\beta} = \Delta\delta_{n\alpha n_0\alpha_0}\delta_{n_0\alpha_0 m\beta} \quad .$$
(3.1)

The molecular levels have been assumed to be nondegenerate.

3.2.3 Conductivity Tensor and Green's Function of a Doped Crystal

The absorption of electromagnetic radiation in doped crystals is best described by the conductivity tensor. The matrix elements of the current operator $j_{n\alpha}$ are generally different for the guest and the host molecules. Therefore, we obtain the following equation instead of (2.43) for the conductivity tensor of a doped crystal:

$$\sigma(\Omega K) = -\frac{1}{\Omega}\frac{1}{vN}\,\text{Im}\left\{\sum_{n\alpha m\beta} G_{n\alpha m\beta}(\Omega)\exp\left[-iK(n_\alpha - m_\beta)\right]\widehat{j_{n\alpha}j^*_{m\beta}}\right\} \quad ,$$
(3.2)

where $G_{n\alpha m\beta}(\Omega)$ is the Green's function of the exciton in the site representation, defined by (2.25) and (2.26). The role of long-range interactions will be considered in Sect.3.2.8.

The matrix elements of the current $j_{n\alpha}$ remain unaltered in the adopted model of isotopic substitution, and therefore $j_{n\alpha}$ is always equal to j_α. As a result, the following general equation is obtained from (3.2):

$$\sigma(\Omega K) = -\frac{1}{\Omega v}\,\text{Im}\{\sum_{\alpha\beta} G_{\alpha\beta}(\Omega K)\widehat{j_\alpha j^*_\beta}\} \quad ,$$
(3.3)

where $G_{\alpha\beta}(\Omega K)$ is the Green's function in the momentum representation. Using (2.44), this expression can be transformed to

$$\sigma(\Omega K) = -\frac{1}{\Omega v}\,\text{Im}\left\{\sum_{\mu\mu'} G_{\mu\mu'}(\Omega K)\widehat{j_\mu j^*_{\mu'}}\right\} \quad ,$$

$$G_{\mu\mu'} = \sum_{\alpha\alpha'} B^{\mu*}_\alpha G_{\alpha\alpha'} B^{\mu'}_{\alpha'} \quad ,$$
(3.4)

where μ and μ' represent the exciton bands.

43

By an appropriate choice of the phase of the wave funcitons one ensure that the products $j_\alpha j_\beta^*$ are real. Hence, the symbol Im in (3.3) only refers to G, i.e., σ is expressed in terms of the spectral function S = - Im {G}:

$$\sigma(\Omega K) = \frac{1}{\Omega v} \sum_{\alpha\beta} S_{\alpha\beta}(\Omega K)\widehat{j_\alpha j_\beta^*} = \frac{1}{\Omega v} \sum_{\mu\mu'} S_{\mu\mu'}(\Omega K)\widehat{j_\mu j_{\mu'}^*} \quad . \qquad (3.5)$$

According to (3.2), the conductivity tensor of a doped crystal (quite analogous to that of a perfect crystal) can be directly expressed in terms of exciton Green's functions. It is precisely in the language of Green's functions that the mathematical techniques of the local perturbation theory, which are briefly described in Appendix B, are most conveniently formulated. Upon replacement of one of the host molecules by a guest molecule, only those matrix elements of the Hamiltonian in the site representation that appear in a limited number of rows and columns (those whose indices either correspond to the guest molecule itself or to its neighbours) usually change substantially. Such perturbations are called degenerate; in solid-state physics they are also often called local. The simplest perturbation of this kind is described by (3.1).

For an arbitrarily degenerate perturbation, regardless of its magnitude, the Green's function of the perturbed Hamiltonian can be expressed in terms of exact closed equations using the Green's function of the unperturbed Hamiltonian. This means that the Green's function of a crystal with an isolated impurity centre can be expressed in terms of the Green's function of the perfect crystal. Appendix B contains a general expression (B.7) for the Green's function of a crystal with an arbitrarily degenerate perturbation.

Let us use a simple expression (B.18) for G as applied to the perturbation given in (3.1). Using the notation of this section this expression has the following form:

$$G_{\mathbf{n}\alpha m\beta}(\Omega) = G^0_{\mathbf{n}\alpha m\beta}(\Omega) + \frac{\Delta}{1 - \Delta G_0(\Omega)} \, G^0_{\mathbf{n}\alpha \mathbf{n}_0\alpha_0}(\Omega)G^0_{\mathbf{n}_0\alpha_0 m\beta}(\Omega) \quad . \qquad (3.6)$$

The function $G_0(\Omega)$ is a diagonal element of the matrix $G^0_{\mathbf{n}\alpha m\beta}$. In crystals with symmetrically dependent sublattices it is defined by (2.29).

Both the width of the exciton spectrum and the isotopic shift are small compared to the electronic term E_ρ, which represents the centre of gravity of the exciton band. Therefore, it is convenient, as in Sect.2.2, to use E_ρ for the origin of the energy scale and to pass on to the frequency $\omega = \Omega - E_\rho$ and the energy $\varepsilon = E - E_\rho$.

3.2.4 Energy Spectrum

If the isotopic shift Δ is large enough, a discrete level arises for a local exciton. The position of the discrete level is determined by the isolated pole of the Green's function. Therefore, the equation to be used for finding the position of the impurity level is

$$G_0(\varepsilon_i) = \frac{1}{\Delta} \quad , \tag{3.7}$$

or in an explicit form

$$\Delta \int \frac{\rho(\varepsilon)}{\varepsilon_i - \varepsilon} d\varepsilon = 1 \quad , \tag{3.8}$$

where $\rho(\varepsilon)$ is the density of states in the exciton spectrum (2.32). In (3.8), an integration with respect to ε is performed over the entire spectrum of the perfect crystal $[\rho(\varepsilon) \neq 0]$. Conversely, ε_i is outside the spectrum of the perfect crystal in the region $\rho(\varepsilon) = 0$.

From (3.8) it can be seen that the positions of the levels of local excitons in crystals with symmetrically dependent sublattices are entirely determined by the density of states $\rho(\varepsilon)$. They neither depend on the detailed form of the dispersion law $\varepsilon_\mu(\mathbf{k})$ nor on the coefficients $B_\alpha(\mu k)$. Therefore, in principle, (3.8) can be regarded as the equation for finding $\rho(\varepsilon)$ from the experimentally determined dependence of ε_i on Δ. Since (3.8) is an integral equation of the first kind with respect to $\rho(\varepsilon)$, it can only be expected to yield the general shape of the curve $\rho(\varepsilon)$, but not details of it.

For crystals of the benzene- and naphthalene-type the exciton spectrum occupies a single continuous section on the energy axis ε (Sect.2.5). Outside of this region $G_0(\omega)$ is real and its modulus decreases monotonically. Therefore, (3.7) has not more than one solution. For $|\omega| \gg \mathcal{M}$ the following expression is valid:

$$G_0(\omega) = \frac{1}{\omega} [1 + \mu_2/\omega^2 + \ldots] \quad , \quad \mu_2 = \int \varepsilon^2 \rho(\varepsilon) d\varepsilon \quad ; \tag{3.9}$$

here the coefficient μ_2 represents the second moment of the spectrum. Therefore, if $|\Delta| \gtrsim \mathcal{M}$, then

$$\varepsilon_i = \Delta(1 + \mu_2/\Delta^2 + \ldots) \quad . \tag{3.10}$$

The first term in this expansion is taken into account by the oriented gas model. Since the second term in parentheses is positive, $|\varepsilon_i|$ will always be greater than $|\Delta|$, i.e., the local level will be "repelled" by the exciton band. The *repulsion of the impurity levels* reflects their closeness to the exciton band and may be used as a source of information on the structure of the exciton band.

The situation is somewhat more complicated for small values of Δ. In this case, the determining factor is the behaviour of G_0 near the edges of the exciton spectrum, which itself depends on the behaviour of $\rho(\varepsilon)$. To be specific, we shall consider the case in which $\Delta < 0$, i.e., when the levels split off the exciton band downwards.

The behaviour of $\rho(\varepsilon)$ near the edges of the spectrum essentially depends on two factors: the dimensionality of the system and the magnitude of the long-range dipole interactions. This leads to a non-analytical behaviour in the region of small momenta $\mathbf{k} \approx 0$ (Sect.2.2).

If the second factor is neglected, then, near the bottom of the exciton band, $\varepsilon_\mu(\mathbf{k})$ is a quadratic form of the components of the wave vector $\mathbf{k}$. In three-dimensional systems, $\rho(\varepsilon) \propto (\varepsilon - \varepsilon_{min})^{1/2}$. However, one often encounters situations in which an interaction is either dominant within certain planes or along a definite direction. The exciton bands in such crystals are quasi two dimensional or quasi one dimensional, and the density of states near their bottom either approaches a constant value or diverges as $\rho(\varepsilon) \propto (\varepsilon - \varepsilon_{min})^{-1/2}$. Elementary calculations show that the behaviour of $G_0(\omega)$ near the edge of the spectrum (when $\omega < \varepsilon_{min}$) follows the laws (3.11-13):

$$G_0(\omega) \approx -a + b\sqrt{\varepsilon_{min} - \omega} \quad , \quad a \sim 1/\mathcal{M} \quad , \quad b \sim 1/\mathcal{M}^{3/2} \quad ; \tag{3.11}$$

$$G_0(\omega) \approx -a \ln [\mathcal{M}/(\varepsilon_{min} - \omega)] \quad , a \sim 1/\mathcal{M} \quad ; \tag{3.12}$$

$$G_0(\omega) \approx -a/\sqrt{\varepsilon_{min} - \omega} \quad , \quad a \sim 1/\mathcal{M} \quad ; \tag{3.13}$$

for the three-, two- and one-dimensional cases, respectively. In all cases, $a, b > 0$.

In the last two cases $G_0(\omega)$ diverges as $\omega \to \varepsilon_{min}$, so that the level is observed at an arbitrarily small value of $|\Delta|$ [3.14]. The binding energy is of the order of

$$\varepsilon_{bind}^{1-dim} \sim \Delta^2/\mathcal{M} \quad \text{and} \quad \varepsilon_{bind}^{2-dim} \sim \mathcal{M} \exp(-\beta\mathcal{M}/|\Delta|) \quad ,$$

where β is a constant of the order of unity.

In three-dimensional systems the level arises at a sufficiently large value of Δ, i.e., $|\Delta| > |\Delta_{cr}|$, where $|\Delta_{cr}|$ is of the order of $\mathcal{M}$. Later on, it will be useful to know the approximate value of the coefficient that relates Δ_{cr} and $\mathcal{M}$. To obtain it, one can use the model density of states,

$$\rho(\varepsilon) = \frac{2}{\pi\mathcal{M}^2}\sqrt{\mathcal{M}^2 - \varepsilon^2} \quad , \tag{3.14}$$

which shows a correct square root behaviour at the edges of the band. The corresponding function $G_0(\omega)$, which was calculated using (2.30) and (2.32), is equal to

$$G_0(\omega) = \frac{2\omega}{\mathcal{M}^2}\left(1 - \sqrt{1 - \frac{\mathcal{M}^2}{\omega^2}}\right).$$
(3.15)

Figure 3.1 illustrates the behaviour of the functions ρ and $G_0' = \mathrm{Re}\{G_0\}$. From (3.15) it follows that $|\Delta_{cr}| = \mathcal{M}/2$. Later on, we shall use this relationship for semiquantitative estimates.

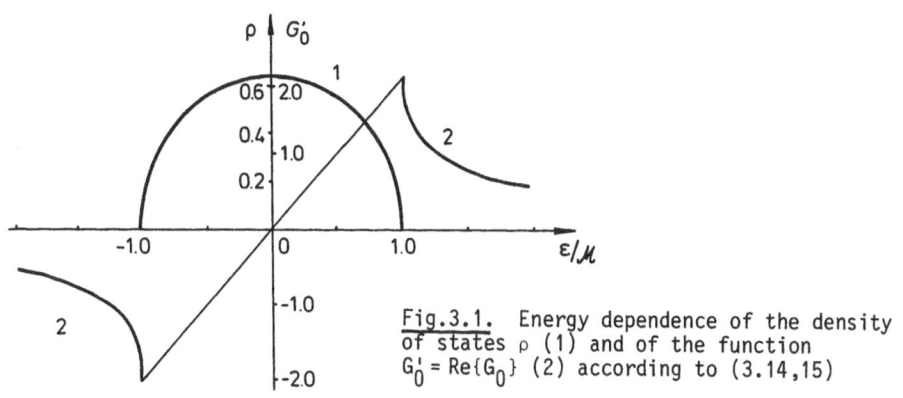

Fig.3.1. Energy dependence of the density of states ρ (1) and of the function $G_0' = \mathrm{Re}\{G_0\}$ (2) according to (3.14,15)

For a comparison, however, it is expedient to consider two additional distributions for $\rho(\varepsilon)$: i) when the density of states is predominantly concentrated near the centre of the band and is described by a δ function,

$$\rho(\varepsilon) = \delta(\varepsilon) \quad , \quad G_0(\omega) = 1/\omega \quad ;$$
(3.16)

then $|\Delta_{cr}| = \mathcal{M}$; ii) when $\rho(\varepsilon) = \text{const.}$ within the band, then $|\Delta_{cr}| = 0$ because of the two-dimensional behaviour of $\rho(\varepsilon)$ at the edges of the band.

We have considered a model with a density distribution that is symmetric relative to the centre of the band. However, it is easy to see that with a non-symmetrical distribution of $\rho(\varepsilon)$, $|\Delta_{cr}|$ is larger and the "repulsion" is smaller at the edge of the band where the density of states is lower.

For strong transitions, the discontinuity in the dispersion law $\varepsilon_\mu(\mathbf{k})$ at the point $\mathbf{k} = 0$ is large and cannot be neglected. This case will be discussed below.

3.2.5 Wave Functions

According to (B.21) and (2.29), the wave function of a local exciton is determined by the following equation:

$$\psi_i(\mathbf{n}\alpha) = G^0_{\mathbf{n}\alpha\mathbf{n}_0\alpha_0}(\varepsilon_i) \Big/ \left| \frac{dG_0}{d\varepsilon_i} \right|^{1/2} . \qquad (3.17)$$

As usual, the magnitude of $\psi_i(\mathbf{n}\alpha)$ determines the amplitude of the excitation on the site $\mathbf{n}\alpha$. Later on, the amplitude of the excitation for the guest molecule,

$$a^2 \equiv \psi_i^2(\mathbf{n}_0\alpha_0) = \frac{1}{\Delta^2} \left| \frac{dG_0(\varepsilon_i)}{d\varepsilon_i} \right|^{-1} = \left| \frac{d}{d\varepsilon_i} \frac{1}{G_0(\varepsilon_i)} \right|^{-1} = \frac{d\varepsilon_i}{d\Delta} , \qquad (3.18)$$

will be of particular importance.

At a large value of $|\Delta|$ the probability a^2 of the excitation of a guest molecule approaches unity, which corresponds to the localization of the exciton on the guest molecule. As $|\Delta|$ decreases, so does a^2, which corresponds to a progressing delocalization of the excitation. For the model density given in (3.14) the dependence of a^2 on ε_i has the following form,

$$a^2 = 2 \sqrt{\varepsilon_i^2 - 1} \left(|\varepsilon_i| - \sqrt{\varepsilon_i^2 - 1} \right) , \qquad (3.19)$$

and is shown in Fig.3.2. An appreciable drop in a^2 only occurs in the immediate vicinity of the exciton band. The difference between a^2 and 1 allows one to determine the degree of excitation delocalization. Nevertheless, the value of a^2 is only weakly dependent on the degree of delocalization due to two reasons: i) the expansion in (3.9) begins with a quadratic term, which rapidly decreases over large distances; ii) the second moment μ_2, which appears in (3.9), is smaller than $\mathcal{M}^2$ for some numerical coefficient. For example, for the model density given in (3.14) $\mu_2 = \mathcal{M}^2/4$.

The higher moments μ_p, which determine the subsequent terms in the expansion of $G_0(\omega)$ in (3.9), are even much smaller than the corresponding

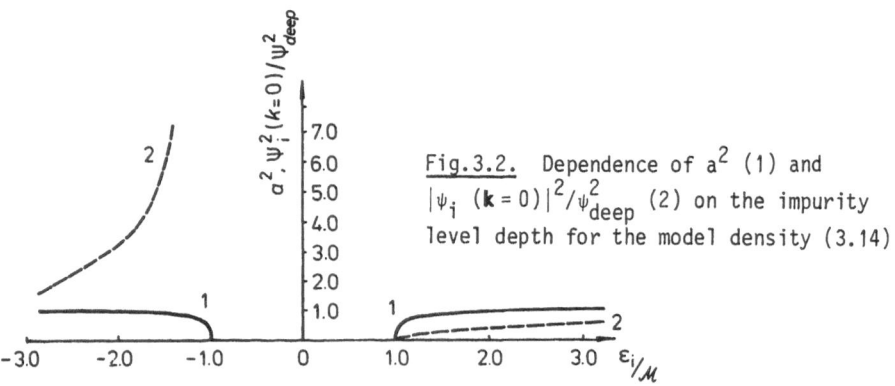

Fig.3.2. Dependence of a^2 (1) and $|\psi_i(\mathbf{k}=0)|^2/\psi_{deep}^2$ (2) on the impurity level depth for the model density (3.14)

powers of $\mathcal{M}$. Thus, only in the immediate vicinity of the exciton band the equality $dG_0/d\omega \approx -1/\Delta^2$ is not valid. As a result, a^2 is always approximately equal to unity and is always the predominant contribution to the normalizing integral, except in the relatively narrow interval immediately following the edge of the exciton band.

Let us now transform the equations into the μk representation according to (2.8) and (3.17). The expression for $|\psi_i|^2$, averaged over the positions α_0 of the guest molecules, is

$$\overline{|\psi_i(\mu\mathbf{k})|^2} = \frac{1}{NZ} \frac{[G_\mu^0(\varepsilon_i\mathbf{k})]^2}{|dG_0/d\varepsilon_i|} \quad , \qquad G_\mu^0(\varepsilon\mathbf{k}) = [\varepsilon - \varepsilon_\mu(\mathbf{k}) + i0]^{-1} \quad . \tag{3.20}$$

For deep centres, i.e., for large values of $|\Delta|$, $|\psi_i(\mu\mathbf{k})|^2 \to 1/NZ \equiv \psi_{deep}^2$, which corresponds to the oriented gas model. If the point $\mathbf{k} = 0$ is at the bottom of the band, then the dependence of $|\psi_i(\mathbf{k}=0)|^2$ on ε_i for the model density determined by (3.14) is

$$NZ|\psi_i(\mathbf{k}=0)|^2 = \frac{|\varepsilon_i - 1|^{1/2}}{2|\varepsilon_i + 1|(|\varepsilon_i| - \sqrt{\varepsilon_i^2 - 1})} \quad . \tag{3.21}$$

A plot of this dependence is given in Fig.3.2. The delocalization of the exciton is evident much more clearly and over a much wider interval of ε_i in $|\psi_i(\mathbf{k}=0)|^2$ than in a^2. The numerator in (3.20) plays the predominant role in the strong dependence of $|\psi_i(\mathbf{k}=0)|^2$ on ε_i.

Thus, the delocalization of the exciton is reflected to varying degrees in various physical quantities.

3.2.6 Discrete Absorption Spectrum

To obtain an explicit expression for the conductivity tensor of a doped crystal, the expression for the Green's function (3.6) must be substituted into (3.2). The first term describes the absorption by a perfect crystal, while the second term describes the entire impurity absorption. Equation (3.6) is written for a single impurity centre. In calculating σ_{imp} the contribution from the second term must be multiplied by the total number of impurity centres in the crystal. It is also necessary to average over α_0, i.e., over the guest molecules that belong to different sublattices.

Using (2.13,27,44) yields

$$\sigma_{imp}(\omega\mathbf{K}) = -\frac{C}{E_\rho V} \sum_\mu \frac{\widehat{\mathbf{j}_\mu \mathbf{j}_\mu^*}}{[\omega - \varepsilon_\mu(\mathbf{K})]^2} \, \text{Im}\left\{\frac{\Delta}{1 - \Delta G_0(\omega)}\right\} \quad , \tag{3.22}$$

where C is the concentration of the impurity, i.e., the fraction of the sites occupied by guest molecules. In the first factor Ω was replaced by E_ρ.

From (3.22) it follows that the intensity distribution in crystals with symmetrically dependent sublattices, as well as the positions of the impurity bands (Sect.3.2.4), are completely determined by the density of states in the energy spectrum of the excitons. The positions of the exciton multiplet bands ε_μ ($\mathbf{k} \rightarrow 0$) only enter the resonance factor of (3.22).

The following calculation of σ_{imp} is slightly different, depending on whether the absorption is associated with the isolated impurity levels or with the continuum. In the first case the imaginary part of $G_0(\omega)$ is exclusively due to the infinitesimal term in the frequency $\omega + i0$, which is only important near the pole of G, i.e., near the frequency $\omega = \varepsilon_i$ that satisfies (3.7). Therefore,

$$\sigma_{imp}(\omega\mathbf{K}) = \frac{\pi C}{E_\rho v} \left\{ \sum_\mu \frac{\widehat{j_\mu j_\mu^*}}{[\varepsilon_i - \varepsilon_\mu(\mathbf{K})]^2} \middle/ \left| \frac{dG_0}{d\varepsilon_i} \right| \right\} \delta(\omega - \varepsilon_i) \quad . \tag{3.23}$$

This equation can be transformed by using (3.20)

$$\sigma_{imp}(\omega\mathbf{K}) = \frac{\pi C}{E_\rho v} NZ \sum_\mu \overline{|\psi_i(\mu\mathbf{K})|^2} \, \widehat{j_\mu j_\mu^*} \delta(\omega - \varepsilon_i) \quad , \tag{3.24}$$

i.e., the impurity absorption can be directly expressed in terms of the magnitude of the wave function ψ_i of the local exciton on the momentum value $\mathbf{K} \approx 0$.

The intensity of an absorption is often characterized by the oscillator strength f of a transition. For anisotropic molecules and impurity centres one must introduce oscillator strengths f_μ that correspond to the individual components of the spectrum. Therefore, it is convenient to transform (3.24) into the oscillator strength language by comparing the oscillator strength of impurity absorption f_{imp} (per guest molecule) with the oscillator strength of intrinsic absorption f (per host molecule). From a comparison of (2.47) and (3.24) it follows that

$$(f_{imp})_\mu = \frac{\overline{|\psi_i(\mu\mathbf{K})|^2}}{\psi_{deep}^2} f_\mu \quad . \tag{3.25}$$

Thus, the change in the intensity of the impurity absorption caused by a partial delocalization of the excitation assumes a particularly simple form in the oscillator strength language.

We now consider the band-intensity equations starting with the polarization ratio. According to (3.23), the conductivity ratio of the two components of the spectrum has the following form:

$$P_{\mu_1/\mu_2} = \frac{(\sigma_{imp})_{\mu_1}}{(\sigma_{imp})_{\mu_2}} = \left(\frac{\varepsilon_i - \varepsilon_{\mu_2}(\mathbf{K})}{\varepsilon_i - \varepsilon_{\mu_1}(\mathbf{K})}\right)^2 P^0_{\mu_1/\mu_2}, \quad P^0_{\mu_1/\mu_2} = \frac{|\mathbf{j}_{\mu_1}|^2}{|\mathbf{j}_{\mu_2}|^2} \quad . \tag{3.26}$$

If $|\varepsilon_i| \gg \mathcal{M}$, the quadratic factor is almost equal to unity, and the polarization ratios for the extrinsic and intrinsic absorption are identical. As ε_i approaches the edge of the exciton band, P_{μ_1/μ_2} changes drastically. If the points $\mathbf{K} \approx 0$, to which the optical transition in the perfect crystal is allowed, are located far from the edge of the exciton band, then P_{μ_1/μ_2} approaches a finite limit. If, however, ε_μ ($\mathbf{K} = 0$) is at the edge of the band, e.g., ε_{μ_1} ($\mathbf{K} = 0$) $= \varepsilon_{min}$, then P_{μ_1/μ_2} diverges as ε_i approaches ε_{min}.

Let us now consider the behaviour of the absolute intensity. When $|\varepsilon_i| \gg \mathcal{M}$, it is easy to simplifiy (3.23) by using (3.20):

$$\sigma_{imp}(\omega) \approx \frac{\pi C}{E_\rho V} \sum_\mu \widehat{\mathbf{j}_\mu \mathbf{j}_\mu^*} \delta(\omega - \varepsilon_i) \quad . \tag{3.27}$$

This equation differs from (2.47) for the conductivity tensor of the perfect crystal only in the factor C. Hence, in this limiting case corresponding to deep centres, the intrinsic and extrinsic absorption per molecule are equal, i.e., $(f_{imp})_\mu = f_\mu$.

As ε_i approaches the edge of a band in a three-dimensional crystal, it follows from (3.11) that $dG_0/d\varepsilon_i \propto |\varepsilon_{bind}|^{-1/2}$.

This increase in the denominator of (3.23) causes a decrease in σ_{imp} ($\propto |\varepsilon_{bind}|^{1/2}$) when the transition to the bottom of the band in the corresponding spectrum component is forbidden. If, however, the transition is allowed, the intensity of the absorption increases with $\varepsilon_{bind}^{-3/2}$ due to the rapid growth of the numerator. As a result, the intensity of the impurity absorption for shallow centres with $\varepsilon_{bind} \ll \mathcal{M}$ differs appreciably from that of the intrinsic absorption (per molecule).

In (3.24), the conductivity tensor is directly expressed in terms of $\psi_i(\mathbf{K} = 0)$. As a result, for the model density given in (3.14) the dependence of the intensity of the impurity bands on ε_{bind} is illustrated by the same Fig.3.2, in which ψ_i^2 ($k = 0$) is plotted.

Thus, as $\varepsilon_i \to \varepsilon_{min}$, the impurity absorption is enhanced if the transition to the bottom of the band is allowed but is suppressed if this transition is forbidden. This gigantic change in absorption as the impurity level approaches the edge of the exciton band (*the Rashba effect* [3.12]) is the most conspicuous manifestation of the delocalization of the exciton.

51

The physical mechanism of this effect is particularly clear from (3.24). As the impurity level approaches the edge of the band, the expansion of ψ_i in the wave functions of the band exciton mainly consists of functions that belong to the narrow vicinity of the bottom of the band. Therefore, if the bottom of the band is at $k \neq 0$, both ψ_i ($\mu K = 0$), and the intensity of the impurity absorption will decrease progressively. Conversely, ψ_i ($\mu K = 0$) will increase continuously if the bottom of the band is at the point $k = 0$. The rate of increase is easy to estimate for a crystal with one molecule in a unit cell. The radius of the local state $r_i \propto \varepsilon_{bind}^{-1/2}$. From the normalization condition it follows that the wave function $\psi_i \sim r_i^{-3/2}$, if r_i is expressed in units of the lattice constants. Therefore,

$$\psi_i \ (K = 0) = \frac{1}{\sqrt{N}} \sum_n \psi_i(n) \propto r_i^{-3/2} r_i^3 \propto \varepsilon_{bind}^{-3/4} \quad . \tag{3.28}$$

This quantity is proportional to the dipole moment of the coherent vibrations in a volume having the order of magnitude of r_i^3. Hence, the intensity of the absorption is proportional to $\varepsilon_{bind}^{-3/2}$, which is in full agreement with the conclusions drawn from (3.23). The gigantic increase in the intensity of the impurity absorption is caused by the inclusion of a progressively increasing number of host molecules into the total coherent excitation; a kind of an "antenna" is formed, as noted in [3.15]. The oscillator strength of the impurity absorption naturally grows at the expense of the intrinsic absorption, which is thus weakened. When $Cr_i^3 \sim 1$, the intensities of extrinsic and intrinsic absorption become comparable[1].

From the above qualitative considerations it follows that the enhancement law, $\sigma_{imp} \propto \varepsilon_{bind}^{-3/2}$, is not restricted to a specific type of interaction but is common to different local excitons in systems with a three-dimensional spectrum. The detailed suppression law depends on the specific form of the Hamiltonian of the impurity centre. The simple law for $P_{\mu_1/\mu_2} (\varepsilon_i)$ [given by (3.26)] is valid for the above model.

Throughout this section we applied the simplest approximation of the Heitler-London method, in which a single excited level of a molecule is con-

---

[1] The phenomenon described here is of quite general nature and may be observed in various types of crystals with the excitons of different nature. Indeed, it has been observed in semiconducting crystals with local Wannier-Mott excitons. It is particularly strong for them because of the large value of the parameter $\mathscr{M}/\varepsilon_{bind}$ ($\sim 10^2 - 10^3$). As a result, the oscillator strength can change by a factor as large as 10^4 [3.16]. More recently, the Rashba effect was observed by BELOUSOV et al. [3.17,18] on phonons in some inorganic molecular crystals and by YEREMENKO et al. [3.19] on magnons in antiferromagnetics.

sidered and the perturbation of the wave function of this state by the crystal field is neglected. As a result, both the contribution to the dielectric permeability of the crystal from all of the other levels and the mixing of intramolecular levels are ignored. Both of these assumptions are invalid for weak transitions (e.g., for the first exciton states of benzene and naphthalene). Therefore, the question arises as to how this affects (3.23).

The answer is that (3.23) remains sufficiently exact [3.20], provided that the quantities G_0 and $\mathbf{j}_\mu$ are expressed in terms of the exact parameters of the exciton spectrum, i.e., G_0 is expressed by (2.32) in terms of the exact density of states $\rho(\varepsilon)$ and the vectors $\mathbf{j}_\mu$ are determined from the dielectric permeability dispersion of the perfect crystal near the exciton resonances. If the spatial dispersion effects are insignificant, $\mathbf{j}_\mu$ can also be determined from the exciton absorption intensity. If (3.23) is interpreted in this way, it becomes possible to determine the impurity absorption spectrum from the energy and optical spectra of the excitons in a perfect crystal.

3.2.7 T Matrix

The positions and the intensities of the impurity bands can conveniently be discussed using a somewhat different language. The T matrix concept is closely related to the Green's function and shall be used later in Chap.4. In the most general case these quantities are related by the following expression:

$$G = G^0 + G^0 T G^0 \quad . \tag{3.29}$$

Isolated poles of the T matrix correspond to discrete spectrum levels. For degenerate perturbations of the general type, the T matrix is determined by (B.16). For the simplest isotopic perturbation of the type given in (3.1) and using (3.6), the T matrix is given by

$$T_{\mathbf{n}\alpha\mathbf{m}\beta}(\omega) = T(\omega)\delta_{\mathbf{n}\alpha\mathbf{n}_0\alpha_0}\delta_{\mathbf{n}_0\alpha_0\mathbf{m}\beta} \quad , \quad T(\omega) = \frac{\Delta}{1 - \Delta G_0(\omega)} \quad . \tag{3.30}$$

The properties of the impurity absorption which were established above can be easily formulated in terms of this language. Over a wide range of ε_i equal to $\mathscr{M}$ in the order of magnitude, the predominant contribution to the variation in σ_{imp} is made by the factor $(G_\mu^0)^2$. However, in the immediate vicinity of the edge of the exciton band the dependence of the residue of the T matrix on ε_i also begins to play an important role.

3.2.8 The Effect of Long-Range Interactions

The exact form of the exciton dispersion law depends on whether or not the long-range dipole-dipole interactions are included in the Hamiltonian

(Sect.2.2). The difference in the spectra of mechanical and Coulomb excitons, which is relatively small for weak transitions, becomes significant for strong transitions. As a result, the following question arises: What meaning must be attached to the disperions law $\varepsilon_\mu(\mathbf{k})$ that appears in the different factors of (3.22) and in the subsequent equations? We shall only give a ready-made answer [3.20]. The factors $G_\mu(\omega\mathbf{K}) = [\omega - \varepsilon_\mu(\mathbf{K})]^{-1}$ contain the dispersion law for mechanical excitons, which has a clear-cut meaning for long wavelengths, as noted in Sect.2.2. At the same time, the last factor in (3.22), as well as all of the expressions that include an integration over the Brillouin zone, contain the dispersion law for Coulomb excitons. A similar meaning must be given to the separate factors in (3.17,20).

To understand the changes that occur in the behaviour of the impurity absorption, it is assumed that the optical transition is polarized along one of the crystallographic axes. Then the density of states has the following form, depending on whether the point $\mathbf{K} = 0$ is at the bottom or top of the band:

$$\rho(\varepsilon) \propto (\varepsilon - \varepsilon_{min}) \quad \text{or} \quad \rho(\varepsilon) \propto (\varepsilon_{max} - \varepsilon)^{3/2} \quad , \qquad (3.31)$$

respectively. These laws correspond to "four-" [3.21] and "five-dimensional" behaviour, respectively. In both cases $G_0(\omega)$ is finite at the edges of the band. Therefore, there exists a finite $|\Delta_{cr}|$. For a four-dimensional $\rho(\varepsilon)$, an enhancement or suppression of absorption takes place according to the following laws:

$$\sigma_{imp} \propto [\varepsilon_{bind}^2 \ln (\mathcal{M}/\varepsilon_{bind})]^{-1} \quad \text{and} \quad \sigma_{imp} \propto [\ln (\mathcal{M}/\varepsilon_{bind})]^{-1} \quad . \quad (3.32)$$

For a five-dimensional $\rho(\varepsilon)$, the absorption approaches a finite limit in all of the components of the spectrum.

Thus, for strong transitions the shape of the law governing the variation in impurity absorptions depends on whether the band is approached from the top or bottom.

3.2.9 Induced Absorption in the Exciton Continuum

Whether or not a discrete level for a local exciton arises, the perturbing potential of the impurity always lifts the selection rule with respect to the momentum and induces absorption over the entire exciton continuum [3.22-24]. This absorption is described by (3.22). But, it must be remembered that within the exciton band $G_0(\omega) = G_0'(\omega) - \pi i \rho(\omega)$ is complex as it is seen from from (2.30-32). From (3.22) it follows that

$$\sigma_{ind}(\omega\mathbf{K}) = \frac{\pi C}{E_\rho v} \frac{\rho(\omega)}{[(1/\Delta) - G_0'(\omega)]^2 + [\pi\rho(\omega)]^2} \sum_\mu \frac{\widehat{\mathbf{j}_\mu \mathbf{j}_\mu^*}}{[\omega - \varepsilon_\mu(\mathbf{K})]^2} \quad . \qquad (3.33)$$

The intensity of the induced absorption is particularly large near the exciton absorption bands because of the last factor. The resulting divergence is cut off by the finite free path of excitons caused by their scattering. The spectral distribution of absorption is also determined by the second factor in (3.33) and thus essentially depends on $\rho(\varepsilon)$ and the magnitude of Δ. For example, when Δ is almost equal to Δ_{cr}, i.e., at the edge of the exciton band where $G_0' \approx 1/\Delta$, an absorption due to a virtual level may arise [3.25].

It can be shown that (3.33) which may be considered as an integral equation with respect to $\rho(\omega)$, can be solved exactly (Sect.6.4) and that the function $\rho(\omega)$ can be explicitly expressed in terms of σ_{ind} ($\omega\mathbf{K}$). Therefore, from a purely mathematical point of view, such a method of finding $\rho(\omega)$ through experimental data appears to be very enticing. However, it is difficult to experimentally isolate the impurity component from the background of the exciton-phonon absorption, which is inevitably present in the energy range corresponding to the entire exciton band even in pure crystals. In addition, (3.33) is only valid for a narrow range of low concentrations and rapidly loses its accuracy with increasing concentration. So far σ_{ind} could not be used to determine ρ.

The intensity redistribution between the exciton bands, the impurity bands, and the exciton continuum is outlined in Fig.3.3 for the case in which the exciton continuum lies between the two bands of the exciton doublet. The arrows indicate the redistribution of intensity among the different parts of the spectrum. As the impurity absorption band approaches the edge of the exciton spectrum, its intensity increases in the component $\mu = 1$, where the transition to this edge is allowed, and decreases in the component $\mu = 2$, where the transition to this edge is forbidden.

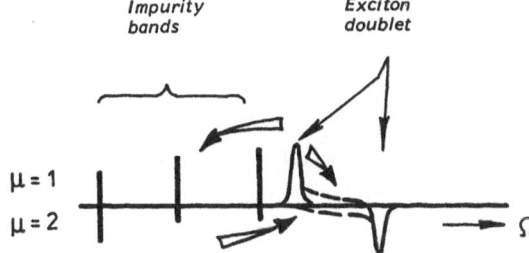

Fig.3.3. Intensity redistribution in the absorption spectrum of a doped crystal

3.2.10 Alternative Approach to Impurity Absorption

The interaction of excitons with an electromagnetic wave leads to the formation of mixed modes, polaritons or light-excitons, which simultaneously contain the energy of the electromagnetic field and of the crystal excitation.

It is the polaritons that are the undamped elementary excitations in the perfect crystal. This problem has already been discussed in Sect.2.4.

If a crystal contains impurities, the polaritons will be scattered by them. If there are no inelastic mechanisms that take place, the scattering only changes the momentum, not the frequency of the polaritons. Following the classical work of PEIERLS [3.26], it can be said that under such conditions the crystal does not behave as an "absorber" (because there is no dissipation of the electromagnetic energy, i.e., it is not converted to heat), but as an ideal "scatterer". The interaction of polaritons with impurities has been discussed by AGRANOVICH [3.27] and others.

Under this approach, the absorption of light by impurities does not exist. Only the elastic scattering of polaritons on impurities takes place. The attenuation of the wave as a result of its scattering is interpreted experimentally as absorption. The scattering cross section is strongly dependent on the frequency. For example, when $\omega \approx \varepsilon_i$ it increases enormously. This corresponds to a resonance scattering of the polaritons by the impurities. If the approach of the impurity level to the edge of the exciton band is accompanied by a progressing increase in the intensity of the impurity band, the coupling of the electronic excitation to the electromagnetic field also increases. As a result, the polariton effects increase even more so.

The scattering of excitons by impurities under such conditions was investigated by SUGAKOV [3.28] and HOPFIELD [3.29]. The obtained results agree with those obtained using the semiclassical approach given above in this section. The width of the impurity scattering peak near $\omega \approx \varepsilon_i$ coincides with the radiative width of the level for the local exciton. A radiative shift in the level also takes place. Scattering occurs at all frequencies, but other than at resonance its intensity is low. In the region of the exciton continuum, taking into account the polariton effects automatically removes the divergence in the induced conductivity near the exciton resonances. Otherwise (3.33) is valid as before.

An appreciable difference between the results of semiclassical considerations and the calculations based on the polariton model could only arise for strong exciton transitions, for which the polariton effects exceed the inevitable side effects associated with the exciton-phonon interaction. Unfortunately, at present practically no experimental data on the shallow levels of local excitons are available for such crystals.

A number of questions concerning the theory of the absorption of light by local excitons are also considered in [3.30-33].

3.3 Exciton Spectra of an Isotopic Impurity

3.3.1 General Characteristics of Monomer Centres

Only two systems of isotopically doped crystals (IDC), namely the mixed crystals of d-benzenes and d-naphthalenes, have been thoroughly investigated. Attempts to study IDC of d-anthracenes have also been made. It was found, however, that the isotopic impurities by which the crystals are doped do not produce a discrete spectrum. Another type of isotopic substitution in aromatic molecules is associated with the replacement of ^{12}C by ^{13}C. Only a few such investigations have been made, and they have not yet yielded any new physical results as compared with deuterosubstitution [3.34,35].

The greatest experimental attraction of the systems of d-isotopes is the possibility of controlling the magnitude of the paramter $\Delta/\mathcal{M}$ by using different, partially substituted molecules. A number of isotopically doped molecules of benzene and naphthalene is available. This makes it possible to produce IDC in which the parameter $\Delta/\mathcal{M}$ can be changed more or less uniformly over a wide range. For IDC of d-naphthalenes, changes in $\Delta/\mathcal{M}$ can be achieved from -1.3 to 1.3 with an average spacing of 0.17. For IDC of d-benzenes this range stretches from -7 to 7 with an average spacing of about 1. With respect to the ranges in the ratio $\Delta/\mathcal{M}$ and, above all, with respect to the magnitude of the spacing average, the IDC of d-naphthalenes are a convenient system for observing the levels of shallow local excitons. In the IDC of d-benzenes the levels of the local excitons are deeper. Therefore, it is no wonder that the IDC of d-naphthalenes first revealed the peculiarities of impurity absorption caused by the Rashba effect [3.13].

Let us consider the quantitative characteristics of these spectra using the model described in the preceding section. We shall compare the calculated and experimental values of: a) the position ε_i of the K_0' band of the impurity absorption (Fig.1.7), b) the ratio of the intensities of the K_0' band, $P_{\mu_1/\mu_2}(\varepsilon_i)$, using two different polarizations of the incident light (μ_1 and μ_2), c) the absolute intensity of the K_0' band, $\sigma_{imp}(\varepsilon_i)$, d) the intensity of the absorption induced in the exciton continuum, $\sigma_{ind}(\omega)$, and e) the amplitude of the excitation for the guest molecule, $a(\varepsilon_i)$.

Before we proceed to the experimental results it is necessary to determine to what degree the assumptions lying at the basis of the model for IDC are fulfilled in the actual experiments. These assumptions are concerned with the independence of the matrix elemens $M_{n\alpha m\beta}$, which represent the interaction between the molecules, and the matrix elements of the current $\mathbf{j}_\alpha$, with respect to isotopic composition, the structural isomorphicity of the isotopic crystals, and the applicability of the low-concentration approximation.

The sensitivity of the matrix elements $M_{n\alpha m\beta}$ to the isotopic composition is reflected in the magnitude of the Davydov splittings. These values are practically constant in isotopically pure crystals of d-naphthalenes. In crystals of d-benzenes they vary within 25% (Table 3.1). The independence of the matrix elements of the current j_α with respect to isotopic composition can be checked by the integral intensity of the 00 transition. In the pure d-isotope crystals of benzene and naphthalene, the intensities for completely deuterated molecules decrease by 15-40% (Table 3.1.). These changes will be taken into consideration below. Decreasing the number of atoms that are replaced weakens the effect considerably.

Table 3.1. Isotopic effect in exciton spectra of crystals of deuterobenzenes and deuteronaphthalenes

Crystal[a]	Position of exciton bands cm^{-1}			Davydov splitting cm^{-1}	Integral intensity of K_0 absorption (arbitrary units)	Polarization ratio[d]
	A_0	B_0	C_0		e	$\mathscr{P}^0_{a/c}$
bd$_0$	37803	37846	37839	40	1.0	0.62[e]
bd$_1$	37841	37880[b]		39	0.98	
bd$_2$	37875	37912[b]		37	0.96	
bd$_3$	37911	37946[b]		35	0.87	
bd$_4$	37945	37978[b]		33	-	
bd$_5$	37982	38014[b]		32	-	
bd$_6$	38017	38050	38045[g]	30	0.63	
	A_0	B_0			f	$\mathscr{P}^0_{b/a}$
nd$_0$	31475[c]	31626[c]		151	1.0	99[f]
α-nd$_1$	31488	31634		146		
$\acute{\epsilon}$-nd$_1$	31490[c]	31642[c]		152	0.78	
α-nd$_4$	31520	31668		148		
$\acute{\epsilon}$-nd$_4$	31549	31697		148		
nd$_8$	31588[c]	31739[c]		151	0.845	75[f]

a) For symbols of isotopic molecules, see Table 1.5
b) Mean frequency of the B_0, C_0-doublet, see [3.39]
c) Band positions in free-mounted crystal specimens [3.40]
d) For the definition of $\mathscr{P}^0_{\mu_1/\mu_2}$ see Sect.3.3.3
e) [3.39]
f) [3.40,41]
g) [3.42]

X-ray and neutron-diffraction studies have not revealed any considerable differences in the structures of nd$_0$ and nd$_8$ naphthalenes [3.36-38], bd$_0$ and bd$_6$ benzenes [3.43,44], and ad$_0$ and ad$_{10}$ anthracenes [3.45,46]. These data suggest that the isotopic compounds are isomorphic and that the molecules of both the solvent and the solute are only weakly deformed upon crystallization.

Experimentally, one has to use impurity concentrations of about 0.01% to determine ε_i and within 1-5% to accurately determine the impurity-band intensities and the polarization ratios. At these concentrations new effects arise, which complicate the structure of a spectrum. A quantitative treatment of the impurity spectra, using the model that assumes that the centres are formed by single molecules, can be made because the K_0' absorption bands are the dominant bands. The low concentration of impurities suggests the absence of any substantial interaction among the centres.

3.3.2 Position of Impurity Bands

It is convenient to measure the distance of the K_0' band from some characteristic peak in the host-crystal spectrum rather than use the value of ε_i that can be determined from the position of the electronic term E_ρ (as a rule, the value of this term is difficult to determine). The origin is usually related to the exciton multiplet. Below, we shall determine the positions of the K_0' bands as distances e_i from the long-wavelength band of the multiplet. The position of this band is given by $\varepsilon = -e_\rho$, where e_ρ is the position of the term.

The function $G_0(e_i)$ is shown in Figs.3.4,5 for the IDC of d-benzenes and d-naphthalenes (the numerical values are given in Tables 3.2,3). The experimental data for $\Delta < 0$ are given in Fig.3.4b and the left-hand side of Fig.3.5, for $\Delta > 0$ in Fig.3.4a and the right-hand side of Fig.3.5, respectively.

Within the framework of the ideal isotopic substitution model, the position of the isotopic impurity level is determined using (3.8). In Fig.3.4a the solid curve represents the function $G_0(e_i)$ which was calculated [3.47] using the density of states ρ (line consisting of dots and dashes) of the host crystal of bd_0, which in turn was determined using the band-to-band vibronic-transitions method [3.48] (Sect.5.2.2). A similar calculation was performed for the systems shown in Fig.3.4b. The calculated functions given in Fig.3.4 take into account the isotopic dependence of the width of the exciton spectra of pure crystals.

Figure 3.5 shows the positions of the impurity bands of the IDC of d-naphthalenes, and the density of states ρ (line consisting of dots and dashes) found by solving the inverse problem numerically [3.49]. The density was chosen so as to obtain an optimum description of the positions of the impurity levels based on (3.8) for the entire accessible range of Δ.

The dashed lines in Figs.3.4,5 illustrate the asymptotic behaviour of $\Delta = e_i - e_\rho$, which holds for large values of $|\Delta|$ [see (3.10)] or when the func-

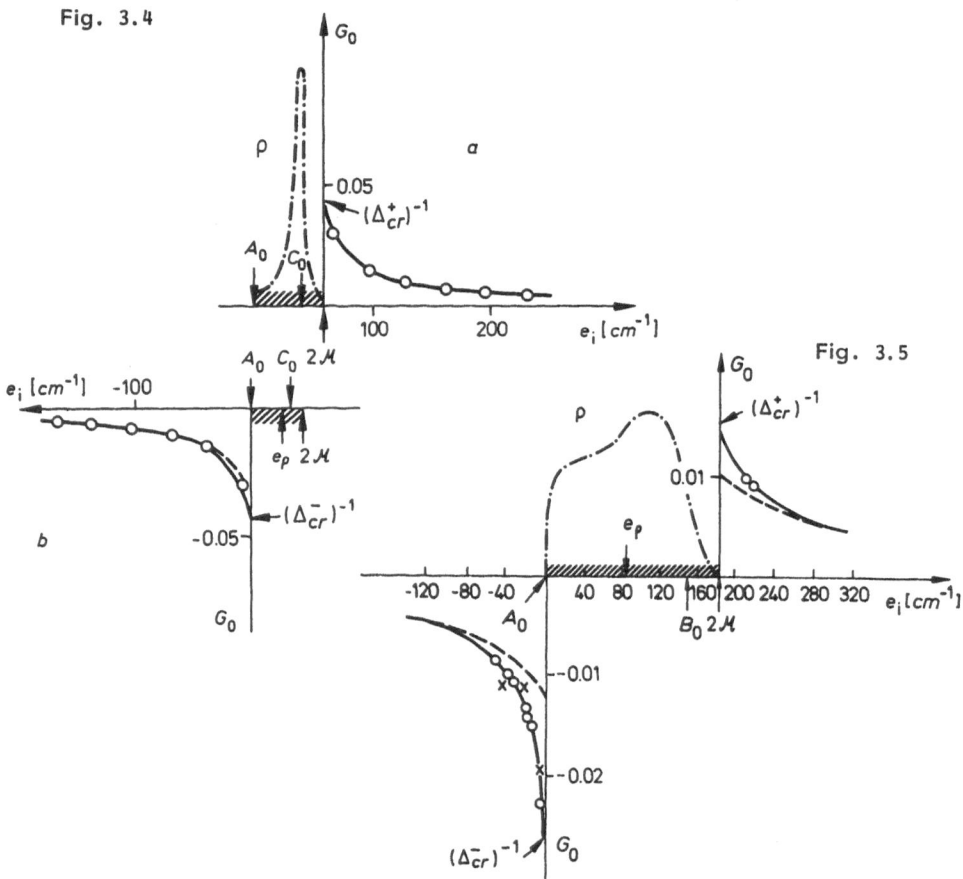

Fig. 3.4

Fig. 3.5

Fig.3.4a,b. Experimental (open circles, Table 3.2) and theoretical (solid curves) dependence between the inverted isotopic shift $G_0 = 1/\Delta$ and the distance e_i of the impurity K_0' bands from the A_0 band of the exciton multiplet in the IDC of d-benzenes [3.47]. The hatched areas correspond to the exciton energy spectra of the crystals of bd_0 and bd_6; A_0, C_0, and e_ρ are the positions of the exciton absorption bands and of the energy spectrum centre of gravity of host crystals, respectively; a) isotopic impurities in the bd_0 host; b) same for the bd_6 host

Fig.3.5. Experimental (open circles [3.50] and crosses [3.51], Table 3.3) and theoretical (solid curves) dependence between the inverted isotopic shift and the distance of the K_0' bands from the A_0 band of the exciton doublet in the IDC of d-naphthalenes. The hatched area corresponds to the exciton energy spectrum of the host crystal ($2\mathcal{M}= 180$ cm^{-1}), A_0 and B_0 are exciton absorption bands

tion ρ resembles a δ-function [see (3.16)], which in turn corresponds to the oriented gas model. The deviation of the experimental values of $1/\Delta(e_i)$ from the dashed lines is a result of the partial delocalization of the excitation.

For the IDC of d-benzenes, a slight deviation of the experimental and calculated values of $G_0(e_i)$ from the asymptotic law only occurs when $|\Delta| \lesssim 50$ cm^{-1}.

Table 3.2. Parameters of K_0^i bands in the IDC spectra of d-benzenes [3.47,52]

Guest[a]	Host[a]	Ω_i [cm^{-1}]	Δ [cm^{-1}]	e_i [cm^{-1}]	$\mathcal{P}_{a/c}$	i_a	i_c
d_0	d_6	37853	-191	-164	0.94	1.35	0.96
d_1	d_6	37884	-160	-133	0.92	1.30	0.90
d_2	d_6	37913	-131	-104	1.12	1.40	0.90
d_3	d_6	37948	-98	-69	1.38	1.80	0.90
d_4	d_6	37979	-67	-38	2.35	2.10	0.60
d_5	d_6	38000/ 38009	-32	-8			
d_6	d_0	38033	+194	+230	0.41	0.71	1.04
d_5	d_0	38001	+163	+198	0.46	0.70	0.90
d_4	d_0	38038	+133	+163	0.33	0.70	1.20
d_3	d_0	37933	+94	+130	0.29	0.40	1.10
d_2	d_0	37903	+61	+100	0.22	0.30	0.80
d_1	d_0	37874	+31	+71	0.17	0.30	1.00

[a] The letter "b" is omitted for simplicity (Table 1.5)

Due to the small total width of the exciton spectrum and the prominent sharpness of the function $\rho(\varepsilon)$, the effect of the delocalization of the impurity excitation is hardly noticeable in the energy spectrum. However, the energy spectrum of the impurity levels in the IDC of d-naphthalenes deviates considerably from $\Delta = e_i - e_\rho$. The value of the repulsion of the impurity level as the exciton band is approached from the low-energy side becomes comparable to Δ.

The experimental dependence of $\Delta(\varepsilon_i)$ on ε_i plotted in coordinates e_i and Δ enables one to determine the position of the centre of gravity of the exciton energy spectrum from the positions of the bands of "far-lying" impurities (i.e., those with large values of $|\Delta|$). This is illustrated in Fig.3.6 for the IDC of d-benzenes.

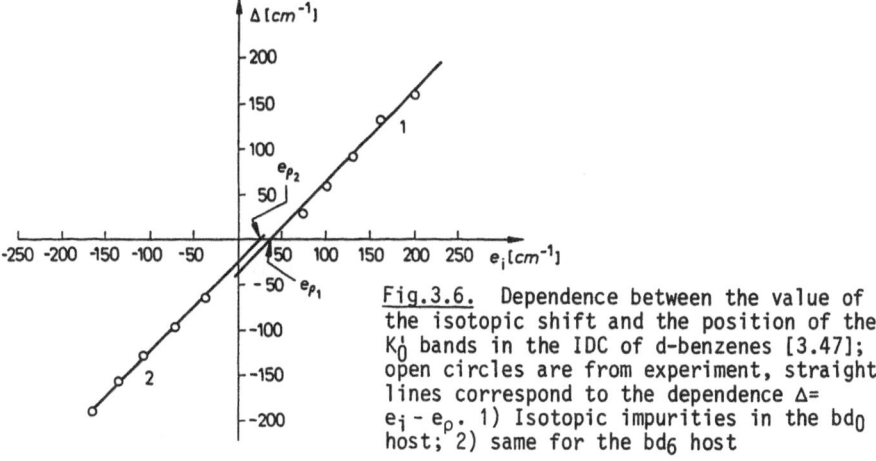

Fig.3.6. Dependence between the value of the isotopic shift and the position of the K_0^i bands in the IDC of d-benzenes [3.47]; open circles are from experiment, straight lines correspond to the dependence $\Delta = e_i - e_\rho$. 1) Isotopic impurities in the bd_0 host; 2) same for the bd_6 host

Table 3.3. Parameter of K'_0 bands in the IDC spectra for d-naphthalenes

Guest[a]	Host[a]	Ω_i [cm^{-1}]		e_i [cm^{-1}]	Δ [cm^{-1}]	$P_{b/a}$	$\mathscr{P}_{b/a}$ [b]			a^2 [b]			i_a
		[3.50]	[3.51]	[3.50]	[3.50]	[3.53]	[3.53]	[3.51]	[3.54]	[3.55]	[3.51]	[3.54]	[3.54]
d_0	d_8	31542	31541.4	-49.6	-118	5.6	3.5	3.8	4.2	0.9	0.83	0.83	0.38
α-d_1	d_8	31551	31549.8	-41.2	-107	4.2	2.6	2.5	3.1	0.7	0.77	0.81	0.52
β-d_1	d_8	31558	31556.1	-34.9	-100	-	-	2.9	-	0.6	0.75	-	-
α-d_2	d_8	-	31557.5	-33.5	-99	-	-	1.8	2.1	-	0.75	0.79	0.54
α-d_3	d_8	-	31565	-26	-88	-	-	1.1	-	-	-	-	-
d_0	β-d_4	31533	31533	-17	-72	2.0	1.25	-	-	-	-	-	-
α-d_4	d_8	31572	31572	-19	-74	-	-	0.83	-	0.5	0.59	-	-
$\alpha\beta$-d_6	d_8	-	-	-18	-74	-	-	-	0.97	-	-	0.59	0.86
β-d_4	d_8	31535	-	-15	-61	1.7	1.05	-	-	-	-	-	-
$\alpha\beta$-d_5	d_8	-	31583	-8	-55	-	-	-	-	-	-	-	-
d_0	α-d_4	31512	-	-8	-44	0.4	0.25	-	0.027	-	-	0.50	3.40
d_8	d_0	31685	-	209	+118	-	-	-	-	-	-	-	-
d_8	α-d_1	31695	-	207	+107	-	-	-	-	-	-	-	-

a) The letter "n" is omitted for simplicity (Table 1.5)

b) The values of the polarization ratios and of a^2 in [3.53,55] have been determined by quantitative photometering of spectra. The mean errors in the determination of $\mathscr{P}_{b/a}$ in [3.51,53,54] are 14, 18, and 10%, respectively, $P_{b/a}$ and $\mathscr{P}_{b/a}$ differ by the ratio $n_b/n_a = 1.6$

The experimental data for the IDC of d-benzenes follow the asymptotic curves because of the smallness of the ratio μ_2/Δ^2 ($\mu_2 \simeq 110$ cm^{-2}) for the entire range of $|\Delta|$ from 30 to 200 cm^{-1}.

For d-naphthalenes ($\mu_2 \simeq 1750$ cm^{-2}), the repulsion is large and is about equal to 14 cm^{-1} even for the deepest impurity ($|\Delta| = 118$ cm^{-1}). Therefore, it has been impossible to determine the position of the centre of gravity of the exciton bands of naphthalene in this way.

Important information on the density of states in the energy spectrum can be derived from the critical values Δ_{cr}^- and Δ_{cr}^+ (Fig.3.1). When $\Delta_{cr}^- < \Delta < \Delta_{cr}^+$, no discrete impurity levels arise.

For the IDC of d-benzenes, the calculated values are $\Delta_{cr}^- = -27$ cm^{-1} and $\Delta_{cr}^+ = 24$ cm^{-1}. Therefore, it is natural that impurity levels can be observed when $\Delta = \pm 30$ cm^{-1} (which corresponds to the substitution of one deuterium or one hydrogen atom in the molecule). However, when one or two ^{12}C atoms of the benzene molecule are replaced by ^{13}C, no impurity bands are visible in the spectrum. For such substitutions, the values of Δ, equal to 4 cm^{-1} and 8 cm^{-1}, respectively, are too small.

For the IDC of d-naphthalenes, the calculated values are $\Delta_{cr}^- = -40$ cm^{-1} and $\Delta_{cr}^+ = 70$ cm^{-1}. It has been experimentally established that the discrete impurity level in the range of negative values of Δ appears at $\Delta = -44$ cm^{-1}, not at $\Delta = -38$ cm^{-1}. For positive values of Δ, the impurity level in the high-frequency region has only been observed when $\Delta \gtrless 107$ cm^{-1}. The differences between the experimental and calculated values of Δ_{cr}^+ may be attributed to the experimental difficulties of detecting the impurity band at the high-frequency edge of the intensive B_0 band of the host crystal. A broadening of the impurity band due to the phonon scattering of the local exciton to the band states is not excluded, either. It must hinder the isolation of the band in the spectrum.

For the IDC of d-benezences, $|\Delta_{cr}^\pm| \approx \mathcal{M}$. As shown in Sect.3.2.4, such a relation holds when the density of states ρ is concentrated in a narrow spectral region.

For an extended symmetrical density of states, the relation $|\Delta_{cr}^\pm| \approx \mathcal{M}/2$ must be valid. For the IDC of d-naphthalenes $\mathcal{M}/2 \approx 45$ cm^{-1}, which practically coincides with the value of $|\Delta_{cr}^-|$. Due to the asymmetry of the ρ-function $|\Delta_{cr}^+| > \mathcal{M}/2$ (Fig.3.19). In accordance with the conclusions of Sect.3.2.4, the asymmetry of the function $\rho(\varepsilon)$ explains the smaller value of $|\Delta_{cr}^-|$ and the stronger repulsion of the impurity levels at the low-frequeny edge of the spectrum. This leads to an asymmetry in the spectra of the "light" ($\Delta < 0$) and "heavy" ($\Delta > 0$) guests.

The broader the exciton energy band, the larger is the value of $|\Delta_{cr}|$.

That is why it has been impossible to detect an impurity level in the IDC of d-anthracenes even in the system ad_0 in ad_{10}, i.e., at the maximum value of $|\Delta| \approx 60$ cm^{-1}. Thus, $|\Delta_{cr}| > |\Delta_{max}|$ and the band of the anthracene crystal is broader than 240 cm^{-1}. For a width of 700 cm^{-1} found by calculation [3.56] (Sect.2.5.3), $|\Delta_{cr}|$ is about 175 cm^{-1}.

3.3.3 Polarization Ratio of the Impurity Bands

The polarization ratio of the intensities of the impurity bands, P_{μ_1/μ_2}, is a measure of the anisotropy of the impurity absorption. In the oriented gas model, P_{μ_1/μ_2} is determined by the orientation of the dipole moment of the optical transition in the guest molecule with respect to the μ_1- and the μ_2-axis of the crystal (Fig.1.2). Since the guest molecules replace the host molecules isomorphically, P_{μ_1/μ_2} in this model is equal to the polarization ratio in the spectrum of the host crystal $P^0_{\mu_1/\mu_2}$. It has already been mentioned in Sect.1.5 that one actually observes a very strong dependence of P_{μ_1/μ_2} on the distance e_i of the impurity band to the low-frequency edge of the intrinsic absorption. This dependence is a direct experimental manifestation of the delocalization of the impurity excitation.

The polarization ratios P_{μ_1/μ_2} were introduced in Sect.3.3.2 as conductivity ratios. However, one usually measures the absorption indices $\varkappa_\mu$, which are determined from the attenuation of the light wave [$\varkappa d = \ln(I_0/I)$ where d is the thickness of the crystal]. Of course, corrections must be made for reflections. The absorption indices are related to the corresponding components of σ_μ by the usual expression, $\sigma_\mu = cn_\mu \varkappa_\mu/2\pi$, where c is the velocity of light in a vacuum, and n_μ is the refraction index. For weak transitions (such as the ones in benzene or naphthalene), the dispersion of n_μ near the exciton bands is small and can thus be neglected. Then instead of P_{μ_1/μ_2} one can consider the polarization ratio $\mathscr{P}_{\mu_1/\mu_2}$:

$$\mathscr{P}_{\mu_1/\mu_2} \equiv (n_{\mu_2}/n_{\mu_1})P_{\mu_1/\mu_2} = \mathscr{K}_{\mu_1}/\mathscr{K}_{\mu_2} \quad , \tag{3.34}$$

where $\mathscr{K}_\mu = \int \varkappa_\mu(\omega)d\omega$ is the integral coefficient of the absorption for the impurity band.

Figures 3.7,8 show that the logarithm of $\mathscr{P}_{\mu_1/\mu_2}$ is almost a linear function of the logarithm of $(\varepsilon_i - \varepsilon_{\mu_2})/(\varepsilon_i - \varepsilon_{\mu_1})$ for both IDC systems. In both cases the slope of the straight line is close to 2, which corresponds to a quadratic function. Such a dependence follows from the theoretical model [see (3.26)], according to which

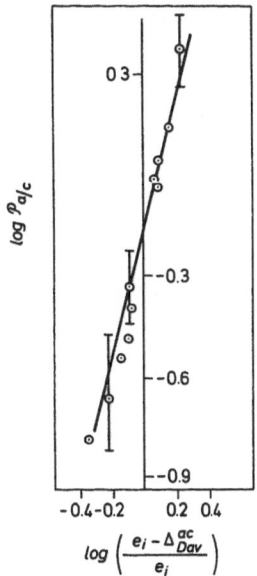 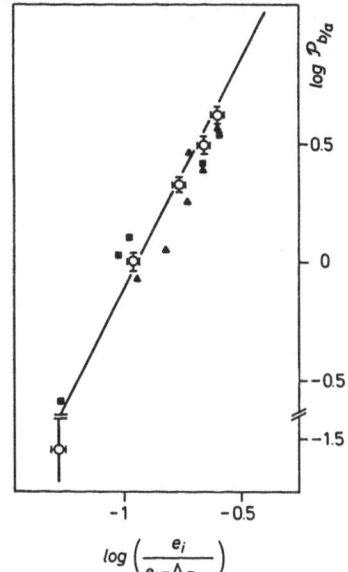

Fig.3.7. Dependence of the polarization ratio of K'_0 bands on their distance from the A_0 band of the exciton multiplet of the host crystal in the IDC of d-benzenes [3.47]

Fig.3.8. Dependence of the polarization ratio of K'_0 bands on their distances from the A_0 band of the exciton doublet of the host crystal in the IDC of d-naphthalene [3.54]. ▲- [3.51], ■- [3.53], o- [3.54]

$$\mathscr{P}_{\mu_1/\mu_2} = \mathscr{P}^0_{\mu_1/\mu_2}\left[\frac{\varepsilon_i - \varepsilon_{\mu_2}(0)}{\varepsilon_i - \varepsilon_{\mu_1}(0)}\right]^2 = \mathscr{P}^0_{\mu_1/\mu_2}\left[\frac{\varepsilon_i}{\varepsilon_i - \Delta_{Dav}}\right]^2 , \qquad (3.35)$$

where ε_{μ_2} corresponds to the low-frequency component of the exciton multiplet and $\Delta_{Dav} = \varepsilon_{\mu_1} - \varepsilon_{\mu_2}$ is the magnitude of the Davydov splitting. This relation is checked by the quadratic functionality and the numerical value of $\mathscr{P}^0_{\mu_1/\mu_2}$, which is given by the intercept of the straight line drawn through the experimental points with the ordinate axis. The values of $\mathscr{P}^0_{\mu_1/\mu_2}$ for the IDC of d-benzenes and d-naphthalenes are equal to 0.64 and 73, respectively. The former is close to the value $\mathscr{P}^0_{a/c} = 0.62$ for a bd_0 crystal (Table 3.1), while the latter agrees with the value $\mathscr{P}^0_{b/a} = 75$ for an nd_8 crystal.

The quadratic dependence in (3.35) is a result of the assumption that the matrix elements of the current j_α are the same for all of the isotopic components. If, however, one takes into consideration the appreciable differences in the absolute intensities of the exciton absorption in individual isotopic crystals (Table 3.1), one must make the appropriate corrections to (3.26) and (3.35).

Let $\mathbf{j}_\alpha$ and $\mathbf{j}_\alpha + \tilde{\mathbf{j}}_\alpha$ be the matrix elements of the current for the host and guest molecules, respectively. The components of the current along the crystallographic axes will be j_μ and $j_\mu + \tilde{j}_\mu$. Hence,

$$\mathscr{P}_{\mu_1/\mu_2} = \mathscr{P}^{quad}_{\mu_1/\mu_2} \left[\left(1 + \frac{\tilde{j}_{\mu_1}}{j_{\mu_1}} \frac{\varepsilon_i - \varepsilon_{\mu_1}}{\Delta} \right) \Big/ \left(1 + \frac{\tilde{j}_{\mu_2}}{j_{\mu_2}} \frac{\varepsilon_i - \varepsilon_{\mu_2}}{\Delta} \right) \right] \quad , \qquad (3.36)$$

where $\mathscr{P}^{quad}_{\mu_1/\mu_2}$ is described by (3.35).

If we assume that the magnitude of $j_\mu + \tilde{j}_\mu$ is the same as that of the corresponding pure crystal, then the correction factor may reach the value 0.91 for d-benzenes (IDC of bd_0 in bd_6) and 1.4 for d-naphthalenes (IDC of nd_0 in nd_8). For d-benzenes this correction leads to a 10 per cent difference in $\mathscr{P}_{\mu_1/\mu_2}$, which is less than the experimental error. The 40% deviation for d-naphthalenes exceeds the experimental error. But here it has not been observed, either. This evidently must mean that the determination of $\tilde{j}_\mu$ from the spectra of pure crystals is not always valid for weak transitions. The difference in the intensities of the electronic transitions in gaseous and solid naphthalene is large. Hence, the values of the matrix elements $\mathbf{j}_\alpha$ do not only depend on the properties of the molecule itself, but also on the strength of the crystal field. Therefore, it is quite possible that the j_μ for different isotopic molecules in the same matrix are closer in magnitude to one another than the corresponding values for pure crystals. Under such conditions the correction factor in (3.36) approaches unity. This also means that in considering the absolute intensities of the impurity bands one can neglect the isotopic change in the absorption band intensities that is observed in pure isotopic crystals.

3.3.4 The Integral Intensities of Impurity Bands

The integral intensities of impurity absorption bands have been measured in the a and c components of the IDC spectra of d-benzenes and in the $\perp \mathbf{b}$ component of the IDC spectra of d-naphthalenes [3.54]. The relative absorption intensities per molecule, $i_\mu = (\sigma_{imp})_\mu / \sigma_\mu = (f_{imp})_\mu / f_\mu$, were measured experimentally, where σ_{imp} and σ are the conductivities in the impurity and the exciton absorption bands, respectively. The values of i_μ are listed in Tables 3.2,3.

The functions $i_a(e_i)$ and $i_c(e_i)$ for the IDC of d-benzenes are shown in Fig.3.9. The left-hand side describes impurity systems in which $\Delta < 0$, while the right-hand side describes systems in which $\Delta > 0$. The behaviour of the intensity of the impurity bands differs substantially for the different spectrum components, depending on whether the absorption bands approach the allowed or the forbidden exciton transition.

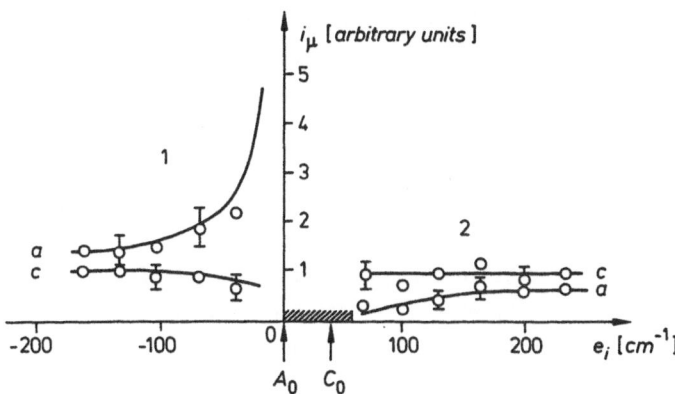

<u>Fig.3.9.</u> Experimental (open circles) and theoretical (solid lines) dependence between the integral intensity of the K_0' bands and their distance from the A_0 band of the host crystal exciton absorption in the IDC of d-benzenes [3.47]; a and c correspond to the two-light polarization. 1) isotopic impurities in a bd_6 host; 2) same for a bd_0 host

   In the ideal isotopic substitution model, the behaviour of σ_{imp} is described by (3.23). The parameter i_μ has the following form:

$$i_\mu = \frac{\Delta^2 a^2}{(\varepsilon_i - \varepsilon_\mu)^2} = \left| \frac{dG_0(\varepsilon_i)}{d\varepsilon_i} (\varepsilon_i - \varepsilon_\mu)^2 \right|^{-1} \quad , \qquad (3.37)$$

where a is the amplitude of the excitation for the impurity centre [see (3.18)]. The solid lines in Fig.3.9 represent the calculated values of i_μ. Similar results are observed for the IDC of d-naphthalene; they are shown in Fig.3.10. The figure exhibits a pronounced Rashba effect (Fig.3.2).

   The example of IDC of d-benzenes quite clearly shows that the various parameters of the optical spectra differ greatly in their sensitivity to the de-

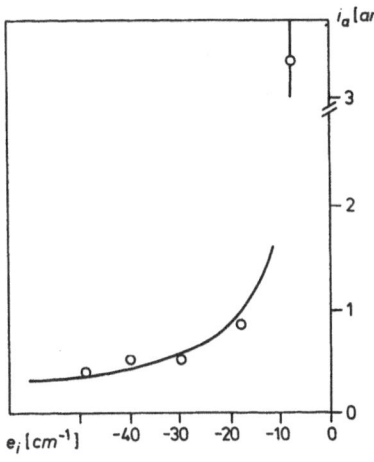

<u>Fig.3.10.</u> Experimental (open circles) and theoretical (solid line) dependence between the integral intensity of the K_0' bands and their distance from the A_0 band of the exciton doublet of the host crystal in the IDC of d-naphthalenes; $\perp$ **b** component of the spectrum [3.54]

localization of the impurity excitation. For practically all of the systems
investigated the function $e_i(\Delta)$ coincides with the asymptotic curve, so that
the impurity level is not repelled from the exciton spectrum. Hence,
$|dG_0/d\varepsilon| \approx 1/\Delta$ and $a^2 \approx 1$. At the same time the intensities of the impurity
bands in the a and c components of the spectra are described by the following
expressions, $i_a = \Delta^2/e_i^2$ and $i_c = \Delta^2/(e_i - \Delta_{Dav}^{ac})^2$. Although the parameter $\Delta/\mathcal{M}$ is
not small, the values of $|e_i|$ are appreciably less than those of $|\Delta|$, while
$|e_i - \Delta_{Dav}^{ac}| \approx \Delta$. This explains the drastic change in the intensity of the impurity absorption for the a component when $\Delta < 0$. The functions $i_\mu(e_i)$ shown in
Figs.3.9,10 actually display the dependence of the square of the wave function
of the exciton $|\psi_i(\mu K)|^2$ on e_i [see (3.20)] and demonstrate the high sensi-
tivity of the wave function to the delocalization of the impurity excitation.

3.3.5 Spectral Distribution of the Intensity of Impurity Absorption

The shape of the impurity absorption bands essentially depends on whether the
absorption is outside or inside the exciton spectrum. As a rule, outside the
exciton spectrum the K_0' bands have a Lorentzian shape at low-impurity concentrations. This is a result of the weak interaction between the impurity
state and the external phonons. The integral intensity of the bands is described by (3.23).

Of great interest is the absorption inside the exciton continuum which is
due to the introduction of the impurity. A quantitative description is possible for two limiting cases: i) for very low concentration of impurity and
ii) for large (comparable) concentrations of the components. In the former
case the absorption is described by (3.33). This case will be considered in
this section. Mixed crystals of isotopic compounds with high component concentrations are discussed in Chap.4.

A change in absorption within the exciton spectrum has only been observed
[3.47] with relatively thick IDC of d-benzenes. In the IDC of d-naphthalenes,
the attempt was unsuccessful [3.54].

Figure 3.11 shows the absorption spectra of a number of IDC of d-benzenes
in the region of the A_0 band of the exciton triplet. The asymmetric broadening
of the absorption band towards the exciton continuum increases in crystals
doped by heavier impurities. The intensity of the induced absorption is (3.33)

$$\sigma_{ind}^\mu(\omega) = \frac{\pi C}{E_\rho v} \frac{\Delta^2}{e_i^2} \sigma_0(\omega) |j_\mu|^2 \quad , \qquad \text{where} \qquad (3.38)$$

$$\sigma_0(\omega) = \frac{\rho(\omega)}{[1 - \Delta G_0'(\omega)]^2 + [\pi\Delta\rho(\omega)]^2} \quad . \qquad (3.39)$$

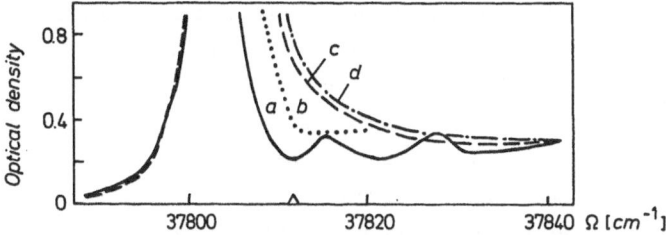

Fig.3.11. Absorption at the edges of the A_0 band of the exciton triplet of a bd_0 host (T = 4.3 K, d = 20 μm); impurity concentration 10%; the light is polarized along the a-direction of the crystal [3.47]. (a) bd_0 crystal, (b), (c), (d) bd_0 crystals doped with bd_1, bd_3, and bd_6, respectively

Since the function $\rho(\varepsilon)$ is known for the benzene crystal, $\sigma_0(\omega)$ can be calculated. In Fig.3.12 the solid line represents the intensity of the induced absorption [calculated using (3.38)]. The conductivity σ_{ind}^{μ} was calculated for a point in the continuous spectrum designated by a triangle on the Ω-axis in Fig.3.11. The filled circles indicate the values calculated for the d-isotope guests in a bd_0 host listed in Table 3.2, while the open circles represent the experimental values. It is worth noting that the effect of the impurities on the continuous absorption spectrum at first increases with increasing $|\Delta|$ but then reaches a limiting value, in contrast to the effect of the exciton band on the impurity absorption which decreases under the same conditions.

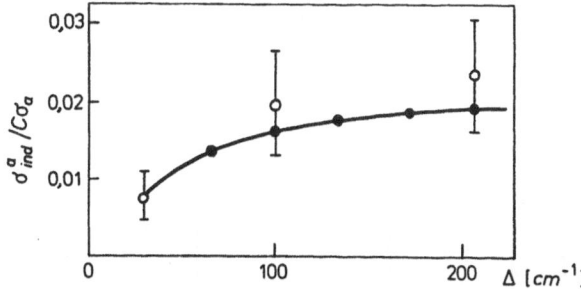

Fig.3.12. Dependence between the induced-absorption intensity in the exciton band of a bd_0 host and the isotopic shift value of the guests [3.47]; solid curve: calculation by (3.38); σ_a: integral intensity of the A_0 band of the exciton multiplet of the host; $\sigma_\mu(\Omega) = \sigma_\mu \delta(\Omega - E_\mu)$ [see also (2.47)]

3.3.6 The Excitation Amplitude on a Guest Molecule

The excitation amplitude on a guest molecule $\psi_i(\mathbf{n}_0\alpha_0)$, [see (3.18)] is an important parameter of an impurity centre. BROUDE et al. [3.55] suggested using the vibronic fluorescence of a local exciton to measure this quantity. The method is based on the fact that two types of vibronic transitions are possible from a local exciton level to the vibrational sublevels of the ground electronic state of the guest and host molecules. The dispersion of the high-frequency optical phonons can be neglected. The vibrational frequencies of the isotopic molecules can differ considerably, so that the vibronic fluorescence spectrum of the local exciton must show the doublet structure (Fig.3.13). The intensities of these transitions i_g and i_h define the ratio $\eta = i_g/(i_g + i_h)$. These quantities are readily expressed in terms of the amplitude a:

$$i_g = \alpha a^2 \quad ; \quad i_h = \alpha(1 - a^2) \quad ; \quad \eta = a^2 \quad . \tag{3.40}$$

Here α is a constant. Thus, a^2 can be directly determined from the relative intensity of the two fluorescence bands. On the other hand, a^2 can be calculated using (3.18) if $\rho(\epsilon)$ is known. It can also be determined independently by differentiating the experimentally determined dependence $\epsilon_i = \epsilon_i(\Delta)$.

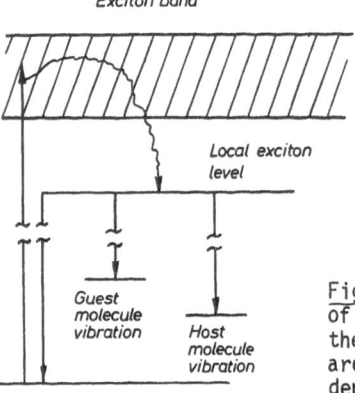

Exciton band

Local exciton level

Guest molecule vibration

Host molecule vibration

Ground state

Fig.3.13. Scheme of IDC fluorescence when impurity excitation is delocalized

Fig.3.14. Patterns of the fluorescence spectra of the IDC of d-naphthalenes in the region of the vibronic transition $00 - \nu_{17}$ (host nd_8, guests are indicated in the figures) [3.51]; the asteriks denote vibronic transitions involving b_{1g} vibration of the host ($\nu_{17} = 496$ cm^{-1})

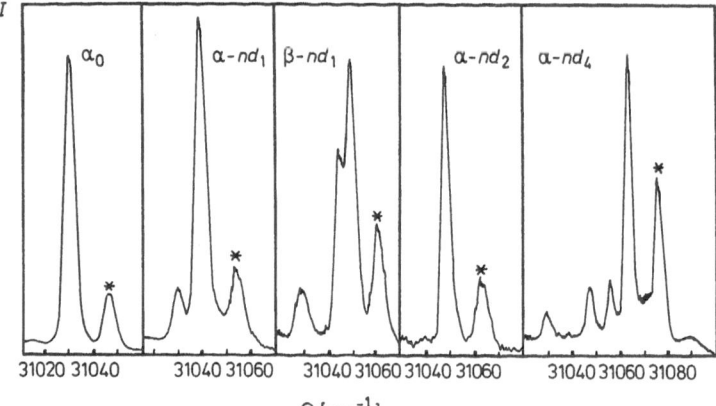

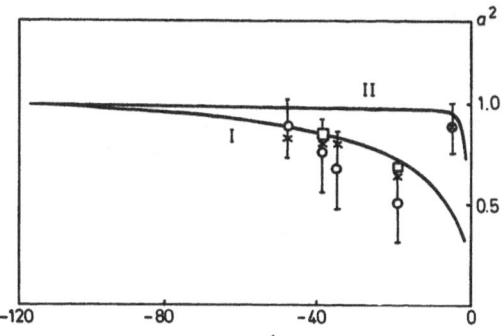

Fig.3.15. Dependence of a² on the distance of the K_0^i bands from the A_0 band of the host-crystal exciton multiplet for the IDC of d-naphthalenes (I) and d-benzenes (II); solid curves: calculation; o and ×: experimental data on vibronic fluorescence spectra according to [3.51] and [3.55], respectively; □: results of graphical differentiation of the experimental dependence $1/\Delta(e_i)$ (Fig.3.5) [3.3]

Figure 3.14 displays certain regions of the vibronic fluorescence spectra of a number of IDC of d-naphthalenes [3.51]. A doublet structure is clearly revealed. The experimental values of a^2 are given in Table 3.3. Figure 3.15 illustrates the function $a^2(e_i)$ for d-naphthalenes. If we take into account the relatively low accuracy of the values of a^2, we can see that the date obtained by different authors are in fair agreement with one another. The same figure shows the calculated values of $a^2(e_i)$ that were obtained using (3.18) and the function $\rho(\varepsilon)$ shown in Fig.3.5. It is seen that the deviation of a^2 from unity, due to the delocalization of the excitation, has been reliably established over the entire accessible range of e_i.

The calculated values of $a^2(e_i)$ have been plotted in Fig.3.15 for d-benzenes as well. The values of a^2 for values of $|e_i|$ up to several cm^{-1} differ negligibly from unity since the function $\rho(\varepsilon)$ is very large near the centre of gravity of the exciton bands and is small near their edges. The first experimental point at which the deviation of a^2 from unity becomes significant corresponds to IDC of bd_0 in bd_1 (see also Sect.5.3.2).

3.3.7 The Structure of Exciton Bands as Determined from IDC Spectra

The above quantitative analysis of the experimental data shows that the model of a monomer impurity centre discussed in Sect.3.2 accurately describes the electronic spectra of such centres. Thus, we can now consider the inverse problem, i.e., determine the parameters of the exciton bands from the IDC spectra.

First of all, we turn our attention to the conclusion drawn from a qualitative analysis of the spectra. The very fact of existence or absence of a
discrete impurity level is very important. If an impurity level exists, the
corresponding value of $|\Delta| > |\Delta_{cr}|$. The quantity $|\Delta_{cr}|$ is related to the width
of the exciton spectrum, $2\mathcal{M}$, by the following expression: $2|\Delta_{cr}| \approx \mathcal{M}$. If
$|\Delta| \approx |\Delta_{cr}|$ (the minimum value of $|\Delta|$ at which the K_0' level ceases to split
off, can be directly established experimentally), then $2\mathcal{M} \approx 4|\Delta|$. If no im-
purity level exists, then $2\mathcal{M} \gtrsim 4|\Delta|$, which gives the lower bound for the ex-
citon spectrum width.

Furthermore, some information is obtained from the observed change in the
intensity of the K_0' bands when $|e_i|$ decreases. The increase in intensity
clearly shows that the K_0' level approaches the exciton state with $\mathbf{k} = 0$. If,
however, the intensity of the K_0' bands decreases near the intrinsic absorption,
the state with $\mathbf{k} = 0$ is located deep within the exciton band (Fig.3.3).

The main source of quantitative information on the distribution of the density of states $\rho(\varepsilon)$ in the energy spectrum of an exciton is the function
$\varepsilon_i = \varepsilon_i(\Delta)$. Measurements made in the near-asymptotic range of ε_i enable one to
determine the position of the centre of gravity and the second moment of the
density of states μ_2 [using (3.10)]. The ratio $\mu_2/\mathcal{M}^2$ indicates to what extent the density of states is concentrated near the centre of gravity. An
estimate of $\mathcal{M}$ can be made if μ_2 is known. An analysis of the function $\varepsilon_i(\Delta)$
enables one to reconstruct $\rho(\varepsilon)$ to a certain extent. This is how the function
$\rho(\varepsilon)$, shown in Fig.3.5, was found for naphthalene [3.49]. The function $\rho(\varepsilon)$
can also be used to find the matrix elements of the intermolecular interaction
that determines the dispersion law $\varepsilon(\mathbf{k})$ [3.51] and to determine the position
of the electronic term.

The data of this section can be used to completely describe the singlet absorption of isotopic monomer centres in molecular crystals. The triplet absorption by an isotopic guest molecule is of a similar nature. It should, however,
be noted that the effects of the delocalization of the impurity excitation are
very small for local triplet excitons, since the magnitude of the parameter
$\Delta/\mathcal{M}$ (e.g., for the IDC of d-naphthalenes [3.51]) is much greater than unity
even for a monatomic substitution. As a result, the triplet levels of the isotopic centres are usually deep. Very recently shallow local triplet excitons
have been discovered and investigated in antracene [3.81].

The general features of the isotopic impurity absorption of organic molecular crystals can also be observed with other classes of solids, e.g., with IDC
of NH_4Cl and sodium, barium, and potassium nitrites containing the isotopes
^{15}N and ^{18}O [3.57,58] as impurities and with crystals of SiF_4 [3.59]. Treating

the high-frequency phonon branches of these crystals as exciton branches, the authors applied the above model to a study of the spectra and obtained a satisfactory quantitative description. Similar phenomena were observed in the Raman scattering by mixed crystals of $Zn_{1-x}Cd_xS$ [3.60].

3.4 The Exciton Spectra of Aggregates of Isotopic Guest Molecules in IDC

3.4.1 The Models of Pair Centres and Their Basic Features

Absorption bands due to a pair of isotopic guest molecules were observed for the first time in the impurity spectrum of bd_0 in the bd_6 host (Fig.1.7) [3.34]. A similar situation was observed by HANSON [3.61] in the spectrum of nd_0 in the nd_8 host. The arrows in Fig.3.16 identify the bands whose intensity is a superlinear function of the concentration. Hanson attributed such a concentration dependence of these bands to the formation of cluster centres that consist of an aggregate of guest molecules. The simplest type of an *aggregate centre* consists of a pair of molecules (a pair centre). From now on, such centres will also be called dimers.

There are many different types of pair centres. Several models of such centres in IDC of nd_0 in nd_8 are shown in Fig.3.17. The centres consist of a molecule located at the origin and one of the neighbouring molecules. As indi-

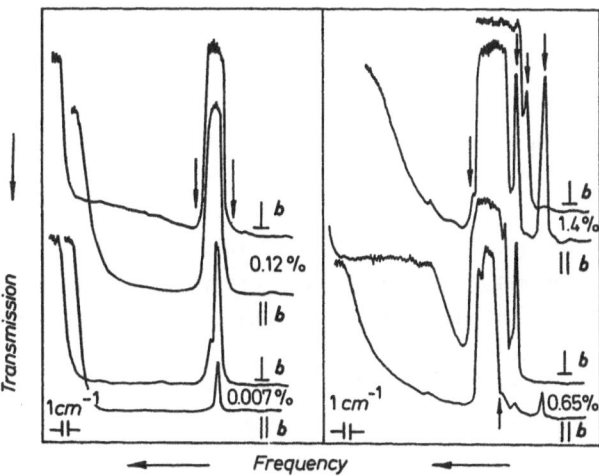

Fig.3.16. Absorption spectra of the IDC of nd_0 in nd_8 at different nd_0 concentrations; to the concentrations 0.007%, 0.12%, 0.65%, and 1.4% correspond thicknesses 1.2, 2.0, 1.0, and 1.6 mm, respectively [3.61]; intensive absorption in the high-frequency region corresponds to the exciton absorption of the host crystal

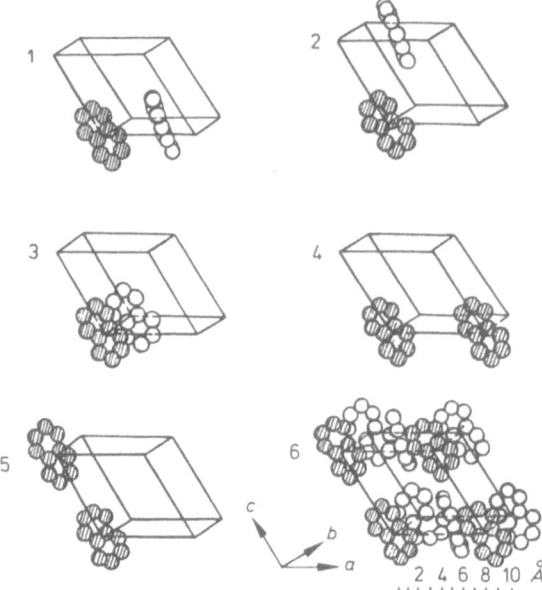

Fig.3.17. Models of nd_0 pair centres: distances between the centres of the molecules at T = 296 K are 1) 5.11 Å, 2) 7.87 Å, 3) 6.00 Å, 4) 8.24 Å, and 5) 8.66 Å; 6) the position of the naphthalene molecules in the unit cell [3.7,37]

cated in Sect.2.5, short-range interaction dominates in an electronic inter-action that is responsible for the exciton spectrum of the naphthalene crystal. This suggests that the types of pair centres shown in Fig.3.17 adequately de-scribe the most important configurations and allows one to neglect the centre-to-centre interactions in the low-impurity concentration range.

The observation of discrete K_0'' bands of pair centres, which are shifted relative to the K_0' band of the monomer centre (and also of the bands of more complex aggregate centres), requires an appreciable interaction within the pair centres for the monomer level to be split into two. This interaction is different for different pair centres, e.g., for naphthalene it is a maximum for the first case (Fig.3.17). It follows that the K_0'' bands of different pair centres appear at different positions in absorption spectrum. This leads to the fine structure of the spectra given in Fig.3.16. Therefore, an analysis of the positions of the K_0'' bands may become a method of determining the ener-gies of the interactions of pairs of molecules, which are needed for the de-rivation of the exciton dispersion law.

The dipole moment for optical transitions in a deep pair centre, to which the oriented gas model can be applied, is determined by the sum or difference of the dipole moments of the transitions in individual molecules. Since the

74

transition moments of the translationally equivalent (TE) molecules (of the type 3-5 in Fig.3.17) that make up the pair centre are parallel, an optical transition is only allowed to the even states of the pair centre. As a result, the spectrum consists of a single band. The polarization in this band is determined by the projections of the dipole moment of the molecule onto the crystallographic axes. The above qualitative conclusions are also valid for shallow centres because they can actually be made from the symmetry arguments. However, the polarization ratio may change considerably due to the Rashba effect.

Translationally non-equivalent (TNE) molecules of the type 1-2 shown in Fig.3.17 form pair centres in which transitions are allowed to both sublevels. In a naphthalene crystal, one of the transitions is polarized along the b-axis, while the other one is polarized in the ac-plane ($\perp$ **b**). This permits the observation of the doublets of the K_0'' bands with sharp polarizations in mutually perpendicular directions.

3.4.2 The General Properties of the Energy Spectra of Pair Centres

The equation that describes the energy spectrum of a pair centre formed by two identical isotopic guest molecules can be written according to the conventional procedure of degenerate perturbation theory (Appendix B). Let the guest molecules be located a distance **R** apart at the sites $n\alpha$ and $m\beta$. The isotopic shift is equal to Δ. Hence, the equation that describes the energy levels has the following form:

$$G_0(\varepsilon) \pm G_R^0(\varepsilon) = 1/\Delta \quad , \quad G_R^0 \equiv G_{n\alpha m\beta}^0 \quad . \tag{3.41}$$

Since the perturbation is of the second rank, no more than two discrete levels arise. It has already been noted in the previous section that the levels can be considered to be a result of the splitting of the levels of the monomer centres. The magnitude of the effective interaction largely depends on **R** and is determined by G_R^0, the off-diagonal matrix element of the Green's function. Let us now consider some of the peculiarities in the behaviour of G_R^0, which will be of use later on.

According to (2.27), G_R^0 is determined by the equation

$$G_{n\alpha m\beta}^0(\varepsilon) = \frac{v}{(2\pi)^3} \sum_\mu \int \frac{B_\alpha(\mu k) B_\beta^*(\mu k)}{\varepsilon - \varepsilon_\mu(k)} \exp\left[ik(n_\alpha - m_\beta)\right] d^3k \quad . \tag{3.42}$$

The integration with respect to **k** extends over the entire Brillouin zone. In fact, a general investigation of the expression for G_R^0 is only possible at large values of **R**. In this case, the investigation is reduced to finding the asymptotics of the Fourier components of a definite function of **k**. These asymptotics always depend on the singularities of the integrand function. For large values of **R**, the function $|G_{\mathbf{R}}^0|$ is small, and the equation for the positions of the $\varepsilon^{\pm}$ levels of the pair centre can be derived from (3.41):

$$\varepsilon^{\pm} \approx \varepsilon_i \mp \frac{G_{\mathbf{R}}^0(\varepsilon_i)}{dG_0/d\varepsilon_i} \quad , \tag{3.43}$$

where ε_i is the position of the level of the monomer centre.

First of all we consider the case in which the matrix elements of the excitation transfer, $M_{n\alpha m\beta}$, decrease exponentially with increasing values of $\mathbf{R} = \mathbf{n}_\alpha - \mathbf{m}_\beta$. It corresponds to triplet excitons. The coefficients B_α and $\varepsilon_\mu(\mathbf{k})$ are analytical functions of **k**, and therefore their Fourier components also decrease exponentially with increasing values of **R**. To find the factors that determine this exponent, it is convenient at first to consider deep levels for which $|\Delta| \gg \mathscr{M}$. In this case, one can expand the denominator in (3.42) into a series in powers of $\varepsilon_\mu/\varepsilon$, integrate with respect to **k**, and perform a summation using (2.11) and the orthonormalization condition (2.14). The first term of the expansion is linear in ε_μ and contributes $M_{n\alpha m\beta}/\Delta^2$ to $G_{\mathbf{R}}^0$. It describes a *direct interaction* between guest molecules that is unperturbed by the surrounding solvent molecules. This term decreases with increasing values of **R**:

$$(G_{\mathbf{R}}^0)_{direct} \sim (M/\Delta^2) \exp(-R/d_{at}) \quad , \tag{3.44}$$

where M is the matrix element of the excitation transfer between neighbouring molecules. The values of d_{at} are of the atomic order of magnitude ($d \sim 1$ Å) because the distance at which the magnitude of the wave function of the molecule is vanishingly small is determined by the constituent atomic orbitals. Such a decrease in wave functions with distance proceeds very rapidly, so that a direct interaction is usually only possible between nearest neighbours.

The decrease is slower if the higher terms of the expansion of the denominator are also considered. If (2.11) is successively applied to each factor ε_μ, the following expansion for $G_{\mathbf{R}}^0$ is obtained:

$$G_{n\alpha m\beta}^0(\varepsilon) = \sum_s \sum_{n'\alpha'\dots m'\beta'} (1/\Delta)^{s+2} M_{n\alpha n'\alpha'} M_{n'\alpha'n''\alpha''} \cdots M_{m'\beta'm\beta} \quad , \tag{3.45}$$

where s is equal to the number of intermediate sites in the sequence $n'\alpha'$, $n''\alpha''$, ..., $m'\beta'$. The largest terms are those in which only the nearest neighbours are connected in tandem along the entire chain of sites and the number of links is a minimum. Then, $s \approx R/d$, where d is the intermolecular distance.

76

The largest summands in G_R^0 are of the order of

$$(G_R^0)_{indir} \sim (1/\Delta)|M/\Delta|^S \sim (1/\Delta)\exp[-(R/d)\ln|\Delta/M|] \quad . \tag{3.46}$$

The right-hand side must be multiplied by the number of chains that make comparable contributions. The very significant point is that in organic crystals $d \approx 5 - 10$ Å, which is appreciably larger than d_{at} ($d \sim 5 - 10\ d_{at}$). In actual situations $\ln|\Delta/M| \gtrsim 2$, so that according to (3.46) the decrease in G_R^0 may prove to be much slower than the decrease according to (3.44). The process described by (3.45,46) corresponds to a transfer of excitation between widely spaced guest molecules through a sequence of virtual transitions between closely spaced sites. It actually involves a tunnelling through the exciton band and is called an *indirect interaction*.

Thus far we have considered deep centres. It is of significance, however, that the radius of the state of the local exciton is large ($r_i \gg 1$) for shallow centres with $|\varepsilon_{bind}| \ll \mathcal{M}$. Therefore, over the entire range of $R \gtrsim r_i d$ the indirect interaction between guest molecules through the host molecules is no more exponentially small and therefore may be quite appreciable.

We now return to a discussion of singlet excitons. For them the multipole interactions are dominant since they fall off according to power laws at large distances. As a result, singularities, which are the strongest for dipole-dipole transitions, arise in $\varepsilon_\mu(\mathbf{k})$ at $\mathbf{k} = 0$ (Sect.2.2). Accordingly, the asymptotics of G_R^0 become power ones; for dipole excitons, $G_R^0 \propto R^{-3}$. To verify this, it is easiest of all to return to the discussion of the deep centres: in this case G_R^0 is reduced to $M_{n\alpha m\beta}/\Delta^2$ and substitution into (3.43) yields

$$\varepsilon^\pm - \varepsilon_i \approx \pm M_{n\alpha m\beta} \propto 1/R^3 \quad . \tag{3.47}$$

Thus, we again obtain direct impurity interaction, which in this case amounts to the usual Förster dipole-dipole "resonance" interaction. It is not hard to understand from physical considerations what happens with shallow centres ($r_i \gg 1$) if $R \gg r_i d$, i.e., the ψ functions of the impurity excitations do not overlap. In this case, a dipole-dipole interaction, i.e., the R^{-3} law, is again observed, but the transition dipole moments of single guest molecules will be replaced by the dipole moments that correspond to the formation of local excitons. According to Sect.3.2,3 they differ drastically from the dipole moments of the transitions in separate molecules. As a result, the indirect interaction may change the magnitude of the dipole-dipole interaction between the impurities several tenfold. For smaller distances, i.e., for $R \gtrsim r_i d$, the interaction obeys a more complicated law.

The interaction between widely spaced guest molecules is of importance when discussing the low-temperature migration of energy. On the contrary, only centres for which the magnitude of $\varepsilon^+ - \varepsilon$ is sufficiently large to be resolved experimentally play a role in the impurity absorption spectra. In practice, this restricts the distance between impurities to $R \sim d$. In this range one cannot use asymptotic formulae, so that one must calculate the models for the centres numerically.

3.4.3 Calculation of the Spectra of Pair Centres in Naphthalene and a Comparison with Experiment

The first calculations for such centres were carried out by HANSON [3.61], who assumed a direct interaction between the guest molecules (the oriented gas model). The calculation yielded a series of matrix elements $M_{n\alpha m\beta}$. However, a calculation of the exciton spectrum $\varepsilon(\mathbf{k})$ using these matrix elements resulted in a strong deviation from the density of states shown in Fig.3.5, which agreed with the experimental data on the band-to-band transitions (Sect.5.2) and the spectra of the monomer centres (Sect.3.3.2).

As a result, HONG and KOPELMAN [3.62,63] attempted to treat these experimental data with due regard for the indirect interaction, i.e., they based their calculations on (3.41). Since G_0 is completely determined by the function $\rho(\varepsilon)$, any new information on the energy spectrum of the excitons is exclusively obtained from the function $G_{\mathbf{R}}^0$. The goal was to find a dispersion law that agrees with both the data on the pair centres and the known density of states $\rho(\varepsilon)$. The dispersion law for a naphthalene crystal was chosen in the form (2.51). The quantities M_a, M_b, $M_{\mathbf{a}+\mathbf{c}}$, M_{12}, and M'_{12} are fitting parameters.

Figure 3.18 shows the calculated dependence of the energies ε^- and ε^+ on the isotopic shift for pair centres of the type 1 and 4 in Fig.3.17. The level splitting in the pair centre is equal to twice the interaction energy 2M, only for large values of $|\Delta|$. The positions of the ε^- and ε^+ levels are generally asymmetrical with respect to the level of the monomer centres. The following bands should be visible in the absorption spectrum: one weakly polarized band corresponding to the level ε^- of the TE pair and a doublet of sharply polarized ε^- and ε^+ bands of the TNE pair.

Four weakly polarized bands and one band sharply polarized in the $\perp \mathbf{b}$ direction were observed near the monomer band (Table 3.4). A comparison of the positions of these bands with the calculated positions [3.63] was based on the assignment of the sharply polarized band to the TNE pair with $\mathbf{R} = (\mathbf{a} + \mathbf{b})/2$. The second component of the doublet of the pair centre, polarized in the $\mathbf{b}$

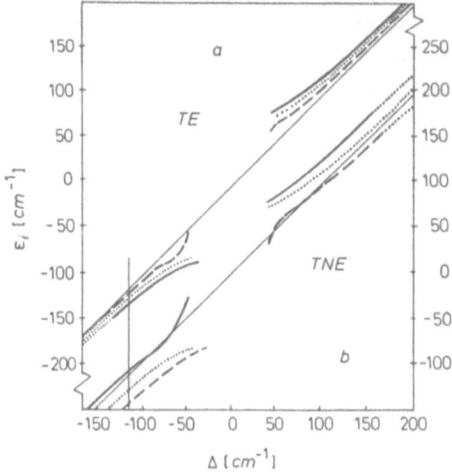

ε_i [cm^{-1}]

Δ [cm^{-1}]

Fig.3.18. Calculated values of the pair centre levels ε^+ and ε^- and of the monomer level ε_i (solid, dashed and dotted curves, respectively) in an nd$_8$ host as a function of the isotopic shift Δ for the TE centre with $R = a$ (a) and the TNE centre with $R = 1/2(a+b)$ (b) [3.63] (calculation was performed with set 1 excitation transfer integrals from Table 3.5; the vertical line corresponds to the IDC of nd$_0$ in nd$_8$

Table 3.4. Comparison of the observed and calculated positions of the bands of pair centres in cm^{-1} [3.65]. The energy, in cm^{-1}, is reckoned from the ε_i level of the monomer centre

	Experiment	Set 1	Set 2	Set 3
	...	21.7 1/2(**a**+**b**)[a]	20.0 1/2(**a**+**b**)	19.0 1/2(**a**+**b**)
	3.7	4.0 **c**	4.0 **a**+**c**	4.1 **a**+**c**
nd$_0$[b]	-3.3	-3.5 **a**+**c**	-3.2 **b**	-3.4 **b**
	-5.1	-5.0 **a**	-5.2 **c**	-4.8 **a**
	-7.9	-8.0 **b**	-7.5 **a**	-7.9 **c**
	-15.3	-15.0 1/2(**a**+**b**)	-14.7 1/2(**a**+**b**)	-14.2 1/2(**a**+**b**)
	18±2	21.5 1/2(**a**+**b**)	20.5 1/2(**a**+**b**)	19.3 1/2(**a**+**b**)
	...	3.8 **c**	3.7 **a**+**c**	3.6 **a**+**c**
α-nd$_1$	-2.2±0.5	-3.7 **a**+**c**	-3.6 **b**	-3.3 **b**
	-4.1±0.5	-5.2 **a**	-5.1 **c**	-4.8 **a**
	...	-7.8 **b**	-7.5 **a**	-7.8 **c**
	-14.1±0.5	-14.5 1/2(**a**+**b**)	-14.5 1/2(**a**+**b**)	-13.7 1/2(**a**+**b**)
	16±2	21.5 1/2(**a**+**b**)	20.6 1/2(**a**+**b**)	19.4 1/2(**a**+**b**)
	...	3.5 **c**	3.5 **a**+**c**	3.5 **a**+**c**
β-nd$_1$	-2.9±0.5	-3.7 **a**+**c**	-3.7 **b**	-3.2 **b**
	-5.0±0.5	-5.5 **a**	-5.0 **c**	-4.8 **a**
	...	-7.7 **b**	-7.5 **a**	-7.5 **c**
	-14±1	-14.0 1/2(**a**+**b**)	-13.8 1/2(**a**+**b**)	-13.5 1/2(**a**+**b**)
	17±2	21.5 1/2(**a**+**b**)	20.6 1/2(**a**+**b**)	19.5 1/2(**a**+**b**)
	3.8±0.5	3.5 **c**	3.5 **a**+**c**	3.5 **a**+**c**
$\alpha\beta$-nd$_2$	-3.5±0.5	-3.7 **a**+**c**	-3.7 **b**	-3.2 **b**
	-5.4±0.5	-5.5 **a**	-5.0 **c**	-4.6 **a**
	...	-7.6 **b**	-7.5 **a**	-7.5 **c**
	-13.4±0.5	-14.0 1/2(**a**+**b**)	-13.7 1/2(**a**+**b**)	-13.0 1/2(**a**+**b**)

a) 1/2(**a**+**b**), etc. indicate the position of the second molecule in the pair
   relative to the first
b) HANSON's data [3.61]

direction, could not be observed in this experiment. It was only observed
with higher impurity concentrations [3.64] (Sect.4.5). The unambiguous identification of the spectrum due to the TNE pair and an additional condition for
finding the magnitude of the Davydov splitting established the values of M_{12}
and M_{12}'. The results are given in Table 3.5.

The matrix elements of the TE interactions could not be found unequivocally
because of the uncertainty in bringing the particular bands and definite models
of centres into correspondence (Table 3.4). Three sets of values, $\{M_i\}$, were
obtained (Table 3.5). They describe the positions of the four bands of the TE
pairs and the shape of the bands of the band-to-band transitions fairly well
(Sect.5.2).

Table 3.5. Integrals of excitation transfer for the B_{3u} state
of naphthalene [3.63] [cm^{-1}]

Position[a]	Set 1[b]	Set 2[b]	Set 3[b]
$1/2(a+b)$[c]	18.0	18.0	18.0
$1/2(a+b)+c$[c]	2.0	1.0	1.0
a[d]	-0.6	-4.3	-1.2
b[d]	-3.9	1.9	1.6
c[d]	6.1	-6.1	-8.9
$a+c$[d]	-3.7	6.0	6.0

a) The position of the second molecule in the pair relative to the first
b) Pairwise interaction
c) TNE pairs
d) TE pairs

The calculated density of states for each of the three sets are shown in
Fig.3.19. OCHS and KOPELMAN [3.65] made one more attempt to improve the choice
of $\{M_i\}$ by analyzing the dependence of the spectra of the pair centres on Δ.
Unfortunately, the dependence proved to be weak, and the uncertainty was not
eliminated. However, they consider the second set as the preferred one.
Table 3.4 summarizes the experimental and calculated positions of the bands
of pair centres in four IDC of d-naphthalenes. It is seen that the three sets
of values, $\{M_i\}$, describe the positions of the bands to within experimental
error.

The low intensity of the K_0'' bands and their superposition upon the intensive
neighbouring K_0' bands of the monomer centres hinder a measurement of their intensity. Therefore, no experimental values of intensity are available. However, as predicted by calculations [3.65], the polarization ratio for the
ε^- band of the TE pair obeys the equation for the monomer impurity centre
(3.26). The expression for the polarization ratio is slightly modified for
the doublet of the ε^+ and ε^- bands of the TNE pair:

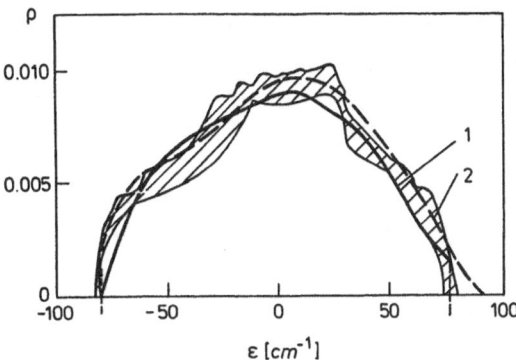

ρ

0.010

0.005

0

-100   -50   0   50   100

$\varepsilon\ [cm^{-1}]$

Fig.3.19. Density of states in the naphthalene-crystal exciton spectrum: 1) result of processing the band-to-band transition band [3.48]; 2) result of fitting the positions of the K_0' levels of the impurity centres [3.49], the dashed area covers all calculated curves for $\rho(\varepsilon)$ with three sets of parameters $\{M_i\}$ in the dispersion law (2.51) (Table 5.3) [3.63]

$$P_{\mu_1/\mu_2} = P^0_{\mu_1/\mu_2}\left(\frac{\varepsilon^- - \varepsilon_{\mu_2}(0)}{\varepsilon^+ - \varepsilon_{\mu_1}(0)}\right)^2 \frac{d\varepsilon^+/d\Delta}{d\varepsilon^-/d\Delta} \ . \tag{3.48}$$

As $\varepsilon^+ - \varepsilon^- \to 0$, (3.48) transforms to (3.26).

3.4.4 Complex Impurity Centres

More complex centres in the IDC of d-naphthalenes with low impurity concentration could only be detected by the fluorescence method [3.66]. Figure 3.20 shows the redistribution of the band intensities in the fluorescence spectrum of IDC of nd_0 in nd_8 that takes place upon changing the concentration of nd_0. The long-wavelength bands associated with the complex aggregate centres become noticeable at higher concentrations of nd_0.

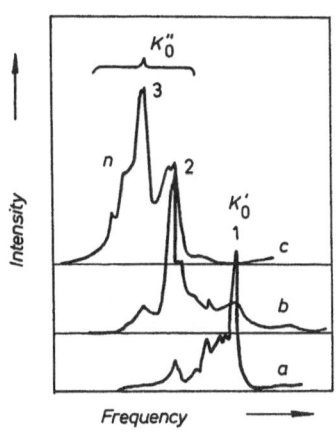

Fig.3.20. Fluorescence spectra of the IDC of nd_0 in nd_8 at different nd_0 concentrations: a) 3%; b) 6%, T = 1.6 K; c) 9%, T = 1.4 K [3.66]. The numerals indicate the bands of centres consisting of one (1), two (2), three (3), and more (n) TNE molecules

81

Aggregate centres that are complexer than pair centres were observed in the absorption spectrum of IDC of nd_0 in nd_8 at nd_0 concentrations that were between 10 - 20% [3.64,67]. At such concentrations the question naturally arises: to what degree is such a centre isolated from the others? The structure of the absorption spectrum associated with the simplest aggregate centres is observed for concentrations of nd_0 up to 50% [3.67]. This suggests that the energy of these centres is hardly perturbed even at such high concentrations (for more details on mixed crystals see Chap.4).

Figure 3.21 shows the absorption spectra of crystals of nd_0 in nd_8 with different concentrations of nd_0. At a concentration of 2.5% the absorption spectrum is dominated by the monomer centre band. On the chosen scale for the absorption coefficient, the K_0'' bands associated with the pair centres are practically inseparable from the background [3.61]. A perceptible broadening of the central band is observed as the concentration increases due to the K_0'' bands of TE centres. Sharply polarized bands appear on both sides of the central band at higher concentrations of nd_0. They are associated with TNE centres of different structures. The behaviour of the spectrum at high impurity concentrations will be discussed in Sect.4.5. For the moment, we shall use

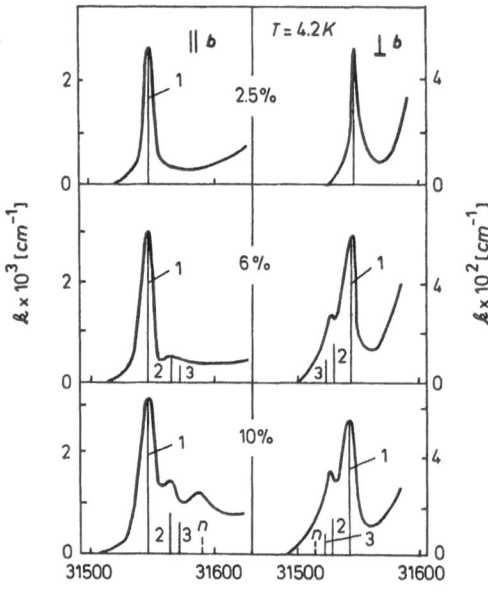

$\Omega \, [cm^{-1}]$

Fig.3.21. Absorption spectra of the IDC of nd_0 in nd_8 at different nd_0 concentrations in polarized light [3.64]. The vertical line segments show the positions and relative intensities of the K_0'' bands of the centres consisting of one (1) molecule and of the K_0'' bands of the centres consisting of two (2) and three (3) molecules; the dashed line segment corresponds to more complicated aggregates (n)

82

the low-concentration approximation to explain the basic experimental features within a simple model used in [3.64].

It has been noted above (Table 3.5) that the dominant interaction in naphthalene crystals is between the nearest TNE molecules in the ab-plane. Therefore, in calculating the naphthalene spectra one can often use the plane-network approximation which implies a network of molecules in a plane that only interact with their nearest neighbours. An interaction with the next-nearest neighbours can be neglected since it is comparable in magnitude to the interlayer interaction. This system is described by a single parameter, the transfer integral M_{12}.

Naturally, such a model noticeably distorts the energy spectrum; for example, the density of states $\rho(\varepsilon)$ becomes symmetrical and shows abrupt changes at the edges of the spectrum. Nevertheless, it is a reasonable one for calculations which do not claim to be highly accurate. In particular, it can be applied to the calculation of the spectrum of aggregate centres. Figure 3.22 shows all of the possible configurations worth considering for centres with three and four guest molecules.

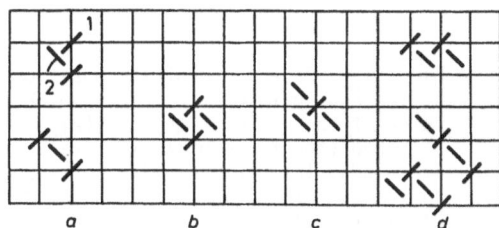

Fig.3.22. Guest aggregates of three and four molecules: numerals 1 and 2 denote TNE molecules

The direct-interaction approximation provides the simplest estimates for the splitting of impurity levels. This approximation yields the best results in describing the splitting of the levels of relatively deep impurities, such as nd_0 in nd_8. In this approximation the energy of the monomer level $\varepsilon_0 = \Delta$. The following results are valid for the positions of the levels and the polarization of the transitions:

a) *Pair centres* (dimers): $\varepsilon_{1,2} = \varepsilon_0 \pm M_{12}$, sharply polarized.

b) Both *trimers* in Fig.3.22a have levels in common: $\varepsilon_1 = 0$, optically forbidden; $\varepsilon_{2,3} = \varepsilon_0 \pm \sqrt{2}\, M_{12}$, weakly polarized, but the degree of depolarization is small.

c) *Tetramers* of the type shown in Fig.3.22b: $\varepsilon_{1,2} = \varepsilon_0$, forbidden; $\varepsilon_{3,4} = \varepsilon_0 \pm 2M_{12}$, sharply polarized.

d) *Tetramers* of the type shown in Fig.3.22c: $\varepsilon_{1,2} = \varepsilon_0$, forbidden; $\varepsilon_{3,4} = \varepsilon_0 \pm \sqrt{3}\, M_{12}$, sharply polarized.

e) All of the *tetramers* of the type shown in Fig.3.22d:

$\varepsilon_{1,2} = \varepsilon_0 \pm (1/2)(1 - \sqrt{5})M_{12}$, forbidden;

$\varepsilon_{3,4} = \varepsilon_0 \pm (1/2)(1 + \sqrt{5})M_{12}$, sharply polarized.

Thus, only the two external bands must be revealed for each of the centres in the spectrum. They must be either sharply polarized or exhibit a considerable preferred polarization. Besides, it is easy to see that all of the levels of the tetramers to which transitions are allowed are closely spaced, namely only $0.4\ M_{12} \approx 8\ cm^{-1}$ apart. As the size of the aggregate centre increases, the levels to which transitions are allowed recede from the monomer band.

The effect of a nearby exciton band, i.e., of an indirect interaction between guest molecules, can also be determined using the plane-network approximation. The calculations of the positions of the levels use the following simplified dispersion law:

$$\varepsilon_\mu(\mathbf{k}) = \pm 4M_{12} \cos (\mathbf{k}a/2) \cos (\mathbf{k}b/2) \quad . \tag{3.49}$$

The results are given in Fig.3.23. They are in qualitative agreement with the results obtained using the direct-interaction approximation.

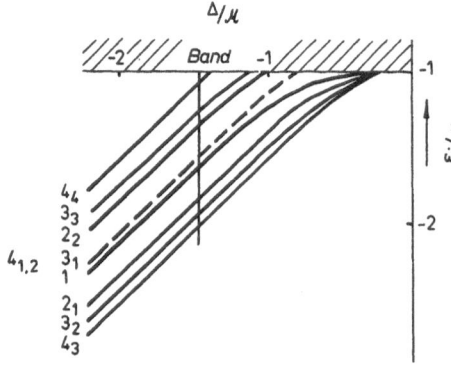

Fig.3.23. Dependence of the position of the aggregate centre levels on the value of the isotopic shift Δ (ε_i is reckoned from the centre of gravity of the band). The vertical line corresponds to the IDC of nd_0 in nd_8: 1) monomer; 2) dimers; 3) trimers; 4) tetramer of type (b) in Fig.3.22. Subscripts 1,2,3, and 4 denote the solutions by analogy with the formulae for various types of the centres (a-e) given at the preceding and this page is; the dashed curve corresponds to the degenerate level, to which transition is forbidden [3.64]

From Fig.3.23 it follows that a weakly polarized band due to monomers in the IDC of nd_0 in nd_8 must be observable. Polarized bands due to dimers, trimers, etc., are arranged on both sides of this band. Since the levels of different aggregate centres are very close together in the low-frequency range, these bands may coalesce into a relatively broad absorption band with a complicated shape. The centre of this broad band will gradually shift towards lower frequencies at higher concentrations. At small values of $|\Delta|$, the high-frequency group of the impurity levels is absent altogether. At large values of $|\Delta|$, the impurity bands polarized along the b-axis must be observable.

As the concentration of the guest molecules increases, the intensity of the impurity absorption is gradually redistributed from the central monomer band to the side bands of the complex centres. They are "embryos" of bands that correspond to excitons associated with the impurity sublattice, i.e., to *impurity exciton* bands. The existence of such bands will be shown in Sect.4.3. One can trace their evolution to a range in which the concentrations of the two isotopic components of the solution are comparable.

The pattern obtained with a random numbers table illustrates the distribution of impurity molecules in a crystal (Fig.3.24). One can see many aggregates that are weakly bound to each other even if the impurity concentration is rather high (17.5%). Such tables can be used to calculate the number of single molecules, pairs and other complexes. The lengths of the vertical bars in Fig.3.21 are a measure of these numbers. They reflect the distribution of the intensity of the impurity absorption in the direct-interaction approximation. The positions of the vertical bars relative to the monomer band correspond to calculations whose results are illustrated in Fig.3.23.

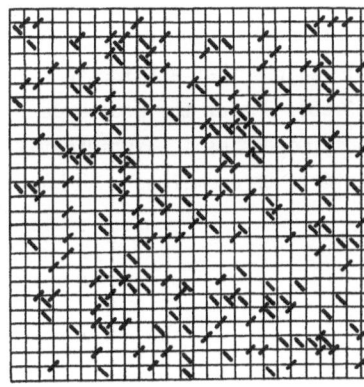

Fig.3.24. Random distribution of guest molecules in the ab-plane of the naphthalene crystal (concentration of $nd_0 = 17.5\%$) [3.64]

Only small differences exist between the calculated and the experimental positions of the bands. It is more difficult to evaluate the degree of agreement between experiment and theory when considering the intensity distribution between the monomer band and those of the aggregate centres. In this case, one cannot even hope to find a quantitative agreement within the framework of the direct-interaction model because the partial delocalization of the impurity states affects the intensities of the bands much more than the positions (Sects.3.2,3).

3.5 Exciton Spectra of Defect Centres

3.5.1 General Characteristics of Impurity Centres

From the experimental point of view, the spectroscopy of defect molecules is
intimately related to the *spectroscopy of real crystals*. The important problem
involved is the presence in the crystal's absorption spectrum of a large number
of weak bands adjoining the long-wavelength edge of the intrinsic absorption
band. More often than not, these bands are highly variable, and their polarizations, intensities, and positions are strongly affected by external factors.
The short-wavelength fluorescence bands at low temperatures coincide with
these bands. A typical absorption and fluorescence spectrum is shown in
Fig.3.25. In most cases, the bands that are several hundred cm^{-1} away from
the origin of the intrinsic absorption can be definitely assigned to certain
chemical impurities. Usually, the bands less than several tens of cm^{-1} away
from the edge of the intrinsic absorption cannot be identified in this way.
At the same time, the decrease in the intensity of these bands and their complete disappearance with increasing chemical purity and structural perfection
of the crystal can be reliably established. This allows to suggest that the
observed spectrum beyond the edge is due to local centres that are formed
in the vicinity of the various crystal defects [3.68]. One of the models describing such a centre implies that a group of host molecules is distorted by
a nearby chemical impurity. Such molecules are said to be *anomalous* [3.68] or
defect molecules [3.69].

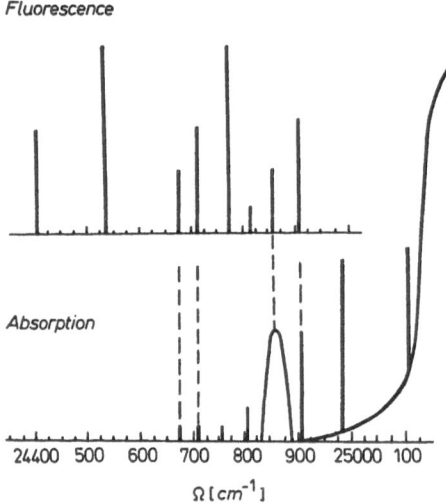

Fluorescence

Absorption

24400 500   600   700   800   900 25000 100

$\Omega [cm^{-1}]$

Fig.3.25. Scheme of the absorption and
fluorescence spectra of the anthracene
crystal at 20 K [3.70]; the two low-frequency bands in the fluorescence spectrum
are vibronic

The earliest investigators of the spectra of defect impurity centres have already observed an anomalous (with respect to the oriented gas model) polarization of the absorption and fluorescence bands. From the theory of absorption of an isotopic impurity given in Sect.3.2, whose qualitative conclusions. are valid for the shallow level of a local exciton of any origin [3.12], it follows that the observed anomaly may be due to a partial delocalization of the exciton associated with the defect molecule [3.68]. As a result, the electronic spectrum of a defect centre must primarily depend on the value of the parameter $\Delta/\mathcal{M}$, where Δ is the change in the electronic excitation energy of a host molecule that is distorted.

A systematic study of defect centres has been carried out by a Kiev group of spectroscopists: OSTAPENKO (SHEREMET), SUGAKOV, and SHPAK. They have published a detailed review of calculations and experimental results [3.5]. We shall now consider those aspects of this study which are related to general problems of the local-exciton spectra and consider the experimental data on the structure of exciton energy bands derived from these spectra. Defect molecules were formed in benzene and naphthalene crystals upon isomorphically replacing the host molecules by five-membered aromatic rings that contain the heteroatoms S, N and O (Fig.3.26a) or by compounds that contain a benzene ring in addition to a similarly constructed five-membered ring (Fig.3.26b). The difference between the electronic excitation energies of these molecules and that of the host considerably exceeds the exciton band width. The various chemical impurities cause different shifts Δ ·in the levels of the defect host molecules.

In the systems investigated, isolated, narrow *defect bands* were observed in the absorption spectra at low-impurity concentrations. This led SUGAKOV [3.71] to propose the picture shown in Fig.3.27 to represent the simplest model for a defect impurity centre. In this model, the asymmetric guest mole-

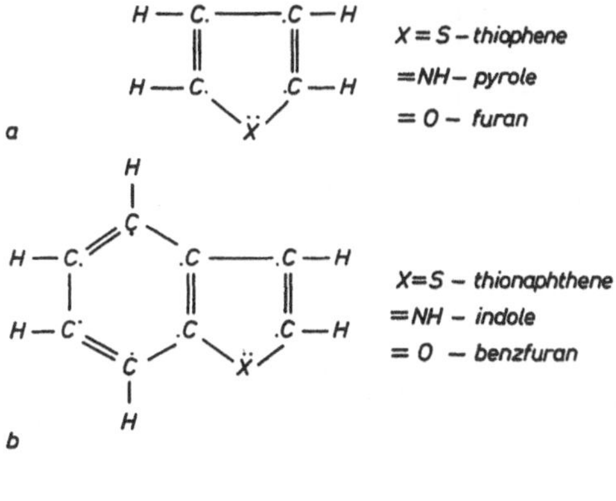

a

X = S – thiophene
=NH – pyrole
= O – furan

b

X=S – thionaphthene
=NH – indole
= O – benzfuran

Fig.3.26a,b. Structural formulae of the chemical guests giving rise to defect molecules in the benzene (a) and naphthalene (b) crystals

cule 2 deform one of the nearest TNE host molecules 3, while the other molecules are left unperturbed. The existence of defect molecules in the vicinity
of guest molecules is naturally a widespread phenomenon. The appearance of a
discrete level for a local exciton is determined by the condition[3] that
$|\Delta| > |\Delta_{cr}|$. The existence or absence of the level depends on the magnitude
of Δ. Defect molecules have also been found in the naphthalene/β-methylnaphtha-
lene systems (so-called X-traps) [3.72] and in the naphthalene/naphthalene-
halide systems [3.73]. However, the only quantitative study of the absorption
spectra of defect centres as a function of Δ has been carried out in [3.5].
The absorption spectra of two impurity systems whose parameters are listed
in Table 3.6, are sketched in Fig.3.28. One observes a large change in the
polarization ratio of the defect absorption near the intrinsic absorption
edge in both systems, i.e., a 10-fold and a 30-fold change for benzene and
naphthalene, respectively (on a 4- and 5.6-fold change in ε_i, respectively).

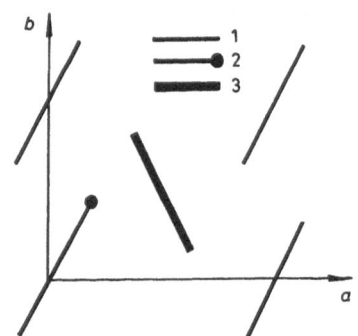

Fig.3.27. Model of an impurity centre with a defect
molecule (L_1 centre) in the ab-plane: 1) ordinary
naphthalene molecule; 2) guest molecule; 3) defect
molecule

Table 3.6. Parameters of K_0^i bands in absorption by defect centres in
benzene and naphthalene crystals [3.5]

Guest	Host	Ω_i [cm^{-1}]	e_i experi-mental [cm^{-1}]	Polariza-tion ratio	Δ [cm^{-1}]	e_i calcu-lated [cm^{-1}]
				$\mathscr{P}_{a/c}$		
Thiophene	Benzene	37759	-45	1.0		
Pyrrole		37777	-28	-		
Furan		37793	-12	0.1		
				$\mathscr{P}_{b/a}$		
Thionaphthene	Naphthalene	31445	-28	1.5	-97	-28
Indole		31462	-14	0.5	-72	-13
Benzfuran		31471	-5	0.05	-56	-4

---

[3] The magnitude of $|\Delta_{cr}|$ is larger than the corresponding value for the iso-
topic impurity in the same crystal; this follows from (3.50) (see following
subsection).

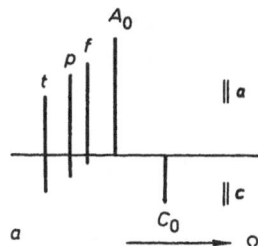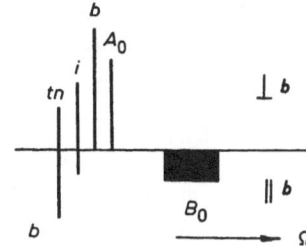

Fig.3.28a,b. Scheme of electronic-absorption spectra of doped crystals of
benzene (a) and naphthalene (b) in a polarized light [3.74-76]: t) thiophene,
p) pyrrole, f) furan, tn) thionaphthene, i) indole, b) benzfuran

3.5.2 The Calculation and Comparison of the Positions of the Energy Levels and the Polarization Ratios of Defect Centres in Naphthalene

The theory of local excitons on defect centres has already been developed in
[3.23,33,71,77]. For a defect centre containing one guest and one defect
molecule a distance **R** apart, the equation that determines the position of
a shallow impurity level associated with a defect molecule (later on called
the *defect level*) has the following form

$$1/\Delta = G_0(\varepsilon_i)[1 - |G_R^0(\varepsilon_i)|^2/G_0^2(\varepsilon_i)] \quad . \tag{3.50}$$

Equation (3.50) can be readily obtained from the general equations (B.7,8).
In deriving (3.50) it was assumed that the level of the guest molecule is
shifted relative to the exciton band by much more than $2\mathcal{M}$. The magnitude of
this shift does not appear in (3.50). Thus, the position of the defect level
is expressed in terms of the same Green's functions as in (3.41).

  Equation (3.50) differs from (3.8), which describes the position of the
monomer level of the isotopic impurity, in the second term containing the
off-diagonal Green's function G_R^0. However, it can be shown that both (3.50)
and (3.8) allow no more than one defect level to be split off each continuous
region of the exciton spectrum. Some difficulty is caused by the necessity of
finding an independent way of determining the magnitude of Δ (which had the
simple meaning of an isotopic shift in the case of an isotopic impurity).
The values of Δ for defect molecules were determined from an analysis of the
spectrum of the defect centre in the vibronic transition region (Sect.6.5.4).
The resulting values are given in Table 3.6.

  Equation (3.50) contains the functions G_0 and G_R^0. The function G_0 may be
expressed in terms of $\rho(\varepsilon)$ and, therefore, is known for the naphthalene
crystal (Fig.3.5). In the model under consideration the defect molecule is

the nearest TNE molecule with respect to the guest molecule. Therefore, $G_{R}^{0} = G_{(a+b)/2}^{0}$. In (2.27) the integrand of this function contains the oscillating factor exp {$i\mathbf{k}(\mathbf{a}+\mathbf{b})/2$}. On the other hand, the integrand of G_0 does not change sign. Therefore, $G_{(a+b)/2}^{0}$ is much smaller than G_0. One can estimate the magnitude of $G_{(a+b)/2}^{0}$ to a lower degree of accuracy using the dispersion law given in (3.49). Taking into account the exact expression for $\rho(\epsilon)$ improves the plane-network approximation (3.49).

It can be seen from Fig.3.29 that the calculated dependence of e_i on Δ and the experimental positions of three defect levels agree fairly well.

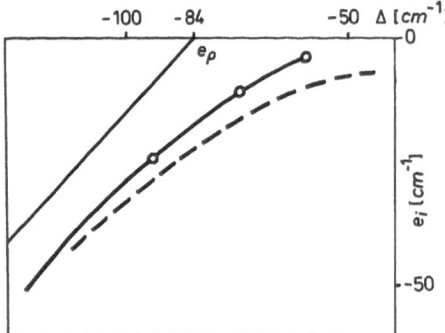

Fig.3.29. Calculated dependences of the position of the defect centre levels (solid line) and an isotopic monomer centre (dashed line) on the value of Δ in the naphthalene host [3.78]: The straight line denotes the position of the defect-molecule term $e = \Delta + e_\rho$ in the oriented gas model (e_ρ is the centre of gravity of the exciton band); the open circles denote the experimental position of the three defect levels

The magnitude of the repulsion of the level (from the term whose position is denoted by a straight line) is greater for the isotopic monomer than for the defect molecules.

In the same model, the polarization ratio for the defect bands is given by the expression

$$\mathscr{P}_{\mu_1/\mu_2} = \left|\frac{\varphi_{\mu_1}}{\varphi_{\mu_2}}\right|^2 \left(\frac{\epsilon_i - \epsilon_{\mu_2}(0)}{\epsilon_i - \epsilon_{\mu_1}(0)}\right)^2 \mathscr{P}_{\mu_1/\mu_2}^0 \quad , \qquad \text{where} \qquad (3.51)$$

$$\varphi_\mu = B_\alpha(\mu,\mathbf{k} = 0) - B_{\alpha_0}(\mu,\mathbf{k} = 0) G_{n\alpha n_0 \alpha_0}^0 / G_0 \quad .$$

The subscripts $n_0\alpha_0$ and $n\alpha$ refer to the guest and the defect molecule, respectively. Due to the factors φ_μ, the change in the polarization ratio as the defect level approaches the exciton level $\mathbf{k} = 0$ obeys a law more complicated than the one describing the change for isotopic centres. Figure 3.30 compares the calculated and experimental dependences of $\mathscr{P}_{b/a}$ on e_i for isotopic monomer centres and for defect centres.

It can be seen from Figs.3.29,30 that, despite of the general qualitative similarity in the behaviour of the spectra, both types of centres also exhibit

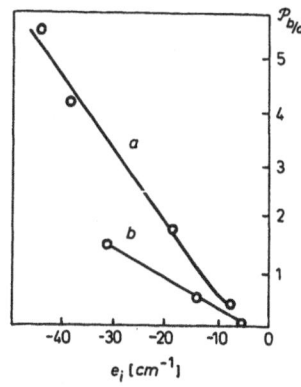

Fig.3.30. Dependence of the polarization ratio $\mathscr{P}_{b/a}$ for the bands of an isotopic monomer center [3.53] (a) and defect centres [3.5] (b) on the distance of K_0' bands from the A_0 band in the naphthalene crystal

considerable differences reflecting the dependence of the spectra on the specific structure of the centre.

3.5.3 Concentration Effects

There is, of course, a finite probability that two, three, or more guest molecules find themselves in a certain volume, even at small impurity concentrations. This gives rise to groups of defect molecules. Aggregates of defect molecules can be found in a crystal, but their structure is more complicated than that of isotopic aggregates [3.79,80]. A few examples of such aggregates are illustrated in Fig.3.31. Dimers, trimers, etc. (L_2, L_3, etc.) are labelled according to the number of defect molecules. Actually, the defect centre L_n contains 2n molecules that differ from the host molecules:

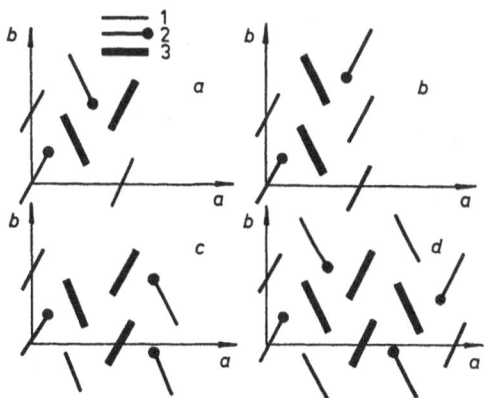

Fig.3.31. Models of aggregate defect centres in the naphthalene crystal [3.5]. Arrangement of the molecules in the ab-plane is shown; a) and b) possible variants of L_2 centres; c) and d) possible variants of L_3 and L_4 centres, respectively; 1) ordinary naphthalene molecule, 2) guest molecule, 3) defect molecule

n guest and n defect molecules. There may also be cases in which two guest molecules deform the same host molecule (a 2L centre). Using this notation, the simplest defect centre considered in Sect.3.5.2 is denoted by L_1.

The calculation of the energy levels and the polarization ratios for more complicated centres is also greatly simplified if the approximation used for L_1 centres is applied. Figure 3.32 compares the calculated and experimental data for crystals of naphthalene containing indole as an impurity. Again, it is the fluorescence spectra that are the most sensitive to the presence of complex centres at low impurity concentrations. The vertical bars in Fig.3.32 denote the averaged calculated positions of the bands of different centres. As the concentration of the impurity increases, the intensity is redistributed in such a way that the fluorescence from deep levels, belonging to large defect centres with complex structures, predominates.

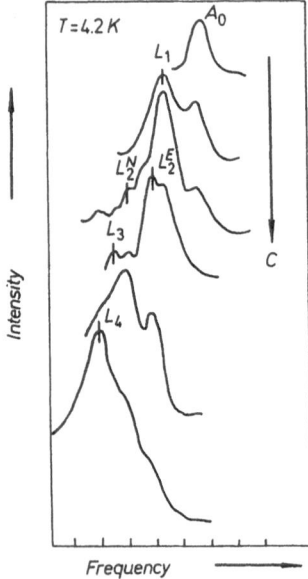

Fig.3.32. Microphotometer tracings of electronic fluorescence spectra of naphthalene crystals with different concentrations of indole guest ($1\% < C < 10\%$) [3.79,80]: A_0 exciton band of the naphthalene crystal, L_2^N and L_2^E correspond to TNE and TE pair centres of defect molecules, respectively

4. Exciton Spectra of Mixed Crystals

This chapter deals with the spectra of mixed crystals, i.e., of binary or complicated crystal solutions with components of comparable concentrations. In crystals such as these it is impossible to discern isolated impurity centres (see Chap.3), and therefore, they must be investigated by completely different methods. The basic available data on the spectra of disordered systems are summarized in this chapter and approximate methods for calculating the spectra are discussed. These methods differ both in the accuracy of the results and in the scope of the information on the Hamiltonian of the perfect crystal that is necessary for these methods to be applied. Two such methods are considered in detail: the average-amplitude approximation and the coherent-potential approximation. Various types of the energy and the optical spectra of disordered systems are analyzed. The experimental data on isotopic mixed crystals of benzene, naphthalene, anthracene, and hexamethylbenzene are also discussed in detail. The absorption spectra of these crystals are used to illustrate the basic types of spectra: one-mode, two-mode, and also some intermediate types. At the same time, the energy spectra of these crystals are classified (with a gap, with a pseudogap, amalgamated bands, etc.). The possibility of establishing the genesis of the exciton multiplets and the exciton bands of perfect crystals by changing the concentration of the components of mixed crystals is demonstrated. The absorption spectra of mixed crystals of deuteronaphthalenes, in which the cluster structure is clearly visible even in equimolecular mixtures, are considered in greater detail. These spectra are compared with the results of calculations.

Finally, the low-temperature kinetics of excitons in mixed crystals in connection with the percolation of singlet and triplet excitons, which has been detected in the fluorescence spectra of several systems of this type, shall be discussed. The interpretation of these results is based on the Anderson localization concept. Such experiments must enable one to investigate various phenomena such as hopping conductivity and transport over the states near the mobility edge under the nonequilibrium conditions determined by the finite lifetime of the exciton.

4.1 Energy Spectra of Mixed Crystals

4.1.1 General Features

The preceding chapter considered the spectra of systems in which individual guest molecules or their simplest aggregates could be independently described by neglecting their interaction. As the concentration increases, the spacing between individual guest molecules and aggregates diminishes, and therefore, their interaction is enhanced. Obviously, both the energy and the optical spectrum of a system are much more complicated in this concentration range, and a satisfactory description requires completely new approaches.

There exists an extensive literature on the spectra of *disordered systems*, including a number of surveys [4.1-6]. Although the surveys generally describe definite types of objects (electrons in disordered systems, nonideal lattice dynamics), the methods and results are essentially the same. We shall focus our attention on problems in which the experimental data on molecular exciton spectra have contributed significantly to existing concepts concerning the properties of disordered systems or have been used for their sufficiently convincing test or illustration.

The basic experimental data refer to the spectra of binary isotopic solutions. Within the framework of the ideal isotopic-substitution model used in Chap.3, such solutions are systems with a *diagonal disorder* described by a discrete parameter. In other words, all of the off-diagonal matrix elements $M_{n\alpha m\beta}$ of the mixed-crystal Hamiltonian are not any different from those of the perfect crystal. Only the diagonal matrix elements $E_{n\alpha} = (E_\rho)_{n\alpha}$ (2.3) are random and can assume two different values E_s (s = 1, 2) depending on which molecule is located at the given site. The isotopic shift Δ is equal to $E_2 - E_1$. In this chapter it is assumed that $E_2 > E_1$; hence, $\Delta > 0$. In accordance with the available experimental data the occupancy of the sites by molecules s = 1, 2 shall be considered to be random. The correlation in their arrangement shall be completely ignored.

It is naturally impossible to obtain a complete quantitative description of such a system. It could also hardly be expected that any one of the approximate methods could accurately describe such a system in detail, considering all of the fine features of the spectrum. However, the purpose of such approximations is to successfully "catch" and describe the gross features of the spectrum in the form of a definite envelope of the true spectrum. As a result, the quantitative analyses of such systems can be made using three approaches.

The first approach is a purely analytical one aimed at establishing the general theorems on the localization of the spectrum and the nature of the wave functions and at elucidating the structure of the spectrum in the neighbourhood of its singular points. Such an approach, which is fascinating in itself and which has already led to a number of fundamental assertions, is also very important in estimating the results obtained by approximate methods.

The second approach involves approximate methods used to determine the fundamental properties but not the details of the spectra of disordered systems.

The third approach involves the direct computation of model systems. The results obtained by this last method have contributed significantly to an understanding of properties of disordered systems, primarily of their energy spectrum.

First we shall consider the spectra of mixed crystals in which one component is present in low concentration. Such crystals contain groups consisting of several guest molecules which are surrounded by host molecules. These entities include the simplest aggregate centres which were discussed in Sect.3.4. Later we shall refer to them as *clusters*, a term widely used in the literature. The cluster concept is obviously somewhat vague, since even singling clusters out is a matter of convention. Of course, the individual clusters interact among themselves. Therefore, we shall define a cluster by taking into account the results to which its existence leads and by proceeding from the character of the related quantum states. We shall include into a cluster those guest and host molecules on which the wave functions of the local excitons are mainly concentrated. For sufficiently deep centres, only the guest molecules themselves may be included in a cluster. However, for shallow centres, the environment surrounding the host molecules, which must be assigned to the cluster, increases progressively. It is particularly important to take into account the host molecules when dealing with the intensities of optical transitions.

The term *mixed crystals* will mainly be applied to solutions whose components are present in comparable high concentrations. The low-concentration approximation is not applicable to such systems. For crystals containing impurities in low concentrations, we shall retain the term *doped crystals*.

4.1.2 Analytical Approach

We shall first of all consider the data on the structure of the energy spectrum which follow from general theoretical considerations. Fundamental results in this field have been obtained by LIFSHITZ [4.1].

According to SAXON-HUNTER's localization theorem [4.7], the mixed crystal does not exhibit a spectrum in the energy ranges where both pure components do not exhibit spectra. On the other hand, as suggested by LIFSHITZ, the exact positions of the spectrum edges for a mixed crystal are independent of the concentration of the components. The spectrum of a mixed crystal spans the entire region in which the spectra of the pure components are found. This follows from the fact that a mixed crystal contains fluctuation regions of individual components, whose dimensions in principle may be arbitrarily large and whose probability of existence is finite. The probability of the formation of large fluctuation regions is naturally exponentially small and, therefore, with a low concentration of one of the components ($C \ll 1$, where C is the fraction of sites occupied by guest molecules) large regions of the spectrum with an exponentially low density of states will exist.

When $C \ll 1$, the impurity-states spectrum is concentrated in narrow regions. If $|\Delta|$ is sufficiently large so that the existence of a monomer impurity level is guaranteed, the position of this level will be that of the simplest singular point. Furthermore, there exist the levels of dimers. They may be classified according to the distance (discrete) between guest molecules (Sect.3.4). As the distances increase, the levels of a dimer converge to the level of the isolated impurity (monomer). Each level of a dimer is the limiting point for the levels of trimers, etc. The number of states that appear in these spectra corresponds to the powers of C (C, C^2, and C^3, ..., respectively). The level distribution in the series is extremely complicated; for example, in an exponentially narrow neighbourhood (of the order of $e^{-1/C}$) of the monomer band, the density of states exhibits a considerable dip.

All of this fine structure is broadened due to additional perturbations, for instance small differences in the value of integrals D, see (2.5), arising from the differences in the interaction of a guest molecule with host molecules and other guest molecules substituting them. An *impurity exciton band* is gradually formed from the whole of this structure with increasing C. Tails with an exponentially low density of states extend from this high density-of-states region and fill the remaining region of the spectrum [4.1].

4.1.3 Approximate and Numerical Methods

As an illustration we shall present results which have been obtained by approximate methods, and also show how their accuracy can be checked through numerical calculations.

Until now the density of states in the energy spectrum of excitons in mixed crystals of nd_0 and nd_8 has been calculated using two different methods. One of these methods, the *coherent-potential approximation* (CPA), is an approximate analytical method [4.6,8,9], which will be considered in more detail in Sect.4.4.1. It enables one to approximate the density of states in a mixed crystal from the density of states, $\rho_0(E)$[1], in the energy spectrum of a one-component crystal. The other method, called the *"negative-factors counting"*, or the DEAN method, involves the direct numerical calculation of the density of eigenvalues from a given Hamiltonian [4.10]. CPA calculations were based on the experimental density of states $\rho_0(E)$ for naphthalene (Fig.3.19, the mean line of the hatched area) [4.11]. The numerical calculations are based on the transfer-integral values, which were determined from the spectra of the aggregate centres. They agree with $\rho_0(E)$ (Sect.3.4). The results are compared in Fig.4.1.

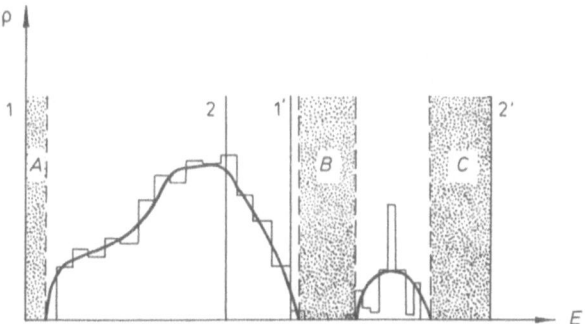

Fig.4.1. Density of states in the energy spectrum of a two-component mixed crystal of nd_0/nd_8; $\Delta/\mathcal{M} = 1.27$, $C_1 = 0.9$, $C_2 = 0.1$ [4.11]: 1 and 1', 2 and 2' are exact edges of the crystal spectra of the first and second components, respectively. Dashed lines) edges of the CPA spectrum determining the positions of the pseudobottom and the pseudotop of the spectrum of each component, respectively. In the A, B, and C regions the density of states is exponentially small; the B region is a pseudogap; the A and C regions are the spectrum "tails" (Sect.4.4). Solid curves) density of states $\rho(\varepsilon)$ in the CPA; histogram - numerical calculation

It is seen that the CPA indicates, as a "spectrum", the sections with a high density of states. The edges of these sections are accurately determined by the CPA. This is true for both the solvent and the solute spectra. The relatively smooth histogram of the first of these spectra is beautifully approximated by the CPA curve. The situation is worse in the region of the solute spectrum, i.e., the density-of-states envelope which can be determined using the CPA differs widely from the highly structured spectrum on the histogram. In this case we are naturally not dealing with fine analytical details, which are lost in the histogram as well, but are rather interested in the main groups of bands of the *cluster structure* (see the discussion of the spectra

---

[1] In this chapter the density of states in a perfect or a mixed crystal is always denoted by ρ_0 and ρ, respectively.

of aggregate centres in Sect.3.4). The accuracy of the CPA and of other similar methods will obviously increase for nearly equimolecular mixtures ($C_1 \approx C_2 \approx 0.5$) and for systems with a considerable contribution from long-range interactions because the fine structure is smeared. On the other hand, this fine structure is clearly revealed in low-dimensional systems. As a result, the accuracy of the CPA is reduced.

From Fig.4.1 it is evident that one must exercise great caution in comparing an experimental spectrum with one calculated using the CPA, since the experimental spectrum (which is closer to the histogram) may prove to exhibit many more structural details than the calculated one.

The densities of state $\rho(E)$ manifest themselves directly in the optical spectra only under special conditions (band-to-band transitions, Sect.5.3). The exciton absorption spectrum differs very strongly from it. In the following section we focus our attention on exciton absorption spectra, while simultaneously discussing the shape of $\rho(E)$.

4.1.4 Choosing Methods to Calculate Spectra

It is necessary to mention one more circumstance, which is very important in the analysis of experimental data. The problem can be formulated from the theoretical point of view as follows: given the Hamiltonian of a disordered system, what is the Green's function or what are the definite correlators? As a result the choice of methods depends exclusively on their efficiency, i.e., on their simplicity and accuracy.

However, the problem is expressed quite differently when applied to the analysis of specific systems, e.g., for molecular excitons. The amount of information on the Hamiltonian of the pure components (which is itself partly derived from the analysis of the impurity spectra!) is very limited, and therefore one must choose those methods for calculating the properties of disordered systems, for which the available information is sufficient. For example, the CPA operates with the density of states in the perfect crystal, $\rho_0(E)$, which may be found by using definite methods (Sects.3.2,3 and 5.2). However, any improvement in the CPA obtained by taking into account the cluster structure requires much more detailed information on the Hamiltonian. This is evident from the equations given in Sect.3.4. Integrals which cannot be expressed in terms of the density of states are encountered in the calculation of any centre containing two or more perturbed molecules. The degree of uncertainty that is still contained in the dispersion law for such a relatively well-studied crystal as naphthalene is evident from the same Sect.3.4. Even a reconstruction of the density of states $\rho_0(E)$ from the experimental data involves great difficulties. Therefore, it is important to have at one's disposal approximate methods which rely on very little spectral information.

It turns out that a fairly good estimate of the basic parameters of the absorption spectrum of a mixed crystal, such as the positions of the centres of gravity of the absorption bands and their intensities, can only be made if the positions of the electronic terms and the components of the exciton multiplets in one-component crystals are known. This approximation (the *average-amplitude approximation*) will be discussed in the following section.

4.2 Average-Amplitude Approximation

4.2.1 Elementary Theory

The approximation used in this simple model was mainly prompted by an experiment carried out in mixed crystals of d-benzenes. The polarized-light absorption spectrum of a crystal of bd_6 with a deuterium content of 81% is shown in Fig.4.2 [4.12]. The structure of this spectrum is complicated in comparison with the simple doublet structure (A_0 and C_0 bands) of the exciton spectrum of a one-component crystal (for example, bd_0, Fig.4.2). The complexity of the spectrum was quite naturally attributed to the fact that the investigated crystal was actually a mixed *multicomponent crystal* containing different isotopic forms of benzene. However, the distribution of the intensity in all of the bands of the components of the spectrum was quite unexpected. It was also difficult to correlate the bands with certain isotopic components. An analysis of the spectrum suggested the following:

1) Both components of the spectrum contain five bands, in accordance with the number of isotopic components of the solution. The bandwidths are comparable to those in a pure crystal.

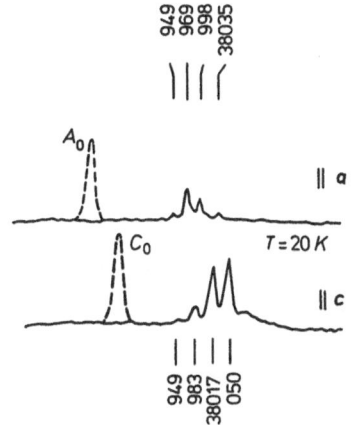

Fig.4.2. Microphotometer tracings of the absorption spectrum of a crystal [4.12] with an isotopic composition: 24% bd_6, 41% bd_5, 24% bd_4, ~5% bd_3, and 6% bd_0; A_0 and C_0 are exciton absorption bands of the bd_0 crystal

2) All of the bands corresponding to isotopic components with a concentration ≳10 are sharply polarized.

3) If we make up exciton doublets from neighbouring bands, their polarization ratio ($\mathscr{P}_{a/c}$) greatly deviates from that of the pure crystal. The magnitude of the Davydov splittings and of the polarization ratios strongly depends on the composition of the mixed crystal. In particular, the magnitude of the splitting increases with the concentration of the corresponding isotopic component.

The existence of distinct doublets of strongly polarized bands suggests the existence in the long-wavelength region ($\mathbf{k} \to 0$) of the exciton states forming *impurity exciton bands*. At the same time, the considerable deviation of the polarization ratios from that in a one-component crystal suggests that the excitation that belongs to a definite impurity exciton band cannot be assigned to a single isotopic component of the mixed crystal. It involves (to a considerable extent) other components as well. Indeed, the mixing of the excited states of different molecules must change the polarization ratio by a mechanism similar to the one considered in Sect.3.2. This is precisely how BROUDE and ONOPRIENKO interpreted their results [4.12]. Their studies laid the foundation for an experimental and theoretical study of the energy and the optical spectra of molecular excitons in mixed crystals.

The equation for Green's function, $G = G^0 + G^0 U G$, in the site representation has the following form

$$G_{n\alpha m\beta}(\Omega) = \frac{1}{\Omega} \delta_{n\alpha m\beta} + \frac{1}{\Omega} E_{n\alpha} G_{n\alpha m\beta}(\Omega) + \frac{1}{\Omega} \sum_{l\gamma} M_{n\alpha l\gamma} G_{l\gamma m\beta}(\Omega) \quad . \qquad (4.1)$$

The energy $E_{n\alpha}$ is reckoned from the ground-state energy. The conductivity tensor is determined by Green's function on the momentum $\mathbf{K} \approx 0$, as in (3.3). As usual, when one calculates the macroscopic quantities, one can take a configurational average, i.e., an average is taken over all of the sites whose probability of being occupied by a molecular species 1, 2, ... is C_1, C_2, ... For a macroscopically homogeneous system, such an averaged Green's function is diagonal in the momentum representation [4.13].

The simplest approximation to the non-averaged Green's function, which corresponds to the physical model described, can be made as follows: $G_{n\alpha m\beta}(\Omega)$ depends exclusively on the sublattice indices α and β, the type of molecules located at the sites $n\alpha$ and $m\beta$ (the indices r and s refer to the components of the solution), and the difference $\mathbf{m} - \mathbf{n}$. Thus, the $G_{n\alpha m\beta}$ are assumed to depend on the average composition of the mixed crystal, but not on the detailed surroundings of the molecules at the sites $n\alpha$ and $m\beta$. Such an approach will be called the *average-amplitude approximation* (AAA); the meaning of this term will be explained below[2].

The quantities $G_{\alpha\beta}^{rs}$ are defined as follows

$$G_{n\alpha m\beta}(\Omega) \rightarrow \begin{cases} G_{\alpha\beta}^{rs}(\Omega,\, n-m) & n\alpha \neq m\beta \quad, \\[2ex] C_s G_{\alpha\alpha}^{ss}(\Omega,\, 0) & n\alpha = m\beta \quad, \end{cases} \tag{4.2}$$

if the site $n\alpha$ contains an r-type molecule and the site $m\beta$ contains an s-type molecule; all of the $G_{\alpha\alpha}^{rs}$ $(\Omega,\, 0) = 0$ at $r \neq s$. Hence, the configurationally averaged Green's function is

$$\langle G_{n\alpha m\beta}(\Omega)\rangle = \sum_{rs} C_r C_s G_{\alpha\beta}^{rs}(\Omega,\, n-m) \quad. \tag{4.3}$$

Transforming (4.3) into the $\mathbf{k}$ representation, we obtain

$$\langle G_{\alpha\beta}(\Omega\mathbf{k})\rangle = \sum_{rs} C_r C_s G_{\alpha\beta}^{rs}(\Omega\mathbf{k}) \quad, \tag{4.4}$$

where

$$G_{\alpha\beta}^{rs}(\Omega\mathbf{k}) = \sum_{n-m} G_{\alpha\beta}^{rs}(\Omega,\, n-m) \exp\,[-i\mathbf{k}(n-m)] \quad. \tag{4.5}$$

If we substitute (4.2) into (4.1) and sum over all the n and m, such that the site $n\alpha$ contains an r-type molecule and the site $m\beta$ contains an s-type molecule, we arrive at the following set of equations for $G_{\alpha\beta}^{rs}$:

$$(\Omega - E_r)\, G_{\alpha\beta}^{rs}\,(\Omega,1) = \frac{1}{C_r}\,\delta_{rs}\,\delta_{\alpha\beta}\delta(1) + \sum_{r'\alpha'1'} C_{r'} M_{\alpha\alpha'}(1')G_{\alpha'\beta}^{r's}(\Omega,\,-1'+1) \quad. \tag{4.6}$$

Finally, transforming (4.6) into the momentum representation, we obtain the following set of equations

$$(\Omega - E_r)G_{\alpha\beta}^{rs}(\Omega\mathbf{k}) = \frac{1}{C_r}\delta_{rs}\delta_{\alpha\beta} + \sum_{r'\alpha'} C_{r'} L_{\alpha\alpha'}(\mathbf{k})G_{\alpha'\beta}^{r's}(\Omega\mathbf{k}) \quad, \tag{4.7}$$

for finding the values of $G_{\alpha\beta}^{rs}$. To solve this system we introduce an auxiliary set of equations,

$$\sum_{s\beta} C_s L_{\alpha\beta} a_{s\beta} = (\mathscr{E} - E_r)a_{r\alpha} \quad. \tag{4.8}$$

From (4.8) it is easy to obtain the orthogonality conditions for the different solutions $a_{s\beta}^\tau$. Imposing the appropriate normalization conditions, we obtain

---

[2] The authors of different papers have used the terms "excitation democracy" [4.14], "average polarizability approximation" [4.15], and "self-consistent-field approximation" [4.16]. The approximation corresponds to the usual average-field type models according to the nature of ansatz (4.2).

$$\sum_{r\alpha} C_r \overset{*\tau'}{a}_{r\alpha} a_{r\alpha}^\tau = \delta_{\tau\tau'} \quad (a) \quad , \quad \sum_\tau \overset{*\tau}{a}_{r\alpha} a_{s\beta}^\tau = \frac{1}{C_s} \delta_{rs} \delta_{\alpha\beta} \quad (b) \quad . \tag{4.9}$$

Using (4.8,9), it can at once be verified that the

$$G_{\alpha\beta}^{rs} = \sum_\tau \frac{a_{r\alpha}^\tau \overset{*\tau}{a}_{s\beta}}{\Omega - \mathscr{E}_\tau} \tag{4.10}$$

are solutions of (4.7), where $\mathscr{E}_\tau = \mathscr{E}_\tau(\mathbf{k})$ are the eigenvalues of (4.8).

A comparison of (4.10) with the usual representation of Green's function in terms of the wave functions of the system [see, for example, (B.9)] shows that the $a_{s\beta}^\tau$ are *excitation amplitudes*, i.e., they determine to what extent the β-type molecules are excited to the quantum state τ with an energy $\mathscr{E}_\tau$.

It is now possible to understand the meaning of the term AAA by comparing (4.2,10). The approximation for Green's functions (4.2) is equivalent to the following approximation for the wave function of the exciton:

$$\psi_{\tau\mathbf{k}}(\mathbf{n}\alpha) \to \frac{1}{\sqrt{N}} a_{r\alpha}^\tau \exp(i\mathbf{k}\mathbf{n}_\alpha) \quad , \tag{4.11}$$

if the site $\mathbf{n}\alpha$ contains an r-type molecule. According to (4.11), all of the r-type molecules at the sites α are assigned the same amplitude $a_{r\alpha}^\tau$.

Since an index r does not appear on the left-hand side of (4.8), we can denote

$$b_\alpha^\tau = (\mathscr{E}_\tau - E_r) a_{r\alpha}^\tau \quad . \tag{4.12}$$

It follows that the amplitudes $a_{r\alpha}^\tau$ obey the *"rule of the lever"*, i.e., the excitation amplitudes for different molecules are inversely proportional to the distance of their terms to the quantum level of the whole system. Using (4.12), we can lower the order of the system (4.8):

$$\sum_\beta L_{\alpha\beta}(\mathbf{k}) b_\beta = \varepsilon b_\alpha \quad , \quad \text{where} \tag{4.13}$$

$$\frac{1}{\varepsilon} = \sum_s \frac{C_s}{\mathscr{E} - E_s} \equiv \phi(\mathscr{E}) \quad . \tag{4.14}$$

The system (4.13) which was obtained here for mixed crystals coincides with a similar system (2.11) which appears in the theory of perfect crystals. Therefore, its eigenvalues are $\varepsilon = \varepsilon_\mu(\mathbf{k})$. They describe the dispersion law of the excitons in the perfect crystal. There is a difference, however, in that a physical meaning must be attached now to the quantities $\mathscr{E}$ that are found from quantities ε by solving (4.14) and determine the positions of the absorption bands. The ε themselves are given no direct physical meaning for mixed crystals.

In the following text we shall label the solutions of the system (4.13) by the index μ, and the different roots of (4.14), corresponding to a given ε_μ, by $\mathscr{E}_{q\mu}$. The subscript q labels the exciton multiplets, while the index μ refers to the components of a multiplet. Also, $\tau = q\mu$.

In a crystal with symmetrically dependent sublattices, all of the $|b_\alpha^\tau|$ approach the same value as $\mathbf{k} \to 0$. If (4.12) is taken into consideration the normalization condition (4.9a) yields

$$|b_\alpha^\tau| = \left| Z \sum_r \frac{C_r}{(\mathscr{E} - E_r)^2} \right|^{-1/2} \equiv b^\tau \quad , \tag{4.15}$$

where Z is the number of molecules in the unit cell. If the quantities

$$B_\alpha^\tau = \frac{b_\alpha^\tau}{\sqrt{Z}\, b^\tau} \tag{4.16}$$

are introduced, they will obey the equations

$$\sum_\beta L_{\alpha\beta} B_\beta = \epsilon B_\alpha \quad , \quad \sum_\alpha |B_\alpha|^2 = 1 \quad , \tag{4.17}$$

which fully coincide with the equations for a perfect crystal. Therefore, the B_α do not depend on the subscript q.

From (4.14) the following expression is obtained for each component of the spectrum, i.e., for any value of μ we have

$$\sum_q \mathscr{E}_{q\mu} = \epsilon_\mu + \sum_s E_s \quad . \tag{4.18}$$

This relation can easily be obtained if (4.14) is considered to be an equation in $\mathscr{E}$ and the sum of its roots is calculated. From (4.18) it can be seen that in the AAA the sum of the Davydov splittings for different exciton multiplets is independent of the concentrations C_s and is equal to the splitting in the perfect crystal.

We shall now consider the band intensities. Integrating the imaginary part of (4.4) with respect to Ω and taking into account (4.9,10), we can convince ourselves that the integral absorption in each of the components of the spectrum is independent of C_s. The intensity distribution is determined by Green's function $\langle G_{\alpha\beta} \rangle$. Inserting (4.10,12,15,16) into (4.4) and taking into account (4.14) we obtain

$$\langle G_{\alpha\beta} \rangle = \sum_{q\mu} \frac{1}{\Omega - \mathscr{E}_{q\mu}} \frac{B_\alpha^\mu \, \overset{*\mu}{B}_\beta / \epsilon_\mu^2}{\sum_s \dfrac{C_s}{(\mathscr{E}_{q\mu} - E_s)^2}} \quad . \tag{4.19}$$

The $\mathscr{E}_{q\mu}$ entering (4.19) are determined from (4.14) in terms of ϵ_μ. The conductivity tensor is

$$\sigma(\Omega) = -\frac{1}{\Omega v} \sum_{\alpha\beta} <G_{\alpha\beta}> \widehat{\mathbf{j}_\alpha \mathbf{j}_\beta^*} = -\frac{1}{\Omega v} \sum_{q\mu} \frac{1}{\Omega - \mathscr{E}_{q\mu}} \frac{\widehat{\mathbf{j}_\mu \mathbf{j}_\mu^*}}{\varepsilon_\mu^2 \sum_s \frac{C_s}{(\mathscr{E}_{q\mu} - E_s)^2}} \qquad (4.20)$$

Thus, within the framework of elementary equations, the AAA makes it possible to describe the positions and intensities of all of the absorption bands of a mixed crystal exclusively in terms of a) the positions of the electronic terms E_s, b) the positions of the bands of the exciton multiplet ε_μ of perfect crystals, and c) the concentrations C_s of the individual components of the mixture. Thus, the AAA is useful in classifying the spectra of mixed crystals. The resulting equations are symmetrical with respect to the components of the mixed crystal. In addition, the AAA exactly describes the first four moments of the absorption spectrum.

Since the poles of (4.20) are on the real axis, all of the resulting absorption bands are infinitely narrow. This is because the AAA does not take into account the exciton scattering. Another shortcoming in the theory becomes evident if one of the C_s tends to zero. Then, according to (4.14), the corresponding $\mathscr{E}_q \to E_q$, i.e., the AAA does not describe the "repulsion" of the impurity levels in dilute solutions (Sect.3.2). Hence, the approximation cannot be used in the low-concentration region.

The dependence of the right-hand side of (4.14) on $\mathscr{E}$ is shown in Fig.4.3 for mixed crystals of two components. The figure also illustrates the graphical solution of (4.14) for $\Delta/\mathscr{M} \gg 1$ (I) and $\Delta/\mathscr{M} \ll 1$ (II) ($\mathscr{M}$ is the half-width of the exciton band of a one-component crystal). In either case, the number of roots of the equation for each component of the spectrum is equal to the number of isotopic components of the mixed crystal. The positions of the roots $\mathscr{E}_{q\mu}$ with respect to the values of E_q and the distance between them for each isotopic component essentially depends on the parameter $\Delta/\mathscr{M}$.

I) $\Delta/\mathscr{M} \gg 1$. From Fig.4.3 it can be seen that in this case all of the $\mathscr{E}_{q\mu} \approx E_q$. Taking into account (4.14), expression (4.19) for Green's function can be maximally simplified to

$$<G_{\alpha\beta}> = \sum_s \frac{C_s}{\Omega - E_s} \delta_{\alpha\beta} \qquad (4.21)$$

This expression is the exact Green's function in the *molecular limit*, i.e., in the limit of non-interacting molecules. The polarization ratio is the same for all of the bands and coincides with that for the exciton multiplet of the one-component crystal.

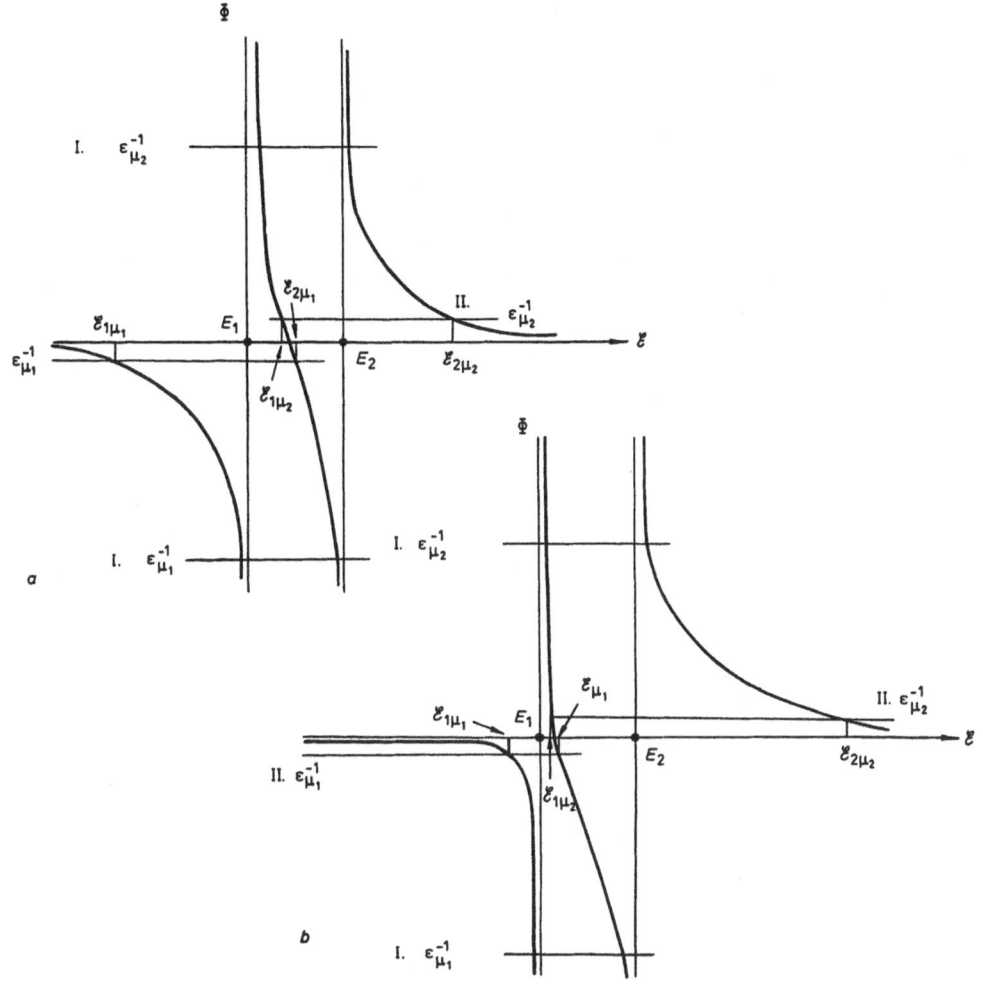

Fig.4.3a,b. Dependence of $\Phi(\mathscr{E})$ on $\mathscr{E}$ for a two-component crystal: a) $C_1 = C_2 = 0.5$; b) $C_1 = 0.2$, $C_2 = 0.8$; I) $\Delta/\mathscr{M} \gg 1$, II) $\Delta/\mathscr{M} \ll 1$; ε_μ) position of the μth component of the exciton multiplet of the perfect crystal measured from the electronic term; E_1 and E_2) values of electronic terms in both isotopic components; $\mathscr{E}_{q\mu}$) roots of (4.14); it is assumed that the electronic term of the perfect crystal is located between the exciton-multiplet components (ε_{μ_1} and ε_{μ_2} have opposite signs)

II) $\Delta/\mathscr{M} \ll 1$. In this case, both the left- and right-hand side external roots of (4.14) are substantially separated from the others (see Fig.4.3) and are approximately equal to

$$\mathscr{E}_\mu^{ext} \approx \varepsilon_\mu + \sum_s C_s E_s \quad . \tag{4.22}$$

In their vicinity, Green's function is equal to

$$\langle G_{\alpha\beta} \rangle \approx \frac{B_\alpha^\mu \overset{*}{B}{}_\beta^\mu}{\Omega - \mathscr{E}_\mu^{ext}} \qquad (4.23)$$

Equation (4.22) is the well-known expression for the energy levels of the *"virtual" crystal* [4.17]. It leads to the correct result at small values of Δ [4.6]. From (4.23) it is seen [taking account of (4.20)] that the oscillator strength of these bands is approximately equal to the total oscillator strength of the intramolecular transitions. The residues of Green's function, which are proportional to the small parameter $(\Delta/\mathscr{M})^2$, correspond to the roots $\mathscr{E}_\mu^{int}$ that lie between the terms E_q. Hence, these bands cannot be seen at small values of Δ. This situation will be called the *amalgamated-band limit*.

The two cases described above are limiting cases. They are useful in following the change in the spectrum of a two-component crystal as a function of the parameter $\Delta/\mathscr{M}$. From Fig.4.3 it can be seen how the magnitude of the exciton splitting $\Delta_q = \mathscr{E}_{q\mu_2} - \mathscr{E}_{q\mu_1}$ for each of the isotopic components increases with decreasing value of $\Delta/\mathscr{M}$. The distance between the external bands of the different isotopic components increases, while the distance between the internal ones decreases. At the same time, the intensity of the internal bands fall off according to (4.19,20), while that of the external bands is enhanced. When $\Delta/\mathscr{M} \ll 1$, the external bands form a multiplet typical of a virtual crystal.

It will be interesting to compare these conclusions with the pattern of an incipient exciton doublet appearing from an impurity band. This phenomenon was described in Sect.3.4.4 in which the spectra of IDC of nd_0 in nd_8 were discussed on the basis of Figs.3.21 and 3.23. The intensity of the two bands with a complicated shape increases with increasing impurity concentration due to a weakening of the monomer band and a strengthening of the bands of the dimers, trimers and the complexer aggregate centres which merge into these two bands. The low frequency band is predominantly polarized perpendicular to **b** and, therefore, is enhanced due to the Rashba effect. On the other hand, the high-frequency band is predominantly polarized parallel to **b** and is weakened for the same reason. The former band is an "external" one, while the latter band is an "internal" one. Therefore, the resulting absorption patterns can be satisfactorily described using either the low-concentration approximation or the AAA if the concentration of one of the components is approximately 10%.

Let us consider the concentration change in the gross features of the absorption spectrum of a mixed crystal as exemplified by a two-component mixture. In this case, (4.14) has the following form

$$\mathscr{E}_{q\mu} - E_0 = \frac{\varepsilon_\mu}{2} \pm \frac{1}{2} \sqrt{\varepsilon_\mu^2 + \Delta^2 - 2\varepsilon_\mu \Delta (C_1 - C_2)} \qquad . \qquad (4.24)$$

The index q can assume one of two values (according to the number of isotopic components) depending on the sign in front of the square root; here $E_0 = (E_1 + E_2)/2$. The values of the terms E_1 and E_2 and, hence, E_0 and Δ may generally depend on the composition of the mixed crystal (*solvent effect*). The magnitude of Δ often remains fairly constant [4.18]. This simplifies a qualitative analysis of the solutions of (4.24).

For each component of the spectrum (i.e,, for a definite value of μ) the curves $\mathscr{E}_{q\mu} - E_0$ as a function of C_1 are branches of the same parabola, whose axis is parallel to the C-axis (provided that Δ is concentration-independent). The parabola always intersects the straight lines $C_1 = 1$ and $C_1 = 0$ at the points $\varepsilon_\mu - \Delta/2$ and $\Delta/2$, and $-\Delta/2$ and $\varepsilon_\mu + \Delta/2$, respectively. When considering spectra for a definite polarization of the incident light, the subscript μ shall be omitted. It will be shown below that the behaviour of the bands of the mixed crystal depends on the mutual arrangement of these four points.

Figure 4.4 shows some characteristic curves for various ratios of ε to Δ, which actually corresponds to different values of the parameter $\Delta/\mathscr{M}$. A change in the sign of $(\varepsilon - \Delta)$ changes the mutual arrangement of the points at each end of the concentration interval and the position of the vertex of the parabola. When $\varepsilon = \Delta$, the vertex is located at the point $C_1 = 1$. When $|\varepsilon| \ll \Delta$ and $|\varepsilon| \gg \Delta$, the vertex moves unboundedly to the right. As a result, in both of these limiting cases the parts of the parabola that lie in the interval $0 < C_1 < 1$ are practically linear. Also the dependence of the positions of the bands on the concentration has to be linear for both isotopic components. The points $\varepsilon \pm \Delta/2$ represent the positions of the exciton absorption bands for both one-

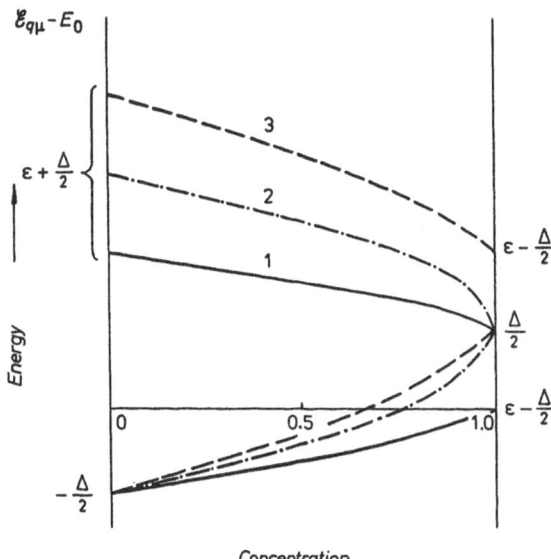

Fig.4.4. Concentration dependence of $\mathscr{E}_{q\mu}$ according to (4.24) for various ratios between $\varepsilon_\mu = \varepsilon > 0$ and Δ (one-spectrum component or a mixed crystal with one molecule in the unit cell [4.18]):
1) $\varepsilon = 0.5\,\Delta$, 2) $\varepsilon = \Delta$, 3) $\varepsilon = 1.5\,\Delta$

component crystals. The terms of these crystals are located at the points $\pm\Delta/2$.

The behaviour of the absorption bands of a mixed crystal is described by the type of concentration dependence of their positions in the spectrum, or by the answer to the question: which points at the ends of the concentration interval do the separate branches of the parabola pass through?

When $\varepsilon < \Delta$, the lower branch of the parabola passes through the points $\varepsilon - \Delta/2$ and $-\Delta/2$, while the upper branch passes through the points $\Delta/2$ and $\varepsilon + \Delta/2$. Thus, each branch of the parabola can tentatively be associated with a separate isotopic component. The evolution of the band of the first isotopic component consists of a gradual transition from the position of the term for the guest molecule (which coincides with the position of the level for the monomer centres if the repulsion is neglected, Chap.3) to the position for the component of the exciton multiplet. The reverse picture is true for the second component. Such a behaviour for the bands of a mixed crystal is called a *two-mode* behaviour. The limiting case ($\varepsilon \ll \Delta$) in which the branches of the parabola straighten out on the segment $0 < C_1 < 1$ corresponds to the molecular limit.

When $\varepsilon > \Delta$, the lower branch of the parabola passes through the points $\Delta/2$ and $-\Delta/2$, while the upper branch passes through the points $\varepsilon - \Delta/2$ and $\varepsilon + \Delta/2$. In this case, the bands $\mathscr{E}_{q\mu}$ cannot, even conventionally, be assigned to one of the isotopic components. The intensity of the lower (internal) band is very low in accordance with (4.20). Practically the entire intensity is concentrated in the upper (external) band. The concentration dependence consists of a gradual transition from the position for the exciton band of one of the one-component crystals to the position for the corresponding exciton band of the other. Such a behaviour is called *one-mode* behaviour. The limiting case in which $\varepsilon \gg \Delta$ corresponds to the limit of the amalgamated bands. The concentration dependences of the positions of the bands in the molecular and amalgamated-band limits for a two-component crystal and for two polarizations of the incident light are shown in Fig.4.5.

The borderline case in which $\varepsilon = \Delta$ (Fig.4.4) separates the two- and one-mode behaviours. The vertex of the parabola is at the point $C_1 = 1$; the values of $\varepsilon - \Delta/2$ and $\Delta/2$ coincide. Therefore, it is impossible to clearly establish through which points at the edges of the concentration interval $0 < C_1 < 1$ each branch of the parabola passes through. As a result, the adopted classification of band behaviour becomes meaningless.

The borderline case corresponds to the limiting situation in which the split off monomer impurity level of the sth component is, appears as $C_s \to 0$. Thus, one of the experimental manifestations of the two-mode band behaviour is the pres-

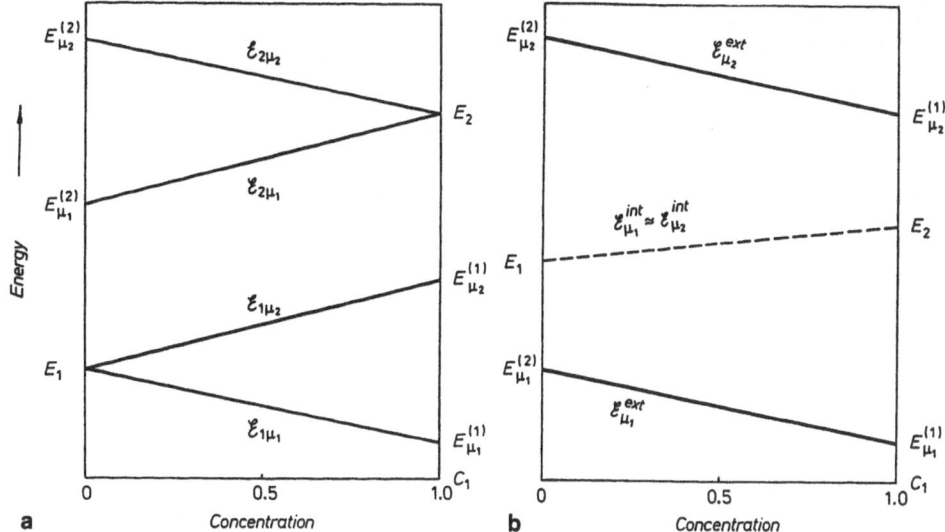

Fig.4.5a,b. Two-mode (a) and one-mode (b) regimes of the concentration dependences of the exciton absorption band positions $\mathcal{E}_{qu}$ in mixed crystals: $E_{\mu}^{(s)} = E_s + \varepsilon_{\mu}$ are the positions of the exciton multiplet components in an s-type perfect crystal

ence of a split-off monomer impurity level for each of the components of a mixed crystal, as $C_s \to 0$. The condition that must be fulfilled for the splitting off of the impurity level is that $|\Delta| > |\Delta_{cr}|$. The values of $\Delta_{cr}^{\pm}$ are close to those of ε_{μ} but still differ from them. This also reflects the inaccuracy of the AAA in the low-concentration range and the vagueness of the definition of the one- and two-mode regimes (Sect.4.4).

The two- or one-mode behaviour of the exciton bands of a mixed crystal implies a definite concentration dependence of their positions in the spectrum. It should be remembered that each band corresponds to the excitation of the molecules of both isotopic components. The collective phenomenon of such an excitation is naturally a maximum in the amalgamated-band limit and a minimum in the molecular limit.

The average-amplitude approximation was developed by BROUDE and RASHBA [4.19] in connection with the interpretation of the experimental data on the spectra of benzene [4.12]. It was later used for a quantitative treatment of the data on solutions of d-naphthalenes [4.20,21] and d-benzenes [4.18]. It was also independently introduced by HUBBARD [4.22] and widely applied to the lattice dynamics (the "*isodisplacements method*") after CHEN, SHOCKLEY and PEARSON published their results [4.23]. A survey can be found in [4.24,25].

4.2.2 Average-Amplitude Approximation and Experimental Results: Mixed Crystals of d-Benzenes

In comparing the results of the AAA with the experimental data one should pay attention to the positions of the bands of the exciton multiplets, $\mathscr{E}_{q\mu}$, and their integral intensities $\sigma_{q\mu}$. It should be noted at once that in the AAA the absorption bands are delta functions. Therefore, according to the AAA the values of $\mathscr{E}_{q\mu}$ should be compared with those for the centres of gravity of the experimental absorption bands. If the spectrum is complicated by a cluster structure, $\mathscr{E}_{q\mu}$ is the overall centre of gravity, and $\sigma_{q\mu}$ is the integral intensity of the group of bands that takes into account the entire cluster structure (Fig.3.21).

We should also mention some of the other limitations of the AAA in the form described in the preceding section. The model is only applicable to isotopically mixed crystals. It is also assumed that the magnitudes of the Davydov splittings in one-component crystals are equal, as are the isotopic shifts Δ for free molecules and those in a crystal.

Just those spectra of mixed crystals which are made up of relatively narrow, isolated absorption bands can be most favourably compared with the results of the AAA. This property is inherent in the absorption spectra of mixed crystals of d-benzenes [4.18]. Therefore, these spectra were used to test the AAA.

As can be seen from Fig.4.2, the absorption spectrum of such a mixed crystal consists of several narrow, distinctly visible bands similar to A_0 and C_0. The bands of different cluster aggregates in crystals of d-benzenes are located in the immediate vicinity of the monomer band (Fig.1.7). Actually, the bands of a mixed crystal are envelopes of a group of bands. They are nearly symmetrical, and therefore, the positions of their maxima can be considered to be the positions of the centres of gravity.

For d-benzenes, however, some other limitations of the AAA are violated. The magnitudes of the Davydov splittings (Table 3.1) and the isotopic shifts for molecules in the free state and in various isotopic solvents (solvent effect) differ noticeably [4.26]. In calculating the spectra of two-component mixed crystals of d-benzenes the differences in the magnitudes of the Davydov splittings are neglected. This introduces a certain error, which will be discussed below. The dependence of the value of Δ on the composition of the matrix can be taken into account by the interpolation formula

$$E_s(C) = E_s^0 + \sum_{s'} D_{ss'} C_{s'} \quad , \qquad (4.25)$$

where E_s^0 is the energy of the molecular term for the vapour, and $D_{ss'}$ is the vapour-crystal shift that results when an s-type molecule is introduced into an s'-type crystal. For the terms of a binary mixed crystal,

$$E_1 = E_1^0 + D_{11}C_1 + D_{12}C_2 \quad ,$$

$$E_2 = E_2^0 + D_{21}C_1 + D_{22}C_2 \quad . \tag{4.26}$$

Hence,

$$\Delta = \Delta_0 + (D_{21} - D_{11})C_1 + (D_{22} - D_{12})C_2 \quad .$$

The concentration dependences $\mathscr{E}_{q\mu}(C)$ have been investigated for three two-component crystals of d-benzenes [4.18,27]: bd_0/bd_6, bd_0/bd_1, and bd_0/bd_2. We shall only consider in detail the first two systems, since the third is a trivial intermediate case.

The experimental results for the a and c components of the absorption spectra of mixed crystals of bd_0 and bd_6 are shown in Fig.4.6. The solid lines represent the values of $\mathscr{E}_{q\mu}(C)$ which were calculated using (4.24). These dependences for each spectrum component represent the branches of the same parabola whose axis is tilted towards the C-axis because of the concentration dependence $\Delta(C)$. To emphasize the genetic relationship between the bands of the impurity excitons and those of the exciton multiplet in the one-component crystal, we shall label them identically, as in one-component crystals, but in script, and with a subscript corresponding to the mixcture component.

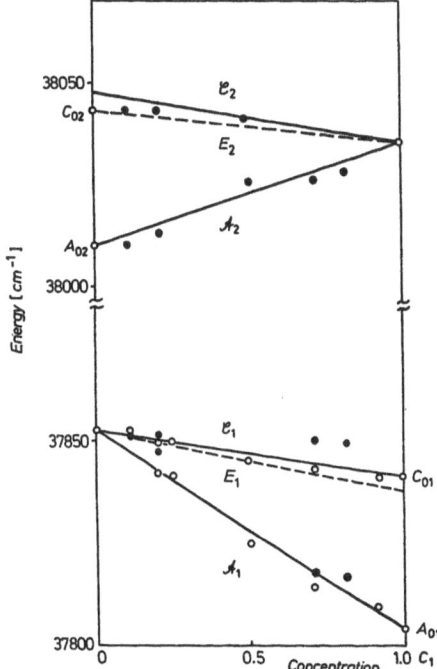

Fig.4.6. Concentration dependence of the exciton absorption band positions $\mathscr{E}_{q\mu}$ of a mixed crystal bd_0/bd_6: A_{0q} and C_{0q} are positions of the A_0 and C_0 bands of the exciton multiplet of the q-th one-component crystal. The $\mathscr{A}_q$ and $\mathscr{C}_q$ bands correspond to $\mathscr{E}_{q\mu}$ with different μ: o and •) experimental data [4.18,27], respectively; solid lines) calculation in the AAA; dashed lines) positions of the electronic terms; C_1) bd_0 concentration

The calculated dependences are practically linear. This is in agreement with the fact that $|\varepsilon| \ll \Delta$ ($\varepsilon_a = -36$ cm^{-1}, $\varepsilon_c = 4$ cm^{-1}, $\Delta \sim 200$ cm^{-1}). Hence, the concentration behaviour of the positions of the absorption bands must be similar to the behaviour in the molecular limit of the two-mode regime (Fig.4.5a). The error due to the difference in the magnitudes of the Davydov splittings in crystals of bd$_6$ and bd$_0$ (Sect.3.3) only manifests itself in the fact that the calculated branch $\mathscr{E}_2$ does not intersect the point that corresponds to the position of the C_0 band of the bd$_6$ crystal.

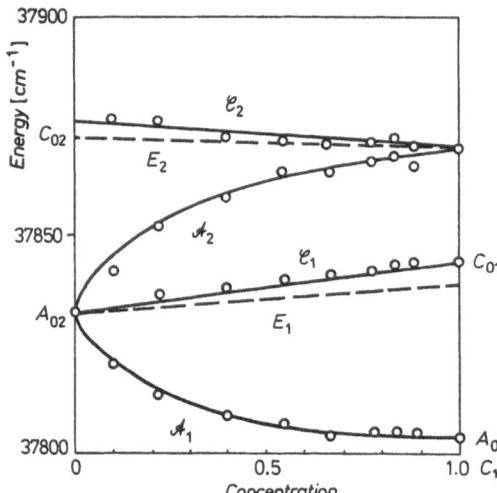

Fig.4.7. Concentration dependence of the exciton absorption band positions $\mathscr{E}_{qu}$ in a mixed crystal bd$_0$/bd$_1$ (see the designations in the caption to Fig.4.6) [4.18]

The data for mixed crystals of bd$_0$ and bd$_1$ are shown in Fig.4.7. The concentration behaviour of the bands is of a two-mode type. The exciton doublets of one-component crystals can be clearly seen to gradually develop from the impurity band of bd$_0$ in bd$_1$ and the band of bd$_1$ in bd$_0$. Four absorption bands are observed over the entire concentration range. As $C_s \to 0$, the impurity bands approach the edges of the exciton bands of the one-component crystals (Fig.3.4 for $|\Delta| = 30$ cm^{-1}). A slight decrease in $|\Delta|$ would make it equal to $|\Delta_{cr}^{\pm}|$ and destroy the two-mode regime.

The solid lines in Fig.4.7 represent the branches of the two parabolas. The calculated dependences $\mathscr{E}_{qu}(C)$ of the two components of the spectrum show quite different behaviours. For the c component, the function $\mathscr{E}_{qu}(C)$ is practically linear over the entire concentration range because $\varepsilon_c \ll \Delta$ ($\Delta = 30$ cm^{-1}). The behaviour of the absorption bands for this component of the spectrum is of a two-mode type with a concentration dependence similar to the one in the molecular limit. For the a component, $|\varepsilon_a| \approx \Delta$, and the behaviour of the band indicates that the regime is close to its breakdown, although

112

remaining two-mode. The large curvature of both branches and the fact that they meet at $C_1 = 0$ suggest this kind of behaviour.

The integral intensities of the absorption bands of a mixed crystal can be expressed in a particularly simple form when the absorption intensity of a one-component crystal is taken to be unity; from (4.19) it follows that

$$I_{q\mu}(C) = \frac{\sigma_{q\mu}(C)}{\sigma_\mu} = \frac{\int \sigma_{q\mu}(\Omega)d\Omega}{\int \sigma_\mu(\Omega)d\Omega} = \frac{\epsilon_\mu^{-2}}{\sum\limits_s \frac{C_s}{(\mathscr{E}_{q\mu} - E_s)^2}} \quad . \tag{4.27}$$

The experimental and calculated results for the intensities for a mixed crystal of bd_0 and bd_6 are shown in Fig.4.8a. A practically linear dependence $I_{q\mu}(C)$ is observed for all four bands. The dependence of the positions of these bands on C (Fig.4.6) is also a linear one. Therefore, the spectra of mixed crystals of bd_0 and bd_6 are an excellent example of the two-mode behaviour under the conditions of the molecular limit.

Quite a different picture is observed for the dependences $I_{q\mu}(C)$ for crystals of bd_0 and bd_1 (Fig.4.8b). The concentration behaviour of $I_{q\mu}$ as well as the positions of the impurity bands differ substantially for the two components of the spectrum. The dependences $I_{q\mu}(C)$ for the c component are almost linear, which is characteristic of the two-mode regime away from the breakdown. The dependences $I_{q\mu}(C)$ for the a component are essentially nonlinear. The weakening of the intensity of the internal bands, which is particularly noticeable for I_{2a}, is conspicuous. Already at $C_1 = 0.1$, the intensity of the "solvent band" decreases by a factor of 2.5 and becomes 1.5 times less than I_{1a}, which

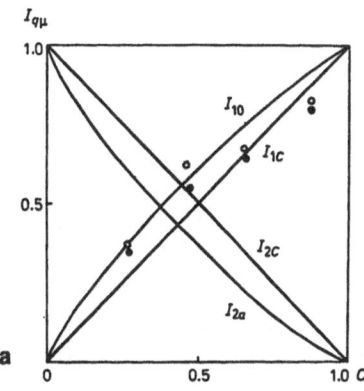
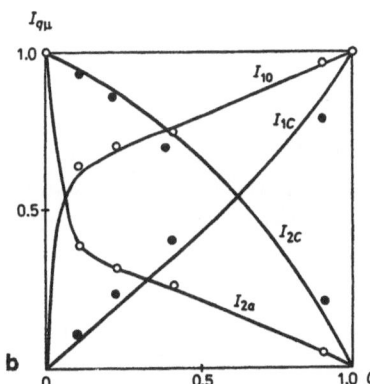

Fig.4.8a,b. Concentration dependence of the integral intensities $I_{q\mu}(C)$ of the exciton bands of mixed crystals: o and •) experimental data from [4.18] for I_{qa} and I_{qc}, respectively; solid lines) calculation in the AAA; C_1) concentration of bd_0; a) mixed crystal bd_0/bd_6; b) mixed crystal bd_0/bd_1

is actually the intensity of the "impurity band". This example illustrates the incipience of a one-mode band behaviour on the verge of the breakdown of the two-mode regime.

4.3 Mixed Crystals of d-Naphthalenes. The Genesis of Exciton Bands

4.3.1 Spectra of Mixed Crystals of d-Naphthalenes

The spectra of mixed crystals of d-naphthalenes have been studied very thoroughly. They have been the subject of attention of many researchers who were interested in both the experimental and the theoretical aspects of mixed-crystal spectroscopy. Until now, the spectra of d-naphthalenes have mainly been used to test new concepts. What makes these spectra so remarkable? Most likely, it is their diversified properties. No approximation, be it the average-amplitude approximation or the coherent-potential approximation, can describe all of the spectral properties of mixed crystals. Essential details were omitted, which stimulated the further development of the theory. Its efficiency was tested by carefully comparing the results with the experimental data.

A qualitative description of the spectra of mixed crystals of d-naphthalenes can best be presented on a three-dimensional "panorama" as shown in Fig.4.9 for a mixed crystal of nd_0 and nd_8. In the foreground one can see the IDC spectrum of nd_0 in nd_8 and in the background the IDC spectrum of nd_8 in nd_0. The main part of the panorama describes the gradual transformation of the spectrum as the concentration of nd_0 increases.

In each vertical section that corresponds to a definite composition of the mixed crystal, the absorption spectrum consists of several bands. This suggests that the spectra are not of a one-mode type, which agrees with the value of the parameter $\Delta/\mathcal{M}$ being equal to 1.3. At the same time, the pattern of the spectrum differs considerably from the simple pattern of the two-mode regime in that a fine structure is observed in the low-frequency range and that the intensity of the absorption in the b component dominates. The origin of this fine structure, discussed in Sect.3.4, is the formation of clusters of various shapes. The cluster structure is only seen in the spectral region of the lighter isotope and is noticeable up to relatively high concentrations ($C_1 \approx 0.5$). The intensity distribution in the spectra is governed by the weakening of the internal bands and a strong dicroism in the exciton absorption of the one-component crystals of d-naphthalenes. This last factor greatly complicates the shape of the spectrum and its interpretation; to say nothing of the fact that at one time it was the main obstacle to make the assignment of the A_0 and B_0 bands as the two components of one exciton doublet.

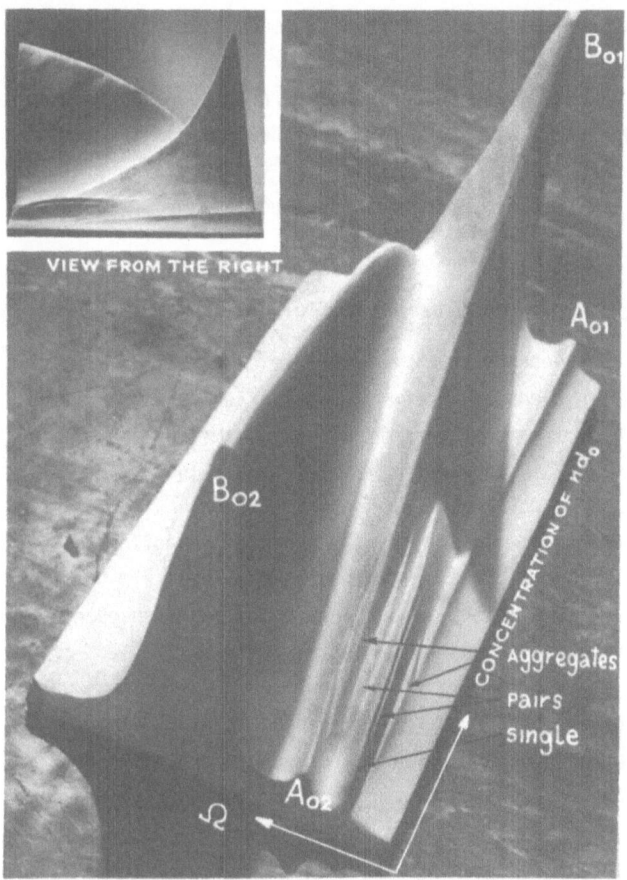

VIEW FROM THE RIGHT

Fig.4.9. Panoramic view of the concentration dependence of the absorption spectrum of a mixed crystal nd_0/nd_8; the right-side view of the spectrum is shown in the insert [4.28]

The spectra of mixed crystals exhibit a high degree of polarization. The most intensive $\mathscr{B}_1$ and $\mathscr{B}_2$ bands, which transform to the B_0 bands of individual components as $C_s \to 1$, are polarized parallel to **b**. A preferred polarization parallel to **b** is also exhibited by the structural spectrum located between the A_0 band of nd_8 and the monomer K_0' band of nd_0 over a wide range of nd_0 concentrations. The weak A_0 band of nd_8 and the weak structural spectrum of nd_0 (located on the low-frequency side of the monomer band of nd_0 and transforming with increasing concentration to the A_0 band of nd_0) are polarized perpendicular to **b**.

Thus, in the spectra of a mixed crystal of nd_0 and nd_8 one can single out four groups of bands which can be paired according to their polarization properties. Therefore, the absorption pattern is of a two-mode type. In this case, the two-mode regime is naturally far from the molecular limit, since the internal bands are greatly weakened. The bands in the perpendicular **b** component are weakened due to the dichroism already present in the initial one-component crystals. An additional weakening due to the mixing of the excitations of the separate isotopic components results in the disappearance of the $\mathcal{A}_2$ band of nd_8 at $C_2 \approx 0.6$.

4.3.2 The Naphthalene Spectra in the Average-Amplitude Approximation

At one time the concentration dependence of the centres of gravity $\mathscr{E}_{q\mu}$ for the four groups of absorption bands of a mixed crystal of nd_0 and nd_8 was the touchstone for the use of the average-amplitude approximation in the quantitative analysis of mixed-crystal spectra [4.21]. The values of ε_μ and Δ are known for the crystals of d-naphthalenes, so that the application of the AAA is a straightforward procedure. In addition, the basic assumptions of the AAA, such as the equality of the Davydov splittings in pure crystals and the equality of the isotopic shifts Δ in the vapour and in any deuteronaphthalene matrix, are satisfactorily fulfilled in these crystals.

Figure 4.10 summarizes the experimental and calculated data for a mixed crystal of nd_0 and nd_8. The $\mathcal{A}_q$ and $\mathcal{B}_q$ bands correspond to a polarization of the incident light perpendicular and parallel to **b**. An appreciable discrepancy between the AAA and experiment only arises near the ends of the con-

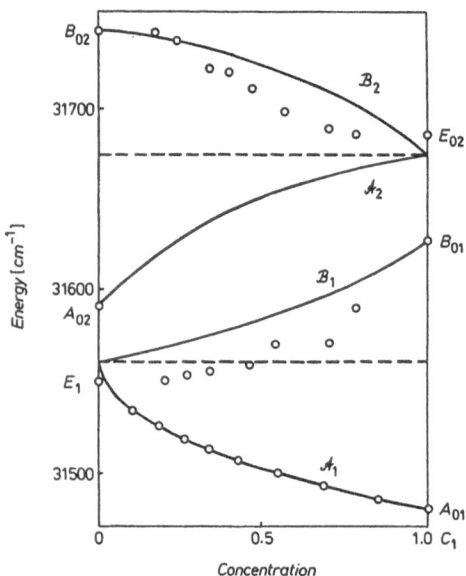

Fig.4.10. Concentration dependence of the exciton-band positions in a mixed crystal $\mathrm{nd}_0/\mathrm{nd}_8$: o) experimental data [4.29]; solid lines) calculation in AAA; dashed lines) electronic term positions; C_1) nd_0 concentration

centration dependence for the minority isotopic component. Both the experimental and calculated dependences $\mathscr{E}_{q\mu}(C)$ in mixed crystals of nd_0 and nd_8
exhibit well-defined two-mode features: doublets of exciton bands ($\mathscr{A}_q$ and
$\mathscr{B}_q$) which transform to the A_0 and B_0 bands of the corresponding one-component
crystals, gradually develop from the impurity bands of nd_0 in nd_8 and nd_8 in
nd_0. This behaviour is also in agreement with a value of $\Delta/\mathscr{M}$ equal to 1.3
and with the fact that $\Delta > |\varepsilon_\mu|$ for both components of the spectrum. In this
case, the two-mode behaviour of the absorption bands is similar to the behaviour for Case 1 in Fig.4.4. The difficulty involved in observing the $\mathscr{A}_2$
band in the component perpendicular to **b** is due to the above-mentioned absorption dichroism in crystals of d-naphthalenes and the weakening of the internal bands.

An essentially different situation is observed in a mixed crystal of nd_0
and $\beta\text{-}nd_4$; it is shown in Fig.4.11. As in the preceding case, the experimental
values of the positions of the $\mathscr{A}_1$ and $\mathscr{B}_1$ bands over the entire concentration
range, $C_1 \lesssim 0.5$, have been determined as the positions of the centres of gravity
of the groups of polarized bands that belong to the cluster structure in the
spectrum of nd_0. Since for this crystal the parameter $\Delta/\mathscr{M} \approx 0.8$, the total intensity of the group of internal bands ($\mathscr{B}_1$) is markedly lower than in a crystal of nd_0 and nd_8. The $\mathscr{A}_2$ band is not observed at all. The positions of the
bands of this crystal agree with the calculated positions over the entire in-
termediate-concentration range. However, this situation is so similar to that
at the border (Fig.4.4, curve 2) that the AAA spectrum and the experimental
spectrum can be assigned to opposite regimes.

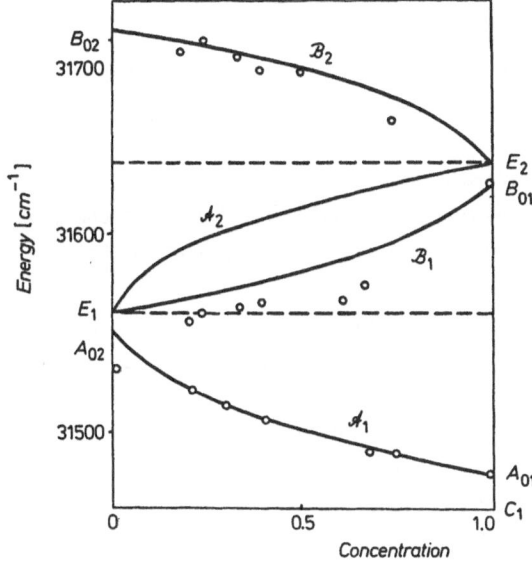

Fig.4.11. Concentration dependence of the exciton-absorption-
band positions in a mixed crys-
tal $nd_0/\beta\text{-}nd_4$: o) experiment
[4.21]; solid lines) calculation
in the AAA; C_1) concentration of
nd_0

In the component of the spectrum perpendicular to **b**, $|\varepsilon_a| > \Delta$. Therefore, according to the AAA calculation, the impurity level for nd_0 in $\beta-nd_4$ does not split off (at $C_1 = 0$ the E_1 level lies slightly above the A_{02} band, Fig.4.11), and the behaviour of the mixed-crystal bands is of a one-mode type. The $\mathcal{A}_1$ branch of the parabola joins the A_0 bands of one-component crystals at the edges of the concentration range. The other branch of this parabola $\mathcal{A}_2$ joins the positions of the terms of both components. Actually, however, a discrete K_0' level exists because of a "repulsion". Therefore, the $\mathcal{A}_1$ branch begins at the level of the monomer nd_0 (at $C_1 \approx 0$) and ends at the position of the A_0 band of $\beta-nd_4$ (at $C_1 = 1$). Hence, from the experimental point of view the regime is of a two-mode type, even though the inner $\mathcal{A}_2$ band is not observed.

For the component of the spectrum parallel to **b**, $\varepsilon_b < \Delta$. Therefore, E_2 lies above B_{01} (Fig.4.11) and the resulting behaviour of the $\mathcal{B}$ bands in the AAA is of a two-mode type, although it is close to border behaviour. However, a splitting off of the impurity level of $\beta-nd_4$ in nd_0 cannot be experimentally observed, since $\Delta < \Delta_{cr}^+$. The position of the external $\mathcal{B}_2$ band changes continuously from that of the B_0 band of one one-component crystal to that of the B_0 band of the other one-component crystal.

This specific situation is naturally due to the fact that $\Delta \approx |\varepsilon_\mu|$ for both components of the spectrum, and therefore, the behaviour of all of the bands is of a borderline type. The above analysis shows that a classification of the regimes in such cases is only a matter of convention.

The two examples of mixed-crystal spectra considered above permit a qualitative description of all of the other possible cases for crystals of d-naphthalenes. For mixed crystals with $\Delta > \Delta_{cr}^+ \approx 107$ cm^{-4} (Sect.3.3) the absorption spectra must have the same general shape as the spectrum of a mixed crystal of nd_0 and nd_8 (Fig.4.10). If $\Delta_{cr}^+ > \Delta > |\Delta_{cr}^-|$ ($\Delta_{cr}^- \approx -44$ cm^{-4}, Sect.3.3), the mixed-crystal spectra must be similar to the spectra of a mixed crystal of nd_0 and and $\beta-nd_4$. The spectra must exhibit a one-mode behaviour of the energies $\mathcal{E}_{q\mu}$ in the component parallel to **b** and a two-mode behaviour in the component perpendicular to **b**. If $\Delta < |\Delta_{cr}^-|$, the two-mode regime will transform to a one-mode regime, so that the spectrum must adopt a form similar to the one for Case 3 in Fig.4.4. Since the values of Δ for binary systems of d-naphthalenes can vary from $\Delta > \Delta_{cr}^+$ to $\Delta < |\Delta_{cr}^-|$ (Table 3.3), any kind of absorption spectra can be obtained. This is what makes mixed crystals of d-naphthalenes so open to experimental investigations. The unique position of the d-naphthalenes in mixed-crystal spectroscopy, as well as in that of doped crystals, is also due to the fact that the value of the parameter $\Delta/\mathcal{M}$ varies from 1.3 to 0.12, which is the most interesting range for investigations. It is also significant that the widths of the exciton bands, $\mathcal{M}$, of one-component crystals are sufficiently large. Accordingly, the cluster splittings are also large and the corresponding structure is easily observed. Examples of the opposite behaviour are the spectra of the mixed crystals of d-benzenes, in which the cluster structure is concentrated within a narrow frequency range because of the relative small values of $\mathcal{M}$. Therefore, this structure can barely be observed in the absorption spectra.

The analysis of the mixed-crystal spectra of d-naphthalenes concludes the consideration of the use of the AAA for the quantitative analysis of mixed-crystal spectra. The series of crystals investigated thus far includes practically all of the cases of interest: the two-mode regime in the molecular limit (mixed crystal of bd_0 and bd_6) and away from it (mixed crystal of nd_0 and nd_8), the intermediate cases of a two-mode regime close to the breakdown (mixed crystal of bd_0 and bd_1) and a broken-down two-mode regime for one component of the spectrum (mixed crystal of nd_0 and β-nd_4). In every one of these cases, the AAA led to a correct description of the concentration dependences and a reasonable estimate of the positions of the centres of gravity of the experimentally observed bands or band groups in the absorption spectra.

4.3.3 The Difficulties Involved in Singling out an Exciton Multiplet

Throughout the book we have repeatedly referred to exciton multiplets which correspond to the pure electronic excitation of molecular crystals. In analyzing the experimental spectra it is often difficult to single out such multiplets.

The main experimental feature that suggests that a certain band belongs to an exciton multiplet is its sharp polarization along one of the principal crystallographic axes or along an axis perpendicular to it. This sharp polarization, however, may not suffice to be able to assign a band to an exciton multiplet, since the absorption spectrum of the crystal may contain sharply polarized impurity or defect bands as a result of the accidental orientation of the guest molecules along one of the crystallographic axes or as a result of the Rashba effect.

An additional quantitative feature which may aid in assigning sharply polarized bands to a multiplet is the ratio of their intensities in the different components of the spectrum. In the Heitler-London approximation, the intensity of an exciton multiplet is determined by the squares of the projections of the dipole moment of the electronic transition in the molecule onto the crystallographic axes. This approximation can be applied to allowed molecular transitions. Generally, however, the magnitude of the corrections to the dipole moment of an electronic transition in a molecule, due to the perturbing crystal field, are comparable to the magnitude of the dipole moment itself. In this case the approximation is not valid. An outstanding example is the polarization ratio $\mathscr{P}_{b/a}$ for the B_0 and A_0 bands of the naphthalene crystal. We shall consider in detail the spectrum of this crystal so as to clearly demonstrate the difficulties involved in singling out an exciton multiplet.

The sharp polarization of the exciton bands of the naphthalene crystal must be observed in directions perpendicular (ac-plane) and parallel to the monoclinic b-axis. The corresponding two components of the spectrum are shown in Fig.1.3. The component parallel to **b** shows a broad intensive B_0 band; it is sharply polarized, and its assignment as the b component of the exciton doublet appears quite natural. In the component perpendicular to **b** of the absorption spectrum of a very pure crystal one can observe a weak, sharply polarized A_0 band. The half-width of the A_0 band is almost 20 times less than that of the B_0 band (at $T = 4.2$ K), while its intensity is 100 times smaller ($\mathscr{P}_{b/a} \approx 100$). In the Heitler-London approximation the value of $\mathscr{P}_{b/a}$ must be equal to 0.24.

If no special precautions are taken in the purification of the crystals, other sharply polarized bands at wavelengths longer than that of the A_0 band and of comparable intensities can also be observed in the absorption spectrum. As a result, the assignment of the A_0 and B_0 bands as an exciton doublet had been the subject of discussions for many years. The problem was only resolved after an independent method of establishing the *genesis of exciton absorption bands* from their evolution in the absorption spectra of mixed crystals was found.

4.3.4 Genesis and Evolution of Exciton Multiplets

Figures 4.6,7 and 10 are "composition-property" diagrams, in which the "property" is the position of an absorption band. They can be used to establish the genesis of exciton bands. The *genesis of multiplets* is completely described by the one-to-one correspondence between the electronic terms of the molecule and the multiplets of the exciton bands of the crystal. The *evolution of a multiplet* implies a controlled variation of the splitting. In a mixed crystal, the composition is the regulator of this variation.

The problem of determining the genesis of a multiplet consists of two stages. At first, one must establish a correspondence between the terms of the free molecule and the molecule in the crystal matrix and then determine a relationship between the latter term and the exciton multiplet of the perfect crystal. Practically all of Chap.3 is dedicated to the solution of the first part of

120

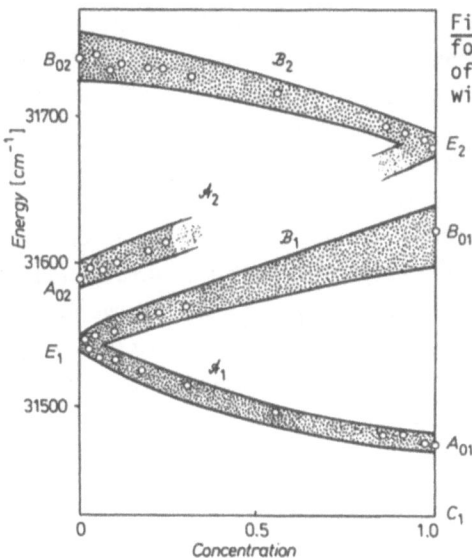

Fig.4.12. "Composition-property" diagram for a mixed crystal nd_0/nd_8: o) positions of the centres of gravity of absorption within the dotted areas [4.29]

the problem and considers in detail the relationship between the observed position of the monomer level of the guest molecule in the crystal matrix and the values of its term in the crystal matrix and in the free state. On the other hand, this section will be dedicated to the second part of the problem.

Figure 4.12 shows the experimental dependence of the positions of sharply polarized absorption bands on the composition of the mixed crystals of nd_0 and nd_8. The shaded areas denote the frequency ranges in which the absorption has a relatively high intensity. It is seen that the A_0-B_0 splitting in one-component crystals gradually diminishes (evolution) with decreasing concentrations of the corresponding component. $\mathcal{A}_1$ and $\mathcal{B}_1$, and also $\mathcal{A}_2$ and $\mathcal{B}_2$, coalesce to form bands of absorption by single impurity molecules (genesis). Thus, Fig.4.12 visualizes the genesis and evolution of an exciton doublet of the A_0 and B_0 bands in a naphthalene crystal.

To establish the genesis of an exciton multiplet in the two-mode absorption regime of a mixed crystal "composition-property" diagrams can be used (Fig.4.5a). In the case of the one-mode regime (for example, the limiting case of amalgamated bands, Fig.4.5b), it is impossible to directly establish the genesis of the exciton multiplet. However, its evolution can be followed by the concentration dependence of the positions of the external absorption bands $\mathcal{E}_\mu^{ext}$ of the spectrum. The concentration dependence and the continuous transformation of these external bands to absorption bands for both components are indications that these bands belong to exciton states.

In the "composition-property" method under consideration, the Davydov splittings are the parameters that are measured experimentally and characterize the "property". But the evolution of exciton energy bands remains the main problem of mixed crystals. The magnitude of the Davydov splittings may be used to estimate the widths of the energy bands if the bands stick together at the boundary of the Brillouin zone (as in crystal lattices of the benzene- and naphthalene type; Sect.2.5). If, in addition, the nearest-neighbour interactions dominate, then the magnitude of the Davydov splitting is usually almost equal to the total width of the exciton spectrum. Therefore, to a large extent, the evolution of the multiplet is also a reflection of the evolution of the entire spectrum. The band-to-band transition method of determining the density of states in the exciton energy spectrum of molecular crystals shall be discussed in Sect.5.2. Its application to mixed crystals (Sect.5.3) yields additional information on the width of the exciton energy spectrum that is otherwise obtained from "composition-property" diagrams.

Above, we have described a method of observing the genesis and evolution of an exciton multiplet for crystals with $Z > 1$. For each electronic molecular term there are Z exciton bands (Sect.2.2). The number of components in an exciton multiplet is determined by selection rules. The presence of a group of strongly polarized bands is usually considered to be an indication of their exciton nature. In a crystal with $Z = 1$ there is no such criterion. Therefore, it is particularly significant that in this case the study of the spectra of isotopically mixed crystals also helps to assign the single intrinsic electronic absorption band to an exciton state.

The pattern of evolution of the exciton states of such crystals is shown in Fig.4.4. The concentration dependence of the position of an absorption band and its motion from the monomer band to the intrinsic absorption band of the one-component crystal (two-mode regime, Case 1 in Fig.4.4) or from the intrinsic absorption band of one isotopic component to that of the other (one-mode regime, Case 3 in Fig.4.4) allows an unambiguous assignment to certain exciton states. By analyzing the observed behaviour of an absorption band versus concentration, it is possible to estimate the magnitude of $2\mathcal{M}$ for a known value of Δ.

4.4 Methods of Calculating Mixed-Crystal Spectra

The general analysis of the spectra of mixed crystals that was made in the preceding two sections yielded a sufficiently complete classification, which could be used to establish the genesis of exciton bands.

We shall now consider the spectra of mixed crystals in greater detail. The theory of these spectra shall be divided into two parts: the first part describes the smoothed contours of the densities of state and the absorption spectra according to the above-mentioned coherent-potential approximation (CPA), while the second part considers their fine structure, which corresponds to definite groups of guest molecules (clusters). In the next section we shall compare the calculated data with the experimental ones.

4.4.1 Coherent-Potential Approximation

There are several equivalent methods leading to the CPA which are quite different technically seen. They have been reviewed in detail [4.6,30]. For our purposes it is more convenient to follow SOVEN's approach [4.9], which was essentially developed in [4.31].

This approach is based on the T matrix concept, which has already been used in Sect.3.2.7 and Appendix B. The conventional equation for Green's function can be written in the form $G = G^0 + GUG^0$, where U is the potential of an isolated defect. Solving it formally, $G = G^0 (1 - UG^0)^{-1}$. Expanding the solution into a series yields

$$G = G^0 + G^0 UG^0 + G^0 UG^0 UG^0 + \dots \quad \text{or} \qquad (4.28)$$

$$G = G^0 + G^0 TG^0 \quad , \quad \text{where} \qquad (4.29)$$

$$T = U + UG^0 U + \dots = (1 - UG^0)^{-1} U = (G^0)^{-1} GU \quad . \qquad (4.30)$$

It is obvious that (4.29) is equivalent to (3.29).

To get a better understanding of the meaning of the T matrix it is instructive to write down the exact expression for the wave function of the system for the range of energies in the continuum:

$$\Psi = \Phi + GU\Phi \quad , \qquad (4.31)$$

where Φ is the unperturbed wave function (i.e., the incident wave) corresponding to the same energy E as Ψ. The validity of (4.31) can readily be verified by ascertaining that from $H_0\Phi = E\Phi$ follows $H\Psi = E\Psi$. Taking into account (4.30) we obtain from (4.31)

$$\Psi = \Phi + G^0 T\Phi \quad . \qquad (4.32)$$

The T matrix transforms the incident wave Φ to a scattered wave. If G^0 is also expanded in terms of the free-motion wave functions [see (B.9)], it becomes evident that the matrix element for the transition between the a and b states is equal to $\langle b|T|a \rangle$. This quantity is called the scattering amplitude (it only differs from the standard definition for the scattering amplitude [4.32] by a numerical factor). If the potential U is small, the first term dominates the series in (4.30) and the exact amplitude is reduced to the Born amplitude $\langle b|U|a\rangle$. The explicit form of the T matrix for an isolated isotopic impurity is given by (3.30).

We shall now consider a binary crystal system with a diagonal disorder. The aim of the coherent-potential method is to introduce an effective auxiliary medium with a regular (i.e., space-periodic) potential. This potential is complex, it is a function of the frequency, and it must be chosen so that it includes a considerable portion of the scattering processes. The Hamiltonian of the effective medium is called the *coherent-potential Hamiltonian*.

The potential of the effective medium is denoted by U. In the CPA, a mixed crystal is treated as a crystal immersed in an effective medium. The random component of the site potential, which causes the scattering, is determined by the values of $E_s - U$, where $E_s = E_1$ or E_2, which are the electronic terms of the components of the mixed crystal. The T matrices, which describe the scattering on an isolated impurity centre in the effective medium, will be denoted by T_1 and T_2 for perturbing potentials equal to $E_1 - U$ and $E_2 - U$, respectively. The principal requirement for an optimum choice of U is that the average values of the T matrix be equal to zero, i.e.,

$$C_1 T_1 + C_2 T_2 = 0 \quad . \tag{4.33}$$

This condition makes the theory self-consistent.

It should be emphasized that the CPA is a one-site approximation. This means that the T matrices appearing in (4.33) describe the scattering on an isolated impurity site. The processes involving an interference of waves which have been scattered from several centres are omitted. Such processes are generally only taken into account in U. This aspect can be explained much better in the language of Feynman diagrams [4.6]; the latter approach has been widely covered in the literature.

Substituting (3.30) into (4.33) yields an expression for the coherent potential $U = U(\Omega)$:

$$\frac{C_1(E_1 - U)}{1 - (E_1 - U)G_0} + \frac{C_2(E_2 - U)}{1 - (E_2 - U)G_0} = 0 \quad . \tag{4.34}$$

It should be emphasized that the term G_0 is Green's function for the effective medium and is, therefore, itself dependent on U:

$$G_0 \equiv G_0(\Omega - U) = \frac{v}{Z(2\pi)^3} \sum_\mu \int \frac{d^3k}{\Omega - U - \varepsilon_\mu(k)} \quad , \tag{4.35}$$

where $\varepsilon_\mu(\mathbf{k})$ is the exciton dispersion law for the perfect crystal. After elementary rearrangements, (4.34) transforms to

$$(E_1 - U)(E_2 - U)G_0 = \bar{E} - U \quad , \qquad (4.36)$$

where

$$\bar{E} = C_1 E_1 + C_2 E_2 \qquad (4.37)$$

is the electronic term of the mixed crystal, that is the centre of gravity of the exciton spectrum. It is convenient to choose the term $\bar{E}$ as the origin and let $\omega = \Omega - \bar{E}$ and $u = U - \bar{E}$.

In the momentum representation the Green's function for a mixed crystal has the following form:

$$G_\mu(\omega\mathbf{k}) = \frac{1}{\omega - u(\omega) - \varepsilon_\mu(\mathbf{k})} \quad . \qquad (4.38)$$

It is diagonal in μ. The spectral density on the zero momentum $S_\mu(\omega, \mathbf{k} = 0) =$ $- \mathrm{Im}\{ G_\mu(\omega, \mathbf{k}=0)\}$ determines the absorpiotn spectrum in accordance with (3.5). At the same time the imaginary part of (4.35) determines the density of states in the exciton spectrum in the usual manner:

$$\rho_{CP}(\omega) = -\frac{1}{\pi} \mathrm{Im}\{G_0(\omega - u(\omega))\} \quad . \qquad (4.39)$$

To distinguish this density from $\rho_0(\varepsilon)$, i.e., the density of states in the perfect crystal, the subscript CP is added, which also indicates the approximation under which it was calculated.

The potential U that is determined from (4.36) is a complex function because G_0 is complex. However, since $\mathbf{k}$ does not explicitly appear in (4.36), the potential $U = U(\omega)$ only depends on the frequency variable. From (4.38) it is seen that U is the self-energy part $\Sigma \equiv U$. Hence the CPA corresponds to an approximation in which Σ does not depend on either the momentum or the subscript μ. This fact was made use of in writing (4.38).

Equation (4.35) is rewritten as follows with due consideration for the independence of U from $\mathbf{k}$:

$$G_0 = \int \frac{\rho_0(\varepsilon)}{\omega - u(\omega) - \varepsilon} d\varepsilon \quad . \qquad (4.40)$$

From (4.40) it is seen that Green's function G_0 and, hence, the potential U and the density of states ρ_{CP} in a mixed crystal are completely determined by the density of states in the one-component crystal. To determine the absorption spectrum of a mixed crystal, i.e., the spectral function S_μ, one must also know $\varepsilon_\mu(\mathbf{k} = 0)$, i.e., the positions of the exciton absorption bands in the one-component crystal.

Since U is a complex function, the coherent potential describes the damping of states with a given momentum. The CPA yields a) the correct limit when the concentration of one of the components tends to zero (isolated guest molecules); b) the correct behaviour in the molecular limit and as $\Delta = E_2 - E_1 \to 0$; and c) the correct values for the first eight moments of the density of states and for the first seven moments of the spectral density (the latter is true, in any case, for the simplest of lattices [4.33,34]).

Since the states are damped, the spectral density $S_\mu(\omega, \mathbf{k})$ is only different from zero where $\mathrm{Im}\{U(\omega)\} \neq 0$. The edges of $S_\mu(\omega, \mathbf{k})$ are the same for all values of $\mathbf{k}$ and coincide with the edges of $\rho_{CP}(\omega) \neq 0$, because U is independent of $\mathbf{k}$.

The general features of an energy spectrum are shown in Fig.4.13 with a density of states given by (3.14). The shaded areas denote the distribution of the density of states in a mixed crystal for different values of the isotopic shift $\Delta/\mathcal{M}$. These areas only cover about half of the region spanned by the mixed-crystal spectrum and correspond to a high density of states, while in the dot-filled areas the density of states is low. These are the "tail" regions; towards the exact edges of the spectrum the density of states falls off exponentially [4.1].

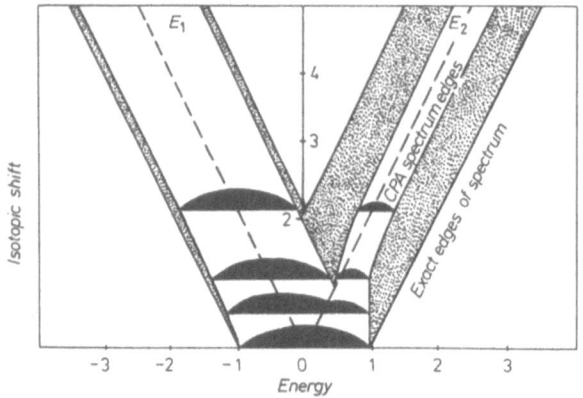

Fig.4.13. Arrangement of the exact spectrum edges of a two-component system and the density-of-states distribution in its energy spectrum in CPA according to [4.35] ($\mathcal{M} = 1$; crystal composition: $C_1 = 0.9$, $C_2 = 0.1$)

If $\Delta/\mathcal{M} \gtrsim 2$, there is a true gap in the spectrum. However, it is seen from Fig.4.13 that a gap appears much earlier in the CPA spectrum, i.e., at $\Delta/\mathcal{M} \approx 1$. The appearance of a gap in the CPA spectrum which is absent in the exact spectrum is an indication of the existence of an extended region with a low (although nonzero!) density of states, i.e., a *pseudogap*. The appearance of a pseudogap (in contrast to a true gap!) depends on the concentrations of

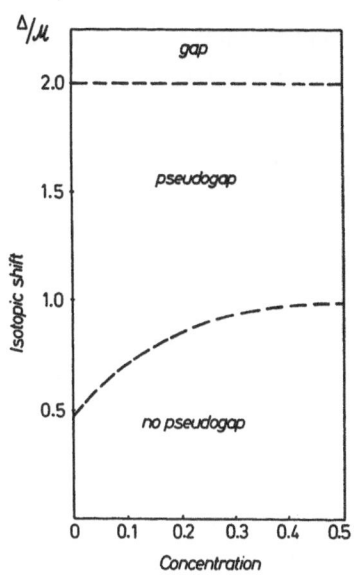

Fig.4.14. Conditions for the existence of a pseudogap in the spectrum of a binary mixed crystal in the CPA according to [4.31]; calculation for the model density of states

the components. This can be seen in Fig.4.14. For a low concentration of one of the components the condition for the opening of a pseudogap coincides with the condition for the splitting-off of a discrete impurity level, i.e., $\Delta_{pg} = \Delta_{cr}$. For a density of states given by (3.14), a decrease in concentration from $C = 0.5$ to $C \rightarrow 0$ reduces the value of Δ_{pg} from $\mathcal{M}$ to $\mathcal{M}/2$ (Sect.3.2.4).

Moving away from the edges of the high density of states into the pseudo-gap or the tails, the density decreases extremely rapidly. The levels located in the middle of long tails and particularly near the exact edges of the spectrum appear in regions of extended spatial fluctuations of a special type whose probability is monstrously small. Therefore, in many experiments an important role is played by the relatively narrow regions near the edges of the high density of states and not by the exact edges of the spectrum. The lower edge of this kind will be called the *pseudobottom* of the spectrum (Sect.4.1.3).

The very appearance of Fig.4.13 clearly shows that even after the pseudo-gap disappears, a dip in the density of states ρ_{CP} of a mixed crystal must exist in a certain energy interval. For $C = 0.5$ and the model density of states ρ_0, the dip only disappears at $\Delta \approx \mathcal{M}/2$. The regime corresponding to $\mathcal{M}/2 < \Delta < \mathcal{M}$ is often called the regime of *incipient pseudogap*. At lower values of Δ the curve ρ_{CP} gradually approaches the curve ρ_0. Of course, the borders of individual regimes cannot be determined precisely (as in Sects.4.2.2,4.3.2). They also depend on the shape of the function ρ_0 for the one-component crystal. The very possibility of introducing some of the regimes arises due to the existence of a small numerical factor in the expression for the second moment

μ_2^0 of the function ρ_0 (Sect.3.2.4), e.g., for the model density $\mu_2^0 = \mathcal{M}^2/4$. At $C \sim 0.5$, an incipient pseudogap corresponds to the parameter range

$$\mathcal{M} \gtrsim \Delta \gtrsim \sqrt{\mu_2^0}.$$

It is convenient to use the following classification of the energy spectra based on the distribution of the density of states in a mixed crystal [4.8,36].

1) For a two-band regime, $\Delta > 2\mathcal{M}$ and the bands are separated by a true gap.

2) For a pseudo-two-band regime, $2\mathcal{M} > \Delta \gtrsim \mathcal{M}$ and the bands are separated by a pseudogap over the entire concentration range.

3) For an incipient pseudogap, $\mathcal{M} \gtrsim \Delta \gtrsim \sqrt{\mu_2^0}$. At a low concentration of a minority component a pseudogap may exist. At $C \sim 0.5$, there is a dip in the density of states.

4) For a one-band regime, $\sqrt{\mu_2^0} \gtrsim \Delta$.

There are no clearly defined borders between these regimes[3]. Similarly, the determination of the edges of the pseudogap and the values of the isotopic shift, Δ_{pg}, at which it disappears cannot be determined rigorously and depends on the approximation used to calculate spectra. The shape of the density of states in all four regimes is illustrated in Fig.4.15.

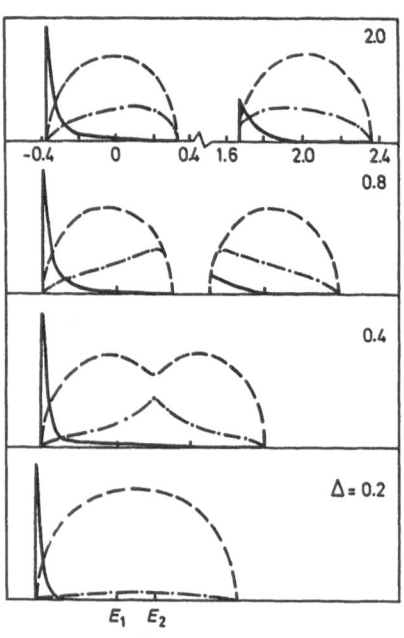

Fig.4.15. Density of states ρ_{cp} (dashed line), optical absorption spectrum $S(\omega, k = 0)$ (solid line), and $\mathrm{Im}\{U(\omega)\}$ (dot-and-dash line) for $C_1 = C_2 = 0.5$ [4.8]; calculations for the model density of states (3.14); $2\mathcal{M} = 1$

---

[3] The terms "separated bands", "persistent case", "incipient gap", and "amalgamation limit" are also often used.

ONODERA and TOYOZAWA [4.8] were the first to use the CPA to calculate the absorption spectrum, i.e., the spectral function $S(\omega, \mathbf{k} = 0)$. The results are given by the solid lines in Fig.4.15. The calculations were based on a model of a crystal whose absorption spectrum exhibits a single exciton band that corresponds to a transition to the bottom of the exciton energy band. Such a model corresponds either to a crystal with one molecule per primitive cell or to the spectrum of a crystal with a more complicated cell, but in one component of the spectrum only.

All of the curves $S(\omega, \mathbf{k} = 0)$ are much narrower than the corresponding curves $\rho_{CP}(\omega) \neq 0$. This suggests that the absorption bands must, as a rule, be relatively narrow and explains why the AAA (Sects.4.2,3) yields satisfactory results in describing an absorption spectrum. For comparable concentrations of components and at $\Delta \sim \mathcal{M}$, there are naturally no small "letter" parameters, and therefore, the narrowness of the bands must be attributed to the existence of some small numerical factors in this problem.

It is also important to compare the classification of spectra according to the distribution of the density of states with the classification according to the behaviour of the spectra (one- and two-mode behaviour, Sect.4.2). It is clear that both the two- and the pseudo-two-band regimes exhibit two-mode behaviour, i.e., each of the bands can be assigned to one of the isotopic components of the mixed crystal. This is also clearly seen in Fig.4.15 for $\Delta / \mathcal{M} = 4.0$ and 1.6. Incidentally, these figures also illustrate other regularities already known to us from the AAA. Thus, the high-frequency band, which is "internal" (in the sense of Sect.4.2), is strongly weakened with respect to the low-frequency ("external") band. The weakening is more pronounced for smaller values of $\Delta / \mathcal{M}$. The one-band regime exhibits one-mode behaviour, i.e., the strong absorption band is always located at the low-frequency edge of the CPA spectrum (Fig.4.15, $\Delta / \mathcal{M} = 0.4$). This absorption band moves together with the edge when the concentration changes. It moves continuously between the exciton bands of one-component crystals. The incipient-pseudogap regime is intermediate. The detailed behaviour of the absorption bands depends on whether or not the pseudogap is formed when $C_1 \to 0$ and, if it does form, on its region of existence. For example, for a model density and $\Delta / \mathcal{M} = 0.8$ the pseudogap exists at $C_1 \lesssim 0.06$. In the absorption spectrum calculated in [4.8] two bands can be clearly seen for concentrations up to $C_1 = 0.125$. The intensity of the low-frequency band, arising from the impurity band (it is the "external" one) rapidly increases and already dominates the spectrum when $C_1 = 0.125$. With a further increase in C_1 the low-frequency band is smoothly transformed to the exciton absorption band of the one-component crystal. The behaviour of this band is of the two-mode type. The motion of the high-frequency ("internal") band cannot be followed because it broadens and weakens rapidly. In the concentration range where $C_1 \sim C_2$ the high-frequency band is not visible; for example, it is absent in Fig.4.15 ($\Delta/\mathcal{M} = 0.8$). It is possible, however, that this is a result of an excessive smoothing of the structure in the CPA, which has already been noted in Sect.4.1. According to the AAA, the intensities of the two bands must differ by about one order of magnitude. A preciser result can only be obtained through numerical calculations with specific models. These have not yet been carried out.

Some curves that correspond to the imaginary part of U (or Σ) are shown in Fig.4.15. They characterize the spectral dependence of the damping of states with $\mathbf{k} = 0$.

The analytical equations for the absorption spectrum are very unwieldy in the CPA, even if the model density given in (3.14) is used for $\rho_0(\varepsilon)$. Simple limiting formulae have been obtained for large and small values of $\Delta/\mathcal{M}$.

1) $\Delta/\mathcal{M} \gg 1$: two-band regime.

In this case, one can determine the centres of gravity $\Omega_{r\mu}$ of the absorption bands that correspond to the rth isotopic component and the widths of these bands $Y_{r\mu}$ (determined by their second moment):

$$\Omega_{r\mu} \approx E_r + C_r \varepsilon_\mu \quad , \tag{4.41}$$

$$Y_{r\mu} \approx (C_1 C_2 \mu_2^0)^{1/2} \quad . \tag{4.42}$$

Equation (4.41) coincides with the limiting linear law (Sect.4.2.1). It is readily derived from (4.24). It is significant, however, that the half-widths $Y_{r\mu}$ at $C_1 \sim C_2 \sim 0.5$ are not small. For the model density, $Y < \mathcal{M}/4$. This may result in a considerable shift of the maxima of the absorption bands relative to their centres of gravity. For example, for the model density (3.14) represented by the upper curve in Fig.4.15 the maxima of both absorption bands are close to the low energy edges of the CPA spectrum. The spectral density exhibits high-frequency tails which shift the centres of gravity to higher frequencies. In the model-density approximation the position of the pseudobottom is given by

$$E_{pb}(C_1) \approx E_1 - \mathcal{M}\sqrt{C_1}. \tag{4.43}$$

For small values of C_1, this expression differs noticeably from $\Omega \approx E_1 - \mathcal{M}C_1$, which follows from (4.41).

2) $\Delta/\mathcal{M} \ll 1$: amalgamated-bands regime.

The location of the centre of gravity of the entire spectrum and the magnitude of its half-width are given by

130

$$\Omega_\mu = C_1 E_1 + C_2 E_2 + \varepsilon_\mu \quad , \tag{4.44}$$

$$Y_u = \sqrt{C_1 C_2}\, \Delta \quad . \tag{4.45}$$

The expression for Ω_μ is equivalent to the limiting equation for the position of the "external" band that follows from the AAA (4.22). The width Y_μ is small compared to $\mathcal{M}$. This agrees with the fact that the high-frequency wing of this band in the lower curve of Fig.4.15 ($\Delta/\mathcal{M}$ = 0.4) is considerably shorter than the high-frequency wing in the upper curve ($\Delta/\mathcal{M}$ = 4.0). The maximum of this band is very close to the pseudobottom. Therefore, the shift of the pseudo-bottom relative to the centre of gravity must be moderate. The position of the pseudobottom can be estimated by using the equation

$$E_{pb} \approx C_1 E_1 + C_2 E_2 \dot{-} \mathcal{M} - \frac{2\Delta^2}{\mathcal{M}} C_1 C_2 \quad , \tag{4.46}$$

which is valid for the model density given in (3.14).

Table 4.1 lists the parameters of the most frequently encountered binary systems of d-benzenes, d-naphthalenes and d-antracenes and classifies them according to the above-described types of energy and absorption spectra. The table is similar to the one given in [4.37]. A more detailed analysis of naphthalene spectra using the CPA will be found in the following section.

Table 4.1. Types of exciton states of binary molecular crystals

Mixture components	Energy spectrum width $2\mathcal{M}$ [cm^{-1}]	Second moment of energy spectrum μ_2^0 [(cm^{-1})^2]	Isotopic shift Δ [cm^{-1}]	Type of mixed-crystal states
bd$_0$/bd$_6$	60	110	200	Two-band, two-mode regime, there is a gap
bd$_0$/bd$_1$	60	110	35	Pseudo-two-band, two-mode regime,
nd$_0$/nd$_8$	180	1750	115	there is a pseudogap
nd$_0$/ß-nd$_4$	180	1750	78	Regime intermediate between two-
nd$_0$/α-nd$_4$	180	1750	44	and one-band. Incipient pseudo-gap
nd$_0$/ß-nd$_1$	180	1750	19	One-band, one-mode regime, there
ad$_0$/ad$_{10}$	≈ 500	-	59	is no pseudogap
nd$_0$/ß-nd$_1$ (triplet excitons)	11	9	19	Two-band, two-mode regime, there is a gap

4.4.2 Application of the Coherent-Potential Approximation to the Spectra of d-Naphthalenes

The CPA was applied to molecular crystals with a complex unit cell by DUBOVSKY and KONOBEYEV [4.38], HOSHEN and JORTNER [4.14,36], and HONG and ROBINSON [4.39,40]. Their attention was focussed on the naphthalene spectra. In [4.38] $\rho_{CP}(\varepsilon)$ and the absorption spectrum were calculated using the model density of states given in (3.14). In the other works [4.39,40] both quantities were calculated using the density $\rho_0(\varepsilon)$ which was determined experimentally by COLSON et al. [4.41] (Sect.5.2, Fig.5.7).

Figure 4.16 describes the pseudo-two-band regime of the system nd_0/nd_8 when there is no gap. However, a pseudogap exists over the entire concentration range. The absorption spectrum is of a two-mode type. Of the two absorption bands shown within each of the areas in which $\rho_{CP}(\varepsilon) \neq 0$, the low-frequency band corresponds to the $\mathscr{A}$ band while the high-frequency band corresponds to the $\mathscr{B}$ band. The intensities of the internal bands are much lower than those of the external bands over the entire high-concentration range. This observation is in full agreement with the conclusions that were drawn from the average-

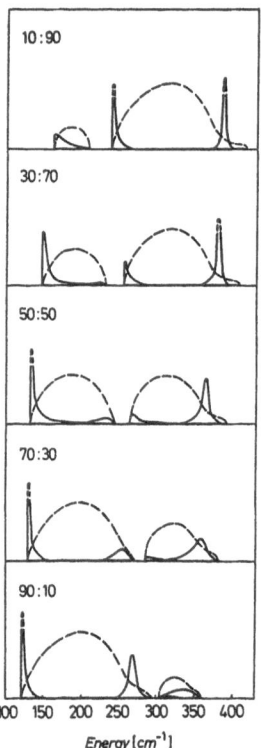

Fig.4.16. Density of states (dashed lines) and the optical absorption spectrum (solid lines) of mixed crystals nd_0/nd_8 calculated in CPA [4.39] ($\Delta = 115$ cm^{-1}; the concentration ratios $C_1(nd_0):C_2(nd_8)$ are indicated; the intensities are so normalized that the A_0 and B_0 band intensities in a one-component crystal coincide)

amplitude approximation. The widths of the individual bands strongly depend on the concentration. As a rule the widths of the internal bands exceed those of the external ones. A trend towards the domination of the width of the external $\mathscr{B}$ band over that of the external $\mathscr{A}$ band can also be observed. The reason is that in a one-component crystal the A_0 band lies at the edge of the energy spectrum, while the B_0 band is displaced relative to its other edge and lies on the weak continuum (Fig.5.7).

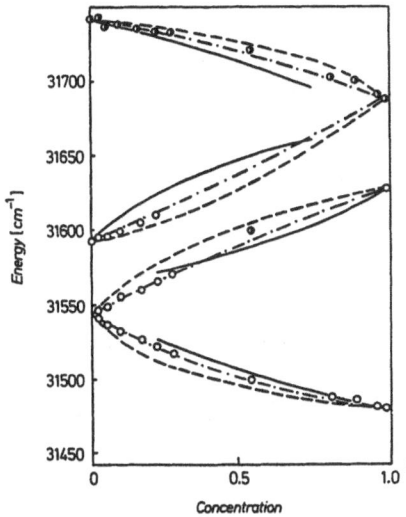

Fig.4.17. Concentration dependence of the exciton band absorption intensity of nd_0/nd_8; calculated data of [4.29]: solid lines) band position according to the AAA; dot-and-dash lines) positions of the band centres of gravity according to the CPA; dashed lines) positions of band maxima according to the CPA; experimental data of [4.42]; o) positions of the centres of gravity of the cluster-structure bands; •) approximate position of the centres of gravity of those bands for which a complete absorption curve has not been obtained; C_1) nd_0 concentration

Quantitative measurements have thus far only been carried out on the system nd_0/nd_8 [4.29,42]. Therefore, the experimental results for this system were compared with the results obtained with the CPA and the AAA. The data on the positions of the bands are summarized in Fig.4.17. The agreement between the CPA and the experimental results is, on the whole, very good. At the same time this confirms the sufficient accuracy of the function $\rho_0(\varepsilon)$ (Fig.3.5), on which the calculations were based [4.29]. The results of the AAA, in turn, agree fairly well with the positions of the centres of gravity that were calculated using the CPA (with the exception of the low-concentration range in which the AAA cannot be used).

Two spectra of the incipient-pseudogap type, that differ noticeably, are shown in Figs.4.18,19. The spectra of the nd_0/β-nd_4 system (Fig.4.18) resemble two-band spectra more than the spectra of the nd_0/α-nd_4 system do. A pseudogap is observed at $C_1 \lesssim 0.1$. At the other end of the spectrum, i.e., at small values of C_2, no discrete level arises, and there is no pseudogap. The asymmetry of the function $\rho_0(\varepsilon)$ manifests itself here (Fig.5.7). However, there does exist

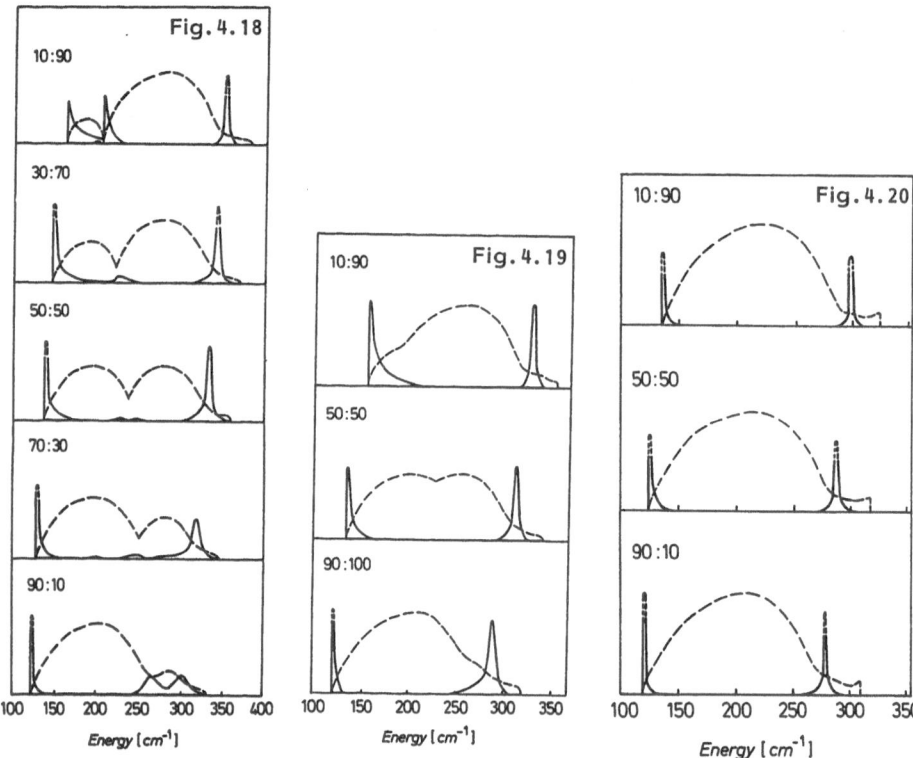

Fig.4.18

10:90

30:70

50:50

70:30

90:10

100   150   200   250   300   350   400
Energy [cm⁻¹]

Fig.4.19

10:90

50:50

90:100

100   150   200   250   300   350
Energy [cm⁻¹]

Fig.4.20

10:90

50:50

90:10

100   150   200   250   300   350
Energy [cm⁻¹]

Fig.4.18. Same as in Fig.4.16, but for $nd_0/\beta-nd_4$ ($\Delta = 74$ cm^{-1} [4.39])

Fig.4.19. Same as in Fig.4.16, but for $nd_0/\alpha-nd_4$ ($\Delta = 51$ cm^{-1} [4.39]. According to [4.43], $\Delta = 44$ cm^{-1}; such a change will bring the regime still closer to being one band)

Fig.4.20. Same as in Fig.4.16, but for $nd_0/\beta-nd_1$ ($\Delta = 21$ cm^{-1} [4.39])

a virtual level (Sect.3.2.9). The high-frequency peak at $C_2 = 0.1$ may be assigned to it. At higher concentrations this peak turns into a high-frequency ("external") $\mathcal{B}$ band. At all concentrations, the function $\rho_{CP}(E)$ has a large dip.

On the other hand, a pseudogap can only be observed in the spectra of $nd_0/\alpha-nd_4$ (Fig.4.19) at minimal concentrations of nd_0 (actually, only in monomer spectra). Over a wide concentration range, the dip in $\rho_{CP}(E)$ is small. Therefore, the regime is almost a one-band one.

For both systems the "internal" absorption bands at $C \approx 0.5$ broaden to such an extent that they are practically invisible. Figure 4.20 corresponds to the one-band and one-mode regime over the entire concentration range.

4.4.3 Shape of the Spectra in the High-Concentration Range

The aim of the CPA calculations is to determine the overall contours of a spectrum. They cannot determine the detailed structure, in particular, the structure associated with clusters. Since the calculated results will be compared with the experimental data on mixed crystals of d-naphthalenes later on, the question arises as to whether the expected cluster structure is sufficiently pronounced. Both Sect.3.4 and Fig.4.9 show that the cluster structure may even be observed at rather high concentrations.

Detailed calculations of the density of states in the spectra of binary mixed crystals of deuteronaphthalenes have been carried out by HONG and KOPEL-MAN [4.11]. The standard negative factors counting method [4.10,44] and the excitation transfer integrals given in Table 3.5 (set 1) were used to determine the density-of-states histograms. The calculations were made for crystals containing 320 to 1280 molecules. The results were compared with the CPA calculations based on the density of states $\rho_0(\varepsilon)$ shown in Fig.5.7. Figure 4.21 illustrates the dependence of the mixed-crystal density of states on the isotopic shift Δ for a constant concentration of components. Although the "impurity" concentration is high, the cluster structure is clearly visible over the entire impurity (in this case, high-frequency) part of the spectrum. This is not so in the upper curve where the regime is close to that of a one-band one. The CPA curve fits snugly on the envelope of the spectrum and determines its width. This is also illustrated in Fig.4.22, in which several spectra of the system nd_0/nd_8 at different concentrations are shown.

Thus, the CPA adequately describes the general contours of the function $\rho(E)$. Under the two-band regime $\rho(E)$ exhibits a distinct cluster structure

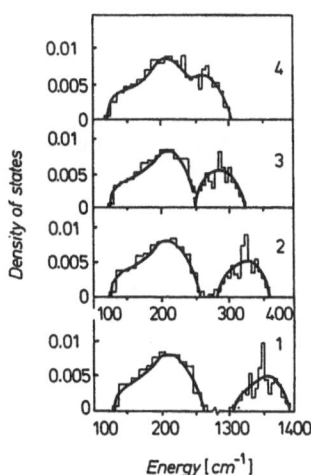

Fig.4.21. Densities of states for mixed crystals of d-naphthalenes with different values of Δ [4.11]; $C_1 = 0.7$ and $C_2 = 0.3$; histograms) numerical calculation; solid curves) CPA; 1) $\Delta = 1150$ cm^{-1}; 2) $\Delta = 115$ cm^{-1} (nd_0/nd_8); 3) $\Delta = 74$ cm^{-1} (nd_0/β-nd_4); 4) $\Delta = 51$ cm^{-1} (nd_0/α-nd_4)

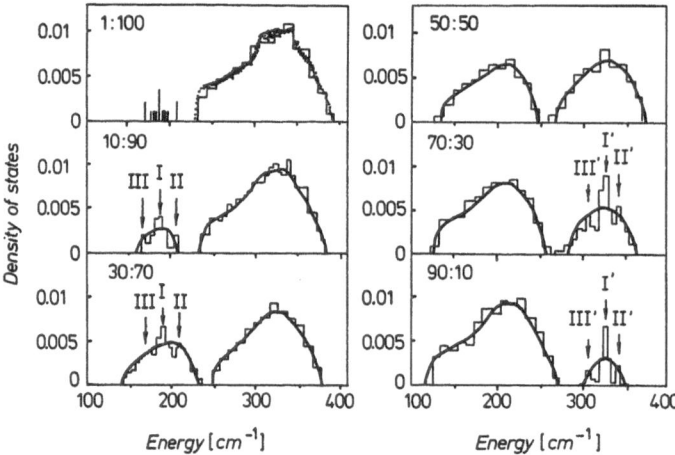

Fig.4.22. Same as for Fig.4.21, but for nd_0/nd_8 [4.11]; the Roman numerals denote definite types of clusters; the concentration ratio $C_1(nd_0):C_2(nd_8)$ is indicated in the figure

almost throughout the concentration range, which is not reflected by the CPA. On the other hand, sharp peaks are particularly conspicuous in experimental curves (of course, only if it is a high-resolution experiment and if extraneous processes, such as an interaction with phonons, do not broaden the spectrum too much). Therefore, one cannot always use the CPA to describe the shape of a spectrum. Experiments which show $\rho(E)$ directly are discussed in Sect.5.3.

It follows from (3.5) that in order to find the absorption spectrum of a mixed crystal, one must calculate the spectral function $S_\mu(E, \mathbf{k} = 0)$. Taking advantage of (3.4) and of the relation

$$S(E) = - \operatorname{Im}\{G(E)\} = - \operatorname{Im}\{(E - H + i0)^{-1}\} = \pi\delta(E - H), \qquad (4.47)$$

where H is the crystal Hamiltonian, we obtain

$$S_\mu(E) \equiv S_\mu(E, \mathbf{k} = 0) = \frac{\pi}{N} \sum_{\mathbf{n}\alpha\mathbf{m}\beta} \overset{*}{B}^\mu_\alpha \langle \mathbf{n}\alpha | \delta(E - H) | \mathbf{m}\beta \rangle B^\mu_\beta \quad . \qquad (4.48)$$

This expression has the structure $\langle X | \delta(E - H) | X \rangle$, where X is a vector whose components are equal to $X_{\mathbf{m}\beta} = B^\mu_\beta$.

Thus, the numerical determination of the absorption spectrum is reduced to finding averages of the type shown in (4.48). A convenient computer method that can be used to calculate such expressions was proposed by Mil'man[4]. The method is based on the approximate representation of the δ function in the form of a Gaussian exponent with a width θ

$$\delta(E - H) = \lim_{\theta \to 0} \delta_\theta(E - H) = \lim_{\theta \to 0} \frac{1}{\theta\sqrt{\pi}} \exp\left[-\left(\frac{E - H}{\theta}\right)^2\right] \quad , \qquad (4.49)$$

---

[4] P.D. Mil'man, cited in [4.16]

and a subsequent Napier factorization of this exponent

$$\delta(E-H) = \lim_{\theta \to 0} \lim_{M \to \infty} \frac{1}{\theta\sqrt{\pi}} \left(1 - \frac{(E-H)^2}{M\theta^2}\right)^M \quad . \tag{4.50}$$

In the computer calculations, both θ and M naturally remain finite. θ is similar to the histogram pitch. In a procedure based on the use of (4.50) all of the calculations involve a repeated application of one and the same matrix,

$$A = 1 - \frac{(E-H)^2}{M\theta^2} \quad ;$$

i.e., the calculations are reduced to determine vectors of the type $A^s X$. Since for short-range interaction the number of nonzero elements in the matrices H and A (expressed in the site representation) as well as the number of components of the vector $A^s X$ is of the order of N, the amount of stored numbers must also be of the order of N [and not N^2, as it is the case for the number of elements of the matrix $(E-H)^{-1}$]. N is the number of unit cells in the crystal. This enables one to carry out calculations for model "samples" containing a large number of molecules.

The *Napier factorization method* (NFM) was applied to the calculation of the absorption spectra of crystals of nd_0/nd_8 [4.16] . The calculations were carried out for "samples" containing up to 4000 molecules. Matrix elements of excitation transfer were taken from Table 3.5 (set 1). The results will be compared with those obtained by experiment (Sect.4.5). For the moment we shall restrict ourselves to a comparison of the results of numerical calculations with those using the CPA.

Figure 4.23 compares the integral absorption intensities as a function of the component concentrations for all four absorption bands: two in the component parallel to **b** and two in the component perpendicular to **b**. The calculations were carried out using three different methods: a numerical one and two approximate ones (AAA and CPA). The relative differences do not exceed 5% of the total spectral intensity. Hence, it is possible to use both the AAA and the CPA to calculate the integral band intensities in two-mode crystals.

We shall now compare the shape of the absorption bands predicted by the numerical method and the CPA. Two questions are to be answered: how accurate is the CPA description of the overall contour of the absorption curve and how pronounced is the cluster structure in the absorption spectrum? The second question will only be considered in the next section. We shall consider the spectrum of a crystal of nd_0 and nd_8 at an nd_0 concentration (C_1) equal to 0.54. For this composition the cluster structure is suppressed to the greatest ex-

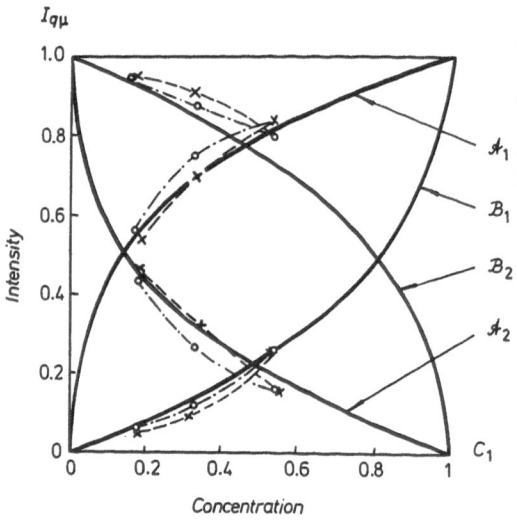

Fig.4.23. Calculation [4.16] of the integral intensity of the exciton absorption bands for nd_0/nd_8 crystals as a function of concentration (in the total-intensity units): solid $\mathscr{A}_1$ curves) AAA; o) CPA; x) numerical method

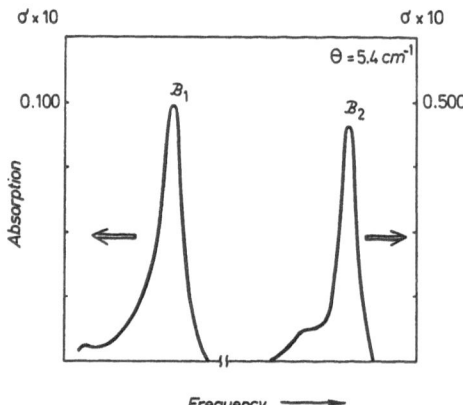

Fig.4.24. $\mathscr{B}$ bands of the absorption spectrum of nd_0/nd_8 crystals at the nd_0 concentration $C_1 = 0.54$ (calculated by NFM [4.16])

tent. The results of the numerical calculations according to the NFM are summarized in Fig.4.24. They should be compared to the results for the spectrum of the same system at $C_1 = C_2 = 0.5$ (Fig.4.16). However, a quantitative comparison is difficult because of the large differences in scales. While in Fig.4.16 the impression is given that the $\mathscr{B}_1$ is highly diffuse, it is evident in Fig.4.24 that both $\mathscr{B}$ bands have comparable half-widths and that their amplitudes only differ by about a factor of five at the most. This is in agreement with the experimental data, according to which both bands are visible in the b component of the spectrum over the entire concentration range (Fig.4.10).

4.5 Mixed-Crystal Spectra. A Comparison of Theory with Experiment

4.5.1 Detailed Analysis of the Absorption and Fluorescence Spectra of a Mixed Crystal

The absorption spectra of a number of mixed crystals were considered in Sects.4.2,3 with the intention of discovering the general qualitative regularities which manifest themselves when the composition of the crystal is changed, e.g., the smooth change in the magnitude of the Davydov splittings of individual components, the borrowing of intensity by the external absorption bands from the internal ones, and establishing the genesis and following the evolution of the exciton bands.

The existing methods that can be used to calculate the spectra describe these regularities equally well. The AAA has an unquestionable advantage at the semiquantitative stage of comparison because of its simplicity. In this section we shall deal with a more detailed analysis of the spectra and compare them with the predicted spectra.

In Sect.4.4 it was possible to visualize with sufficient clarity the structure of the energy spectrum of mixed crystals in the exciton region. Let us now consider the general properties of the optical transitions which are caused by this structure.

Figure 4.25a shows the electronic transitions between the ground and the lower excited state of a mixed crystal which lead to the absorption and fluorescence spectra (Fig.4.25b). We have chosen an energy spectrum with a pseudogap. In this case the absorption spectrum is the most complicated, i.e., two-mode with additional structure due to clusters. After analyzing the structure of this spectrum it will be easy to consider simpler spectra.

The structure of an absorption spectrum has been thoroughly discussed in the preceding sections of this chapter. We shall only recall that a detailed analysis involves a study of the dependence of the shape of the spectrum on the composition of the mixed crystal and the reproduction of this shape by numerical methods.

The fluorescence spectrum has not yet been discussed. Therefore, we shall consider it in greater detail. The fluorescence spectrum of a mixed crystal differs markedly from the absorption spectrum. In a low-temperature fluorescence spectrum the equivalent to an exciton absorption band with a complicated structure is a single narrow exciton band of width $\sim T$ (Fig.4.25). It is assumed that thermodynamic equilibrium has been attained in the excited state. The position of the pseudobottom E_{pb} in the energy spectrum is given by the position of this band. The vibronic fluorescence bands exhibit a doublet structure in contrast to the exciton fluorescence bands because electronic

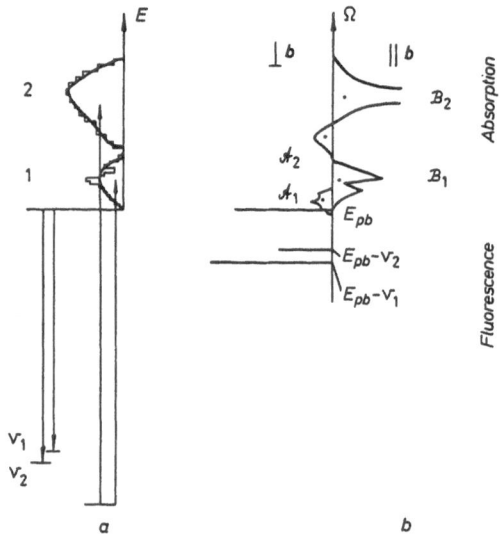

Fig.4.25a,b. Optical transitions (a) and absorption and fluorescence spectra (b) of a mixed crystal. Calculated densities of state and absorption intensities correspond to a crystal of nd_0/nd_8 at $C_1 = 0.2$ and $C_2 = 0.8$ [4.11,16]. 1) and 2) sections where excitation of the first or second isotopic component dominates; ν_1 and ν_2 are frequencies of the respective phonons; $\mathcal{A}$ and $\mathcal{B}$ are exciton absorption bands, the points denote the positions of their centres of gravity; E_{pb}, $E_{pb} - \nu_1$, $E_{pb} - \nu_2$ are the positions of the exciton and vibronic fluorescence bands, respectively

excitations in a mixed crystal are the collective ones and involve the molecules of both components. On the other hand, the vibrational frequencies of these molecules, ν_1 and ν_2, in the ground electronic state are different. The situation closely resembles the one observed under the fluorescence from the shallow level of a local exciton because of the partial delocalization of the impurity excitation (Sect.3.3.6). The difference lies in the fact that the vibrational states of a mixed crystal, as well as the exciton states, transform to two bands, i.e., to those of internal phonons. Therefore, the vibronic transitions under consideration are band-to-band transitions. This allows one to use the temperature dependence of their shape, as will be shown in Sect.5.3, to study the exciton density of states in a mixed crystal. The integral intensities of the vibronic bands with the energies $E_{pb} - \nu_1$ and $E_{pb} - \nu_2$ are determined by the amplitudes of the electronic excitation on the molecules of the components 1 and 2, respectively, in quantum states with an energy close to that of the pseudobottom.

Thus, the low-temperature absorption and fluorescence spectra of mixed crystals yield a variety of experimental parameters which can be compared with theory: 1) the shape of the absorption spectrum, 2) the positions of the centres of gravity in the different components of the absorption spectrum, 3) the position of the pseudobottom in the energy spectrum, and 4) the ratio of the excitation amplitudes for the different isotopic components.

4.5.2 The Experimental Shape of the Absorption Spectrum

Quantitative absorption studies have only been carried out for mixed crystals of nd_0 and nd_8. The absorption coefficients of crystals of different compositions for two different polarizations of the incident light are shown in Fig.4.26. The absorption spectra of mixed crystals are richer in structure than those of one-component crystals. The absorption by the nd_0 component results in at least six bands, whose positions are practically concentration-independent. An increase in the concentration of nd_0 only changes their intensity. This peculiarity can be explained if it is assumed that the individual bands of a structure are due to cluster aggregates of different composition (for example, the 31542 cm^{-1} band is due to the monomer centres, the sharply polarized bands between 31525 and 31562 cm^{-1} are due to pairs of

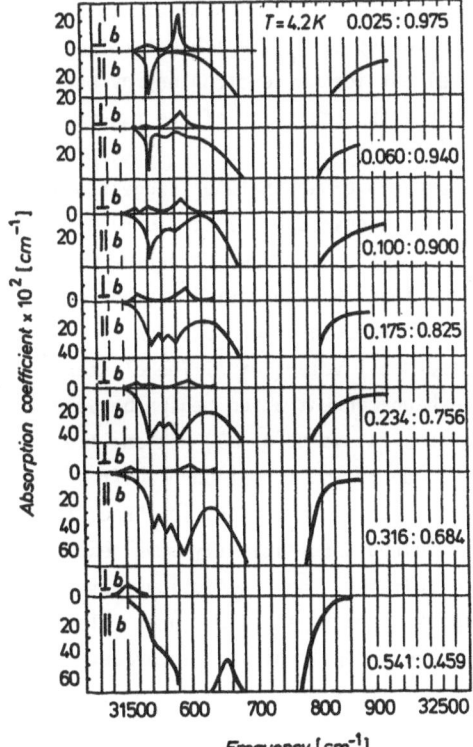

Fig.4.26. Exciton absorption spectra of nd_0/nd_8 (crystal thickness 4 μm or more [4.42]; concentration ratio $C_1(nd_0):C_2(nd_8)$ is indicated)

141

translationally nonequivalent molecules, and the 31508 to 31582 cm^{-1} bands are due to more complicated centres). No cluster structure is visible in the high-frequency absorption spectrum of the component nd$_8$, although a calculation of the density of states provides some implication for the existence of such a structure (Fig.4.22).

SHEKA [4.21] investigated the absorption spectra of a number of binary mixtures of d-naphthalenes in addition to the mixture nd$_0$/nd$_8$. The parameter $\Delta/\mathscr{M}$ varied from 1.3 to 0.12. A common feature of these spectra is the decrease in the intensity of the internal bands with decreasing values of $\Delta/\mathscr{M}$ and the absence of a cluster structure in the high-frequency region of the spectrum. In the low-frequency region, the cluster structure can still be observed at $\Delta = 44$ cm^{-1} (which corresponds to $\Delta/\mathscr{M} \approx 0.5$). At smaller values of Δ it disappears completely. In the same concentration range a change from a two- to a one-mode behaviour is also observed. In the high-frequency region a similar transition takes place at larger isotopic shifts, i.e., at $\Delta \approx 110$ cm^{-1}. The observed "asymmetry" in the spectra of nd$_0$ and nd$_8$ with respect to the observation of the cluster structure is in agreement with the different stabilities of the two-mode regimes. This difference, in turn, is a result of the asymmetry of the energy spectrum of one-component crystals of d-naphthalenes and the difference in the values of $\Delta^{\pm}_{cr}$, which determine the existence of the discrete impurity level outside the exciton energy band. The relationship between the values of $\Delta^{\pm}_{cr}$ and the stability of the two-mode regime is discussed in Sect.4.3.2.

Thus, the experimental data on the mixtures of d-naphthalenes suggest that a necessary condition for the manifestation of a developed cluster structure is the two-mode behaviour of the absorption spectrum. However, experiments with mixed crystals of d-benzenes have shown that this condition may not always be sufficient. The two-mode behaviour of the absorption spectrum is typical of these crystals for both large and small values of Δ. The cluster structure, however, is not observed in the high-concentration region because the excitation transfer integrals (and, hence, the value of $\mathscr{M}$) are small. As a result, the absorption band splittings due to the formation of different cluster centres are small and less than the width of the absorption band. Thus, the observation of the cluster structure requires a sufficiently broad exciton energy band in a one-component crystal. In addition, the interactions between the nearest neighbours (short-range interaction) must play a dominant role. When the contribution from the long-range (e.g., dipole-dipole) interactions is large, the cluster structure is considerably less pronounced.

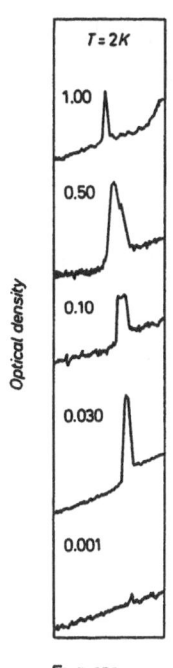

Frequency

Fig.4.27. Low-frequency portion of the exciton absorption spectrum of $hmbd_0/hmbd_{18}$ with varying composition [4.45] (relative $hmbd_0$ concentration is indicated)

The spectra of mixed crystals of $hmbd_0$ and $hmbd_{18}$ shown in Fig.4.27 are interesting examples of the spectra of crystals with a single molecule in the unit cell. For this system the isotopic shift $\Delta = 112$ cm^{-1}, the width of the exciton band of a one-component crystal $2\mathcal{M} = 40$ cm^{-1}, and the parameter $\Delta/\mathcal{M} = 5.6$. The energy spectrum is a two-band one with a gap. The absorption spectrum must be of a two-mode type close to the molecular limit (similarly to bd_0/bd_6, Sect.4.2.2).

The presence of only one molecule in the unit cell simplifies the spectrum, i.e., the exciton spectrum of a mixed crystal exhibits only two bands and the dependence of their centres of gravity on the concentration is described by the branches of a parabola similar to parabola 1 in Fig.4.4. The spectrum in Fig.4.27 illustrates the behaviour of the low-frequency band of the component $hmbd_0$. The observed bands are practically structureless. If the cluster structure is taken into account, the enveloping band becomes slightly asymmetric, with a slower fall-off towards the low-frequency side. The suppression of the cluster structure is evidently due to the small value of $\mathcal{M}$, since the width of the monomer absorption band is considerable.

In mixed crystals of anthracene, ad_0/ad_{10}, the parameter $\Delta/\mathcal{M} \approx 0.30$. The energy spectrum is then a one-band one and the absorption spectrum must be of a one-mode nature. The absorption spectra have not yet been studied, but their one-mode behaviour is seen from the reflection spectra in the exciton absorption region (Fig.4.28). Features I-III in the reflection spectra of Fig.4.28a are attributed to three types of surface excitons [4.46]. The shapes of the reflection bands hardly change as the composition of the mixed crystals is changed. The bands themselves can be found anywhere between the two limiting positions determined by the positions of the reflection bands in one-component crystals.

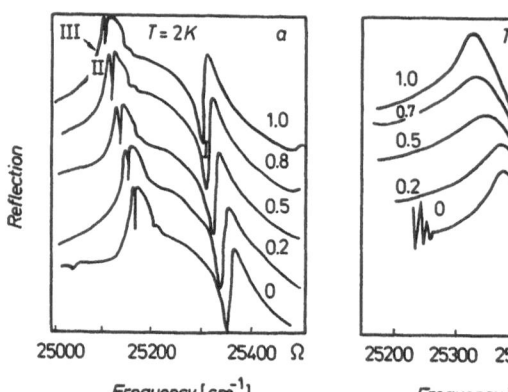

Fig.4.28a,b. Reflection spectra of ad_0/ad_{10} in the region of the exciton band doublet in polarized light [4.46]: a) parallel to **b** and b) perpendicular to **b** (the numerals indicate the ad_0 concentration)

4.5.3 Calculating the Shape of the Absorption Spectrum of a Mixed Crystal

To explain the real structure of mixed-crystal spectra (similar to the structure shown in Fig.4.26 for crystals of nd_0 and nd_8) it is necessary to carry out a numerical calculation of the spectral function S_μ.

The results of a calculation according to the NFM (Sect.4.4.3) along with the experimental data are given in Fig.4.29. The matrix H was constructed for an uncorrelated binary solution with a diagonal disorder. The two diagonal elements differ by 115 cm^{-1}. A dispersion law was adopted in the form expressed in (2.51) with a set 1 of matrix elements from Table 3.5. The values of $\theta = 5.4$ cm^{-1} proved to be sufficient for a manifestation of the cluster structure.

All in all, the calculation correctly describes both the overall shape of the absorption and the cluster structure. This is particularly evident in the low-wavelength region of the spectrum. It is possible to single out the monomer band (1) and the sharply polarized pairs of bands belonging to the dimers (2), trimers (3), and more complicated aggregates (Fig.3.21). The rela-

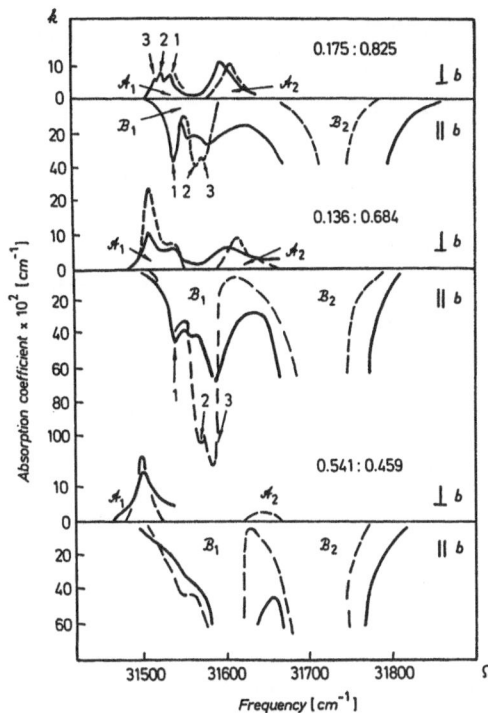

Fig.4.29. Absorption spectra of nd_0/nd_8 in polarized light (the component concentration ratios are indicated): solid curves) experiment [4.42], dashed lines) numerical calculation by NFM [4.16]; 1) monomer band; 2) dimer band; 3) trimer band

tive intensity of the monomer band decreases, while the band intensity for the dimers and more complicated centres increases at higher concentrations of nd_0. One can observe a broadening of these bands due to an interaction between the aggregates of differing structures and an increase in the background intensity due to complicated aggregates.

Despite a complete qualitative agreement, there is, however, an appreciable quantitative discrepancy between the calculated and experimental data. This is partially due to the small "volume" of the model sample, the not-too-small value of θ, and the uncertainties in the dispersion law. A certain contribution is also made by an interaction with phonons.

With respect to the last remark we shall try to establish to what extent the rigid-lattice approximation can be applied to the phenomena under consideration and what computational error is permissible. To this end, we shall refer to Fig.4.29 and consider the widths of the absorption bands shown in this figure.

A comparison of the experimental and calculated data shows that the width of the bands in the component perpendicular to **b** is almost completely due to the disorder. The maxima of the experimental bands in the component parallel to **b** are all below the calculated ones and the experimental bands are noticeably broadened with respect to the calculated ones. This broadening is particularly emphasized by the dip between the $\mathscr{B}_1$ and $\mathscr{B}_2$ bands. The mixed crystal nd_0/nd_8 exhibits a noticeable pseudogap in the energy spectrum (Fig.4.16). Hence, a weak absorption region must appear in the corresponding region of the absorption spectrum. This is actullay observed in the calculated spectra. The experimental spectra exhibit a rather strong absorption in the same region, which is evidently due to an interaction with phonons. The same circumstance is responsible for the discrepancy between the calculated and observed widths of the $\mathscr{B}_2$ band. However, precisely these parameters - the depth and width of the dip between the $\mathscr{B}_1$ and $\mathscr{B}_2$ bands and the width of the $\mathscr{B}_2$ band - are the least sensitive to calculation inaccuracies. Therefore, their deviation from the experimental data suggests the operation of additional broadening mechanisms in the experiment. In concentrated solutions of nd_0 and nd_8, the magnitude of the band broadening due to disorder is comparable to the broadening due to phonon mechanisms and even begins to dominate it. Therefore, the rigid-lattice approximation is certainly applicable to the spectra of concentrated solutions.

The broadening of the exciton absorption bands in the wide range of concentrations was investigated for mixed crystals of $hmbd_0$ and $hmbd_{18}$ [4.45]. Since the cluster structure of the spectra is only weakly defined, it is possible to use the CPA in the calculations. Figure 4.30 shows the concentration dependence of the half-width of the exciton absorption band of $hmbd_0$ in mixed crystals. The half-width of 4 cm^{-1} of the exciton absorption band in one-component crystals is subtracted from the experimental data. This half-width is due to the phonon mechanism. In mixed crystals, the broadening due to disorder is larger than that due to phonons. Therefore, the rigid-lattice approximation

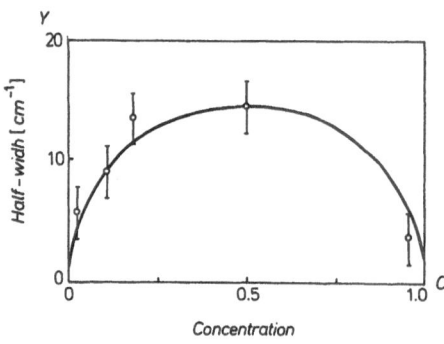

Fig.4.30. Experimental (open circles) and CPA-calculated (solid line) dependence of the half-width of the low-frequency band of $hmbd_0/hmbd_{18}$ on the $hmbd_0$ concentration [4.45]

can be used. Since the ratio $\Delta/\mathcal{M} \approx 5.6$ is still large, the band behaviour is close to the molecular limit and the band half-width is described by (4.42).

4.5.4 Determining the Position of the Pseudobottom of the Exciton-Spectrum

The optical transition scheme in Fig.4.25 shows how the position of the pseudobottom is determined (to within T) by the position of the exciton band in the mixed-crystal fluorescence spectrum. The low-temperature fluorescence spectra of four isotopic systems were studied: $hmbd_0/hmbd_{18}$, bd_0/bd_1, nd_0/nd_8, and ad_0/ad_{10}. These systems represent different types of mixed crystals in accordance with the classification given in Sect.4.4. It is all the more interesting to follow the concentration dependence of the position of the pseudobottom in these systems.

a) Mixed Crystals of $hmbd_0$ and $hmbd_{18}$

The experimental dependence of the position of the pseudobottom of these mixed crystals is illustrated in Fig.4.31. The crystals under investigation have a two-band energy spectrum and their absorption spectrum is two mode (close to the molecular limit). In this case, the position of the pseudobottom is described by (4.43). The centre of gravity of the absorption band is a linear function of the concentration in accordance with (4.41). In Fig.4.31 it is seen that the positions of the pseudobottom and the centre of gravity differ substantially.

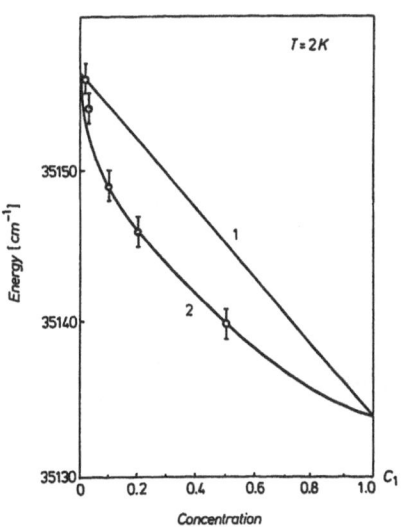

Fig.4.31. Concentration dependence of the centre-of-gravity position of the low-frequency absorption band (1) and of the spectrum pseudobottom (2) in crystals $hmbd_0/hmbd_{18}$; open circles) experimental data on fluorescence, T = 2K; solid line) CPA calculation [4.45]

b) Mixed Crystals of nd_0 and nd_8

The experimental and calculated concentration dependence of the position of
the pseudobottom and the centre-of-gravity of the absorption spectrum in the
region of the $\mathscr{A}_1$ band (Fig.4.25) are shown in Fig.4.32 [4.29]. An appreciable difference in the positions of the pseudobottom and the centre of gravity
can be observed. This has already been noted in Sect.4.4.2 and is attributed
to a slow decrease in absorption towards the high-frequency side. This shifts
the centre of gravity to higher frequencies.

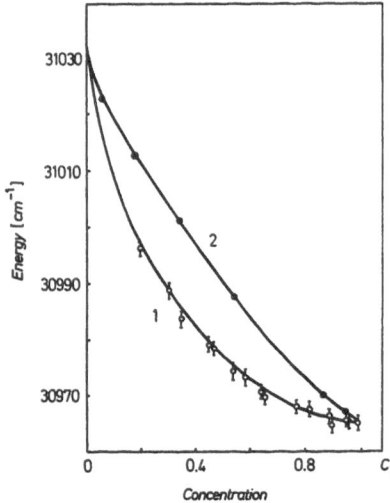

Fig.4.32. Concentration dependence of
the position of the pseudobottom (1)
and the centre of gravity of the A_1
band (2) for nd_0/nd_8; (open circles)
experimental data on fluorescence
($T = 1.6 - 1.8$ K) [4.47], (solid circles)
those on absorption ($T = 4.2$ K) [4.29]
(solid curves) calculation in CPA;
(C_1) nd_0 concentration

c) Mixed Crystals of ad_0 and ad_{10}

The reflection [4.46] and fluorescence [4.48] spectra of these crystals have
been investigated over a wide range of component concentrations. This system
occupies a special place because the experimental data have been analyzed
even though practically no information is available on the energy spectrum
of excitons in a one-component crystal. The structure of the exciton bands
of an anthracene crystal, the density of states in them, the positions of the
electronic terms, and the location of the points with $k = 0$ are still under
discussion. An analysis of the low-temperature fluorescence spectra of mixed
crystals enables one to estimate the half-width $\mathscr{M}$ of the band and, hence,
calculate some other parameters.

The fluorescence spectrum of anthracene (this is also true for mixed crystals of ad_0 and ad_{10}) at low temperatures consists of relatively narrow
($\approx T$) bands. As a result, the exciton fluorescence band determines, to within

the same accuracy as in the two above-discussed systems, the position of the
energy spectrum bottom in a one-component crystal and the position of the
pseudobottom E_{pb} in a mixed crystal.

The concentration dependence of the position of the pseudobottom is shown
in Fig.4.33. The energy spectrum of the mixed crystals of ad_0 and ad_{10} is an
example of a one-band spectrum, which is close to the limit of the amalgamated
bands. In this case, the position of the pseudobottom is described by (4.46),
which can be written in the following form:

$$E_{pb}(C_1) = E_1 - \mathcal{M} + (1 - C_1)\Delta - \frac{2\Delta^2}{\mathcal{M}} C_1 C_2 \quad . \tag{4.51}$$

This expression was obtained for a symmetrical band with a density of states
given by (3.14). Since the values of E_1 and $\mathcal{M}$ are unknown, it is impossible
to calculate $E_{pb}(C_1)$. Assuming that the experimental dependence is given by
(4.51), one can still determine the value of $\mathcal{M}$ by using the deviation of
$E_{pb}(C_1)$ from the linear dependence. The value of $\mathcal{M}$ ($= 230$ cm^{-1}) corresponds
to the experimental position of the pseudobottom for $C_1 = C_2 = 0.5$. The calculated function $E_{pb}(C)$ agrees with the experimental data over the entire concentration range (Fig.4.33a).

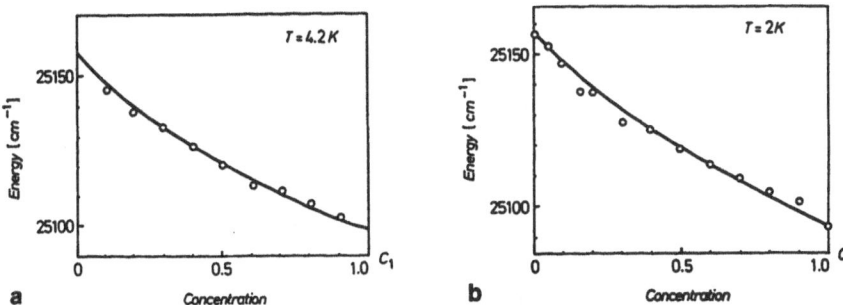

Fig.4.33a,b. Concentration dependence of the position of the energy-spectrum
pseudobottom in ad_0/ad_{10}: (a) open circles) experiment on fluorescence spectra
[4.48]; solid line) calculation by (4.51); (b) open circles) experiment on
crystal reflection spectra: solid line) CPA calculation [4.45]

Setting the centre of gravity of the absorption band of the one-component
crystal equal to the energy of the point $k = 0$ (averaged in a certain way over
the longitudinal-transverse splitting) results in

$$\Omega_\mu = E_1 + \epsilon_\mu \quad . \tag{4.52}$$

For a symmetrical band

$$E_{min} = E_1 - \mathcal{M} \quad . \tag{4.53}$$

Since $\mathcal{M}$ is determined above, E_{min} is known from fluorescence experiments [4.48], and Ω_{μ} is known from absorption experiments [4.50], it is possible to determine the position E_1 of the centre of gravity of the exciton energy band and the positions ε_{μ} of the components of the exciton multiplet relative to the centre of gravity. For an anthracene crystal, $\Omega_b - E_{min} = \varepsilon_{\mu} + \mathcal{M} = 25190 - 25098 = 92$ cm^{-1}. Ω_b denotes the position of the centre of gravity of the long-wavelength b component of the exciton doublet [4.51]. Thus, $\varepsilon_b = -138$ cm^{-1} and, hence, $E_1 = \Omega_b - \varepsilon_b = 25328$ cm^{-1}. Since $\Omega_a = 25440$ cm^{-1} [4.51], $\varepsilon_a = 112$ cm^{-1}. The approximate width $2\mathcal{M}$ of the exciton band determined above is equal to 460 cm^{-1}. Such is the information on the exciton spectrum of a one-component crystal, which can be derived from the fluorescence spectra of mixed crystals.

The position of the pseudobottom in the energy spectrum of mixed crystals of ad_0 and ad_{10} can also be determined from the reflection spectra shown in Fig.4.28 [4.46]. The results are given in Fig.4.33b. For a comparison, the results of a CPA calculation which used the calculated exciton density of states for an anthracene crystal (Fig.2.5) [4.52] with $2\mathcal{M}\approx700$ cm^{-1} are also shown.

4.6 Migration of Excitons and Localization of Exciton States in Mixed Crystals

4.6.1 Kinetics of Excitons

The structure of the energy spectra of excitons and the exciton absorption spectra in isotopically mixed crystals were discussed in the preceding sections of this chapter. It was concluded that the shape of the absorption spectrum over the entire concentration range corresponds to the retention of exciton-like states (the absorption bands remain relatively narrow, the Davydov splitting is retained, etc.). The nature of the exciton-like states essentially depends on the value of the parameter $\Delta/\mathcal{M}$: they are either shared about equally by both isotopic components ($\Delta \gtrsim \mathcal{M}$), or they belong predominantly to one of the components ($\Delta \lesssim \mathcal{M}$).

The image which is associated with the exciton since Frenkel's first work is that of an excitation wave travelling through a crystal. Such a state is delocalized, i.e., its wave function has a finite value everywhere. Whether or not the excitons retain these properties in mixed crystals and whether or not they remain delocalized and retain the ability to transport energy are questions to which answers must be found. To specify the terminology, we emphasize that everywhere in this section the states will be called *localized* or *delocalized*, depending on whether or not their wave functions vanish towards infinity.

The ability of excitons to transport energy is due to their diffusibility. It should be stressed that neither the energy spectra nor the absorption spectra of mixed crystals can directly yield any information on the diffusibility of excitons. Indeed, both the density of states $\rho(E)$ and the conductivity tensor $\sigma(\Omega)$ are determined by the one-particle Green's function of the exciton. The diffusion coefficient is directly related to the two-particle Green's function. It is a well-known fact that these functions cannot be expressed in terms of one another and must be calculated independently. Similarly, one cannot draw any conclusions from the type of the density of states $\rho(E)$ as to whether the wave functions are localized or not.

Information on the kinetics and diffusibility of excitons can be obtained by investigating their migration. Energy migration in itself is a wide field of research and has been considered in a number of monographs and surveys. It is complicated by reabsorption and multiple trapping. We shall only briefly mention several recent experiments on the low-temperature energy migration in crystals of d-naphthalenes and d-benzenes, which are crucial for an understanding of *exciton diffusibility* in mixed crystals.

4.6.2 Low Temperature Exciton Migration in Mixed Crystals of d–Naphthalenes and d–Benzenes

Mixed crystals of nd_0 and nd_8 containing small amounts of β-methylnaphthalene (BMN) were investigated by KOPELMAN et al. The concentration of BMN, C_β, was kept almost constant and close to 10^{-3}. The concentration of nd_0 was varied over a wide range. Their luminescence spectra [below (nd_0/nd_8):BMN] in the region of the first singlet transition (fluorescence) [4.53] and in the region of the first triplet transition (phosphorescence) [4.54] were investigated. The crystals were excited in the region of the intrinsic absorption of naphthalene. Most of the measurements were performed at $T \approx 2$ K.

The luminescence spectrum of the mixed crystal is made up of the bands of nd_0 and BMN. No host emission is observed. This means that the light-produced excitons of the host crystal (nd_8) have enough time to relax to the impurity levels of nd_0 and BMN. The BMN level forms a deep trap. For singlets it is shifted approximately 400 cm^{-1} relative to the nd_0 term. The monomer level of nd_0 is relatively shallow, i.e., for singlets $|\varepsilon_{bind}| = 49$ cm^{-1} (Sect.3.3). In both cases, however, the reverse thermal promotion of the relaxed excitons into the nd_8 band is not possible at $T \approx 2$ K. Therefore, the kinetics of the relaxed excitons is restricted to the subsystem nd_0/BMN.

The simultaneous emission from several types of impurities, including a redistribution of intensity as a result of changes in concentration, is a

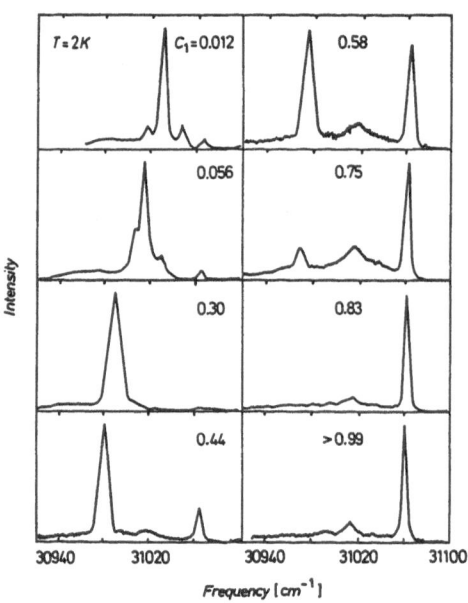

Fig.4.34. Dependence of the fluorescence spectrum of (nd_0/nd_8):BMN on the nd_0 concentration [4.55]; the low-frequency fluorescence peak $(30970 - 31030\ cm^{-1})$ corresponds to the vibronic transition (phonon $\nu_{17} = 509\ cm^{-1}$) in nd_0, and the high-frequency one $(31065\ cm^{-1})$ to the electronic transition in BMN

common phenomenon. However, in this case the concentration dependence of the intensity redistribution proved to be quite unusual.

The fluorescence spectrum of the system $[(nd_0/nd_8)$:BMN] is shown in Fig.4.34. The emission of nd_0 dominates at nd_0 concentrations $C_1 \lesssim 0.3$ (vibronic band, $\nu_{17} = 509\ cm^{-1}$). The BMN band is hardly visible. Due to the large differences in concentrations, i.e., $C_1 \gg C_\beta$, the excitons are mainly trapped on nd_0.

The emission band of nd_0 in the low-concentration region is the result of transitions from the monomer level. In the high-concentration region, this band is the result of transitions from the pseudobottom of the spectrum. As the concentration of nd_0 is increased, it moves to the low-frequency region (Fig.4.32).

The curve corresponding to $C_1 = 0.44$ shows a redistribution of intensity in favour of BMN. It progresses rapidly at higher values of C_1. This is definite proof of an effective mechanism of transferring excitation energy from nd_0 to BMN. It is especially remarkable that the intensity redistribution takes place so abruptly. This can clearly be seen in Fig.4.35, where the ratio of I_β (intensity of the BMN band) to $I = I_\beta + I_1$ (total fluorescence intensity) is plotted as a function of C_1 (I_1 is the intensity of the nd_0 band). One can speak of an abrupt increase in the rate of transfer of excitons from nd_0 to BMN as of the beginning of a *percolation* of excitons through the nd_0 subsystem. It sets in at $C_1 \approx 0.5 - 0.6$.

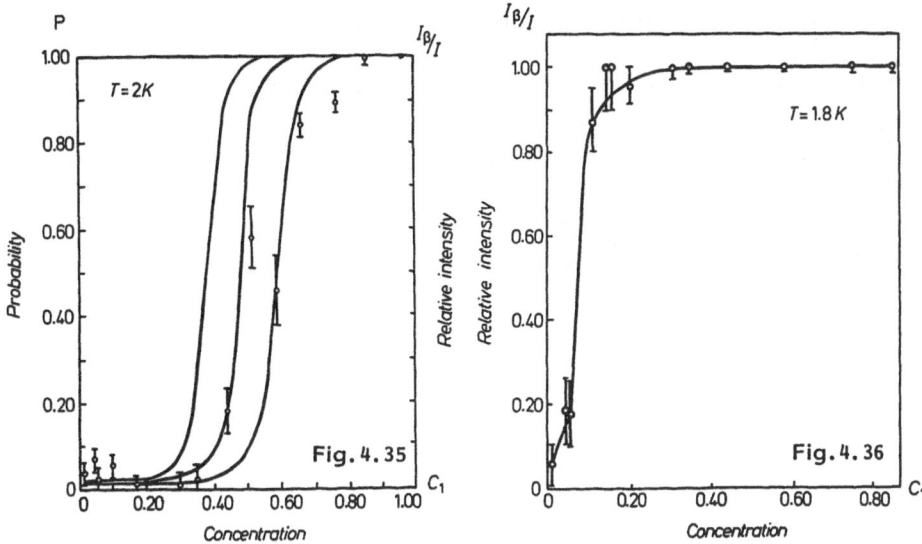

Fig.4.35. Dependence of the relative intensity I_β/I on the nd_0 concentration in the fluorescence spectrum of (nd_0/nd_8):BMN [4.55]: circles) experimental data; curves) probability P (Sect.4.6.4)

Fig.4.36. Dependence of I_β/I on the nd_0 concentration in the phosphorescence spectrum of (nd_0/nd_8):BMN [4.56]

A very similar situation was observed in the phosphorescence spectrum of the same system (Fig.4.36). The difference lies in the fact that the percolation begins at a considerably lower nd_0 concentration, i.e., at $C_1 \approx 0.08$. The difference in threshold concentrations undoubtedly indicates that a percolation through the triplet states does not involve the singlet states, i.e., the percolation takes place after the singlet-triplet conversion. Such a separation in time of the processes in the singlet and the triplet system is a result of the enormous differences in their lifetimes, i.e., $\tau_S \approx 10^{-7}$ s [4.57], while $\tau_T \approx 3$ s [4.58,59].

Similar results for the luminescence spectra of an isotopically mixed crystal of bd_0 and bd_6, which was doped with pyrazine (a trap with an approximate depth of 200 cm^{-1}), were obtained by COLSON et al. [4.60]. They repeat the regularities discussed above with reference to d-naphthalenes. Both the fluorescence and phosphorescence spectra clearly exhibit the percolation effect. It starts to take place for singlets at a considerably higher bd_0 concentration than for triplets.

4.6.3 Localization and Diffusion in Disordered Systems

The question of when an under what conditions elementary excitations are able to spread out over large distances in disordered systems, has been in the centre of attention and under intense development for the past twenty years, especially with respect to the electrical conductivity of amorphous solids and doped semiconductors. The extensive literature is summarized in a number of surveys [4.61-67]. On some problems, there is no concensus of opinion. Often a concensus is not based on rigorous proofs, but on a set of realistic arguments, computer calculations, and an analysis of the experimental results. Nevertheless, the new concepts are very harmonious and explain the extensive experimental results satisfactorily. Below, the main features of these concepts are summarized.

The actual statement of the problem is a result of the classical work of ANDERSON [4.68]. It contained and substantiated an assertion that in a system with a sufficiently high random potential over the entire energy range, which in the perfect crystal belonged to the conduction band, the wave functions undergo a radical reconstruction. All of the delocalized solutions of the plane-wave type disappear and are replaced by localized functions which rapidly decrease towards infinity. Thus, a particle placed in such a state at the instant $t = 0$, at an arbitrary $t > 0$ will remain in a limited region of the order of the radius of the wave function of this state and will not be able to go to infinity. This means that no diffusion is observed in such a system. The qualitative change in the system of the wave functions that is brought about by the random potential is termed the *Anderson transition*.

The development of the Anderson transition is sketched in Fig.4.37. The simplest model is the diagonal-disorder model, which has already been used to describe isotopic solutions. The only difference is that the "isotopic shift" Δ is considered to be a continuous random parameter, distributed over an interval $-W/2 < \Delta < W/2$. Choosing a continuous random parameter eliminates the complications brought about by a cluster structure. It is clear from a general consideration that the critical value of W at which the Anderson transition can take place must be of the order of $W_{cr} \sim \mathscr{M}$. A numerical coefficient of the order of unity is determined by the specific geometry of the lattice. It is highly significant that the localized and delocalized states are not mixed together when $W < W_{cr}$. They are concentrated on certain portions of the energy axis and are separated by demarcation lines called the *mobility edges*. The interval between the mobility edges is named the mobility gap. This concept was advanced by MOTT [4.64].

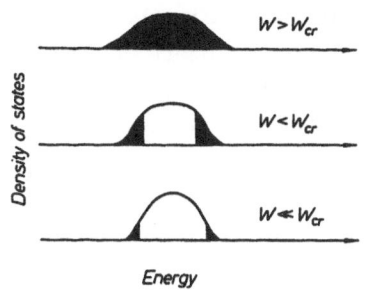

Fig.4.37. Development of the Anderson transition as a function of the value of the random potential W (the area of localized states is blackened)

It was assumed during some period of time that at $T = 0$ the diffusion coefficient $\mathcal{D}$ changes abruptly at the mobility edge from zero in the mobility gap to some finite value outside it. This assumption, however, was a matter of dispute. But the other point of view predominates now. According to it the scaling hypothesis holds near the mobility edge. Therefore, instead of an abrupt change, $\mathcal{D}$ follows the power law $\mathcal{D}(E) \propto |E - E_{m.e.}|^{\gamma}$, where $E_{m.e.}$ is the energy at the mobility edge, E is outside the mobility gap, and the numerical coefficient $\gamma > 0$ is a critical index. Within the delocalized-state region the diffusion mechanism is similar to the usual band mechanism, and $\mathcal{D} \sim v^2 \tau_{fr}$, where v is the velocity and τ_{fr} is the mean free time. $\mathcal{D}$ vanishes at the mobility edge, but increases rapidly away from it. However, in the localized-state region diffusion is only possible through jumps between impurity levels, i.e., it is of the hopping type. The jumps are accompanied by phonon absorption or emission. As a result, $\mathcal{D} \sim l^2/\tau_{jump}$, where l is the length of the jump and τ_{jump} is the average time between jumps. The values of $\mathcal{D}$ for hopping diffusion are usually small (especially at low temperatures). Therefore, the mobility edge separates the regions with large and small diffusion coefficients.

Thus far, the Anderson transition has only been considered to be a function of the parameter $W/\mathcal{M}$. However, in binary mixed crystals, it is more important to consider the Anderson transition as a function of the concentration. This can be done using the *Lifshitz model* - the random arrangement of identical centres. In this model it is assumed that the isotopic shift Δ is sufficiently large, so that discrete impurity levels appear even in the low-concentration range. At low-impurity concentrations ($C_1 \ll 1$), i.e., for large distances between impurity centres, the excitations are only shared by pairs of molecules (dimers). The level splitting is exponentially small (Sect.3.4.2; dipole-dipole interactions are excluded). LIFSHITZ showed that a sharing between trimers is only possible with an almost symmetrical arrangement of all three molecules [4.1]. However, the statistical weight of such configurations is small. The same is true for more complicated clusters. This means that the wave function in each of the quantum states is distributed extremely non-uniformly, even within the same cluster. It is concentrated almost entirely on either one of the monomers or on one of the dimers. Therefore, for $C_1 \ll 1$, all of the impurity states are localized. On the other hand, it is obvious that there is no localization in most parts of the band when $C_1 \approx 1$; therefore, a critical value of C_{loc} must exist for which the localization of the impurity states becomes destroyed near a definite energy value. If we exclude large-radius centres, the problem will lack another parameter with a length dimension, apart from the lattice constant. Therefore, the magnitude of C_{loc} must be of the order of several tenths. It is a function of the lattice geometry and the ratios $M_{n\alpha m\beta}/\Delta$. If only a direct interaction is considered and the well-known Mott criterion for localization in an impurity band ([4.65], Sect.2.7.1) is applied to the lattice, $C_{loc} \approx 1/16$. When $C_1 < C_{loc}$, all of the impurity states are localized.

The following must be emphasized. The entire concept described above was actually developed for electrons. It can only be extended to excitons to the extent that dipole-dipole interactions can be neglected. ANDERSON [4.68] showed that dipole-dipole interactions can destroy the localization. Of course, for large values of $W/\varepsilon_{\ell t}$ the lifetimes of the states will be exponentially large ($\varepsilon_{\ell t}$ is the longitudinal-transverse splitting at $\mathbf{k} = 0$, Sect.2.2). This problem has apparently not yet been studied in greater detail. Of course, dipole-dipole interactions do not exist for triplet excitons. These interactions are very weak for singlet excitons in benzene because the transitions are very weak. It is possible, however, that they may play a significant role in the singlet transitions in naphthalene (Sect.4.6.4).

The approach to the mobility-edge concept described above is based on investigations into the eigenfunctions of the quantum system with a random potential. A similar concept arises in several classical problems, where the classical analog of mobility edge is called the *percolation level* or *percolation threshold*. The percolation concept can be formulated for the flow of a liquid over an irregular relief or through a pore system, or for the passage of a current through a random resistance network. The reader is referred to the relevant reviews [4.69-71]. We shall only state the problem in the most convenient form for discussing the above experimental data.

Imagine an infinite square network made up of identical resistances r, whose nodes are either occupied (then the corresponding resistance terminals are joined together) or vacant (then the resistances at the node are disconnected) (Fig.4.38). The nodes are filled at random. A cluster is formed by a group of nodes that are connected by resistances. It is obvious that if the probability

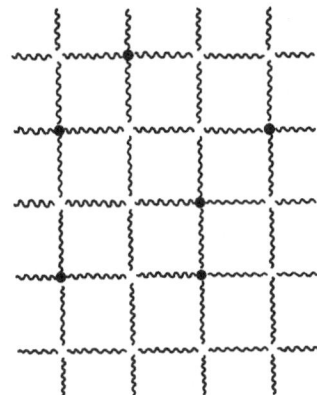

Fig.4.38. Square net with random occupancy of the sites (the site problem); the occupied sites are blackened

C_1 of filling a node is low, the network is dominated by circuit breaks, the clusters are small, and there are no paths for the current. They only appear when $C_1 = C_{per}$, i.e., when a cluster that penetrates the entire network is formed. For a square network, $C_{per} \approx 0.59$ [4.69]. When $C_1 > C_{per}$, the network possesses a finite resistance. The quantity C_{per} is the percolation level or threshold.

It is not hard to understand the relationship between the percolation problem and the exciton migration on a plane network in the nearest-neighbours-interaction approximation (Sect.3.4). If the isotopic shift in a binary mixed crystal is large ($|\Delta| >> \mathcal{M}$), the indirect interaction can be neglected, i.e., the exciton can only visit one type of node. The "accessible" nodes will be regarded as "occupied". Their concentration is denoted by C_1. The conductivity r^{-1} plays the same role as the matrix element M_{12}. It then becomes obvious that a classical diffusion of the excitons is only possible when $C_1 > C_{per}$.

Of course, the real migration problem is a quantum mechanical one and differs essentially in some respects from the classical percolation probelem. First of all, the spectrum contains, in addition to delocalized states, Anderson-localization regions even when $C_1 > C_{per}$ (Fig.4.37). If those states which belong to such a region are of special importance, there will be no diffusion. A direct interaction between widely spaced molecules and an indirect interaction can also open the channel for diffusion, even if $C_1 < C_{per}$. Thus, the diffusion problem cannot be reduced to the classical percolation problem. Nevertheless, any information that is obtained from the much simpler classical percolation problem may prove to be very useful.

In the literature, the appearance of intensive diffusion is often termed percolation regardless of the detailed diffusion mechanism, e.g., the transport by particles in the delocalized-states region. Below, as well as in Sect.4.6.2, the term will be used in this broad sense.

4.6.4 Discussion of Experimental Data

A complete interpretation of the experimental data on migration presented in Sect.4.6.2 has not yet been made. These results will be discussed below on the basis of the scheme proposed by the authors of the original papers (for review, see [4.56,72]) and the general properties of disordered systems which were described in the preceding section.

The distinguishing feature of migration experiments is that an essential role is played by the lifetime τ of the exciton (Sect.4.6.2), i.e., the time which an exciton exists in the crystal before its radiative or non-radiative decay. The problem is whether or not the time τ is sufficiently large to enable the exciton to transfer from an "impurity" (i.e., from the levels of the "low-energy" isotopic component) to a deep trap. The determining factor controlling the transfer of excitons from "impurity" levels to deep traps is the competition between the capture of excitons by traps or by localized impurity states with levels located deep enough below the mobility edge so that no activational transitions into delocalized states (i.e., into states located above the mobility edge) can take place.

To explain the fluorescence of a crystal of $[(nd_0/nd_8):BMN]$, KOPELMAN et al. [4.53,54] used a model according to which an exciton which finds itself on a classical cluster containing at least one trap will necessarily be captured by it. Such a model assumes that transitions between clusters cannot take place within the lifetime τ and that the exciton always manages to reach the trap in the cluster. Under these conditions the emission intensity of the traps is completely independent of τ. It is entirely determined by the probability P that the cluster of nd_0 molecules on which the exciton is trapped contains at least one BMN molecule. The calculated results for a square network and for two other plane networks are shown in Fig.4.35. The agreement between the calculated and experimental values is quite satisfactory. But this agreement does not eliminate the questions associated with the fact that excitons are quantum particles and, therefore, their behaviour must be described by quantum phenomenon - Anderson localization, and not by the classical percolation arguments. Of course, if $C_1 > C_{per}$, the formation of "corridors" of molecules connected by the large matrix element M_{12} promotes a migration. Apparently $C_{per} > C_{loc}$ so that when $C_1 > C_{per}$ the impurity band contains a delocalized-state region. However, a tail of localized states remains below the mobility edge. Therefore, it is necessary to find out what the consequences of this tail are. It would be possible to understand the classical nature of percolation if the inclusion of dipole-dipole interactions can be assumed to suppress the localization near the mobility edge. At the same time, however, we must take into account that the observed percolation threshold for singlets, $C_1 \approx 0.6$, is very high with respect to the threshold for triplets and cannot be explained without making an allowance for localized states which compete with traps in exciton capture. This competition the localized states may be particularly important for singlet excitons because the transfer integrals are large (especially M_{12}). Therefore, the level splittings in the clusters and the deep-lying tails of the density of states are also large. As a result, the distance between the mobility edge and the positions of most of the localized levels at which exciton capture takes place may be much greater than T. Hence, the activational promotion of the exciton to the mobility edge will not take place within the relatively short lifetime of the singlet exciton $\tau \sim 10^{-7}$ s and the migration of excitons already trapped to the localized level will be almost completely suppressed.

A possible clue to the understanding of the situation is as follows. At $C \sim 0.5$ the position of the fluorescence band of nd_0 is close to that of the pseudobottom of the spectrum. This fact has already been discussed in Sect.4.4. As C_1 increases, the cluster structure disappears and the pseudobottom becomes more and more clearly defined. This also results in a narrowing of the fluorescence band in Fig.4.34; at $C_1 = 0.44$ its width is about 10 cm^{-1}. It would seem natural to assume that the mobility edge is within this half-width. At the same time, this width is already comparable to T (T = 2 K). Therefore, an activational percolation to the traps through the delocalized states becomes possible within the lifetime τ of an exciton. From the above point of view the beginning of an intensive trap luminescence at the classical percolation threshold must be regarded in a certain extent as a coincidence. The latter appears apparently due to the fact that the cluster structure is suppressed at approximately the same time when the long "corridors" are formed.

There is also no concensus of opinion in the interpretation of the phosphorescence (Fig.4.36). The initial interpretation [4.53] was based on a model of hopping diffusion between widely spaced impurity centres with indirect pairwise interaction (Sect.3.4.2) [4.73]. Since the transfer integral is small, $M_{12} \approx 1$ cm^{-1}, so is the spread in the levels. At $T \approx 2$ K there are practically no energy restrictions on the jumps between the levels. However, the jump rate given in [4.53] is grossly overestimated. It has also been suggested [4.74] that the suppression of migration at small values of C_1 is due to the Anderson localization in the impurity band because of the diagonal disorder caused by accidental stress and unwanted impurities. Although such a mechanism cannot be excluded, it should be noted that the observed percolation threshold, $C_1 \approx 0.08$, is close to the estimate for $C_{loc} \approx 0.06$ made in Sect.4.6.3. Therefore, it is quite possible that the transition is actually of an Anderson nature and is associated with the appearance of delocalized states in the impurity band. From this standpoint no additional mechanisms of diagonal disorder must be involved in the explanation of this transition. The smallness of M_{12} for triplet excitons which was already mentioned with reference to hopping diffusion ensures a sufficient rate of transitions from the localized states to the mobility edge at $T \approx 2$ K. The above is also true for the luminescence of mixed crystals of deuterobenzenes [4.60].

A detailed understanding of the above-described effects has not yet been achieved. However, a number of hypotheses have been advanced which relate these effects to the general concepts of the theory of diffusion in disordered systems. At the same time, these investigations have stirred an interest in the scientific community and experimental results on similar effects in other substances are now available. For example, a percolation of triplet excitons from the monomers of phenazine-d_0 to the dimers ("deep traps") was observed [4.75] in the low-temperature ($T < 1.33$ K) fluorescence spectra of isotopic solutions of phenazines-d_0 and d_8. Similar phenomena were also observed [4.75] in isotopic solutions of 1,4-dibromonaphthalene, which are particularly interesting because they belong to a group of crystals with a quasi-one-dimensional exciton spectrum.

5. Band-to-Band Transition Spectra

This short chapter is dedicated to the description of a special type of vibronic transitions whose initial and final states are band states. One of these band states contains an electronic exciton; the other contains an internal phonon. Such transitions are called band-to-band transitions. They correspond to an absorption process in which an exciton appears and a phonon disappears and an emission process in which an exciton disappears and a phonon appears. Since the exciton and phonon do not co-exist, there cannot be an interaction between them. As a result, the theory that describes band-to-band transitions is extremely simple and yields convenient formulae which directly relate the shape of the optical spectra with the density of states $\rho(\varepsilon)$ in the energy spectrum of the exciton. The band-to-band absorption spectrum directly yields the "map" of the density $\rho(\varepsilon)$. The emission spectrum differs from $\rho(\varepsilon)$ only in the Gibbs's factor, which describes the exciton distribution in the band. So far, this method has proven very effective in the reconstruction of $\rho(\varepsilon)$ from experimental data. Until now, this has only been possible for pure crystals. However, it may also become possible to reconstruct the partial densities of states in mixed crystals.

5.1 Classification of Vibronic Spectra

Pure electronic spectra correspond to optical transitions in which an electronic exciton either appears or disappears. We shall now consider vibronic spectra in which an exciton transition is accompanied by the emission or absorption of an internal phonon (Sect.1.6). The vibronic spectra of molecular crystals are, of course, closely related to those of molecules. The latter have already been discussed in Chap.1 (Sect.1.2).

The vibronic spectrum is a part of the electronic-vibrational spectrum of a crystal. We will dwell at some length on the factors that made us single out and study vibronic spectra independently.

A general study of the exciton-phonon interaction is, in practice, impossible for the following reasons.

Firstly, the number of branches in the phonon spectrum of organic molecular crystals is extremely large. Even for a relative simple crystal such as benzene the number of vibrational branches is equal to 144 (the number of degrees of freedom per unit cell).

Secondly, the exciton-phonon interaction constants are different for different branches and, in general, depend on both the phonon wave vector and the exciton momentum. The information on the exciton-phonon interaction law is extremely scarce as yet. As a result, the exciton-phonon interaction Hamiltonian is actually unknown[1].

Thirdly, even if this Hamiltonian were known, it would still be impossible to find a general solution that adequately describes the dynamics of an exciton interacting with phonons because the exciton-phonon interaction is generally not weak and nonlinear with respect to nuclear displacements.

Therefore, the exciton-phonon interaction can only be adequately described by choosing simplified but at the same time sufficiently realistic models. In this respect, the vibronic interaction evidently has unique advantages.

Considering internal phonons[2] as a separate group is dictated by the very logic of the existence of molecular crystals. Since intramolecular interactions are much stronger than intermolecular ones, the frequencies of the internal phonons are also greater than those of the external phonons. Usually they differ by about one order of magnitude. The dispersion of internal phonons is related to intermolecular interactions and is small. As a result, the vibrational spectrum is made up of the low-frequency branches of the external phonons (with a high dispersion) and the high-frequency branches of the internal phonons (with a low dispersion) [5.3]. The magnitude of frequencies of the latter are close to those of vibrations in an isolated molecule. Hence, the internal phonons can be considered to be vibrational excitons that arise from intramolecular vibrational modes [5.4].

It is particularly important to emphasize that, due to the domination of intramolecular interactions, the interaction between excitons and internal phonons (i.e., the vibronic interaction) is described by the intramolecular electronic-vibrational coupling constants which, in principle, can be determined from the spectra of isolated molecules. The weak dispersion of the internal phonons and the simple law governing their interaction with excitons simplify considerably a study of vibronic spectra. It should also be added

---

[1] The first determination of the deformation potentials for excitons in naphthalene crystals has recently been reported by OSTAPENKO et al. [5.1,2].

[2] The term "intramolecular phonons" is also often used.

that the large magnitude of the vibrational frequencies often allows one to recognize the spectra corresponding to different phonons and to study their contributions independently. Finally, the high frequency of the vibrations also makes it possible to develop convenient and sufficiently general techniques that can be used to describe the vibronic interaction.

To simplify the task, we do not consider the processes associated with the external phonons, which has always been done in analyzing electronic spectra. In this sense, we shall describe the vibronic spectra using the rigid-lattice approximation.

On the whole, the interaction between excitons and internal phonons has been extensively covered in the literature. Below, attention will be focussed on the relationship between the vibronic spectra and the structure of the exciton bands. The existing methods for the reconstruction of the structure of exciton bands from vibronic spectra shall be discussed and a quantitative treatment of the vibronic spectra shall be given. Proceeding from the presently available information on the structure of exciton bands, a sufficiently comprehensive quantitative comparison between theory and experiment can be carried out for particular cases. It turns out that under certain conditions the concepts and ideas used to describe the vibronic spectra and those used in the theories describing excitons in doped crystals are closely related. Therefore, we shall often refer to Chap.3.

Vibronic spectra essentially involve two distinct experimental situations.

The first situation corresponds to transitions between states, one of which contains an exciton while the other contains an internal phonon. For example, absorption spectra correspond to the disappearance of an internal phonon and the creation of an exciton, while emission spectra correspond to the disappearance of an exciton and the simultaneous creation of a phonon. Such transitions are called band-to-band transitions because both the initial and the final states are band states. Band-to-band transitions are an important source of information on the distribution of the density of states $\rho(\varepsilon)$ in exciton bands.

The second situation corresponds to the simultaneous creation of an exciton and an internal phonon upon absorption of a quantum of light. Since both quasi-particles are simultaneously present in the final state, they interact with one another. The nature and strength of final-state interaction determines both the overall structure and the intensity distribution of the absorption spectrum. It is here that the close relationship of vibronic spectra with those of excitons in a doped crystal arises. Henceforward, it is assumed that the vibronic spectra involve transitions of this type unless it is specified otherwise.

This chapter will consider the spectra of band-to-band transitions, while the vibronic spectra themselves will be discussed in the next chapter.

5.2 Band-to-Band Transitions in Perfect Crystals

5.2.1 Band-to-Band Transition Theory

The absorption spectrum of a crystal is made up of transitions from the ground state at absolute zero (T = 0 K). As the temperature increases, i.e., as the average Planck occupancy numbers $\bar{n}(\nu)$ of phonons with a frequency ν increase, transitions from the vibrational sublevels of the ground electronic state also take place.

Transitions from the phonon band of the ground electronic state to the exciton band are shown in Fig.5.1a. The momentum of the photon, **K**, is small compared to the Brillouin momentum ($\approx \pi/d$), which determines the size of the reciprocal lattice cell, d being the lattice constant. Therefore, the transitions are practically vertical transitions. They are accompanied by the disappearance of a phonon with a momentum **q** and the appearance of an exciton of the same momentum **k** = **q**. Such transitions will be called *band-to-band transitions*.

Usually the width of the phonon band, $2m$, is small, i.e., $m \ll \nu$ [5.3,4]. Only at a temperature T comparable to ν, i.e., when $\bar{n}(\nu)$ becomes significant, is the intensity of the band-to-band absorption not too low. In this case T $\gg m$, and all of the phonons are approximately equally excited. As a result, all of the states of the exciton band are final states, i.e., the band "opens up" and the band-to-band absorption spectrum must map it. The exciton and band-to-band absorption spectra for $\nu > 2\mathcal{M}$ are compared in Fig.5.2a. When $\nu < 2\mathcal{M}$ and thus the band-to-band transitions overlap with the exciton absorp-

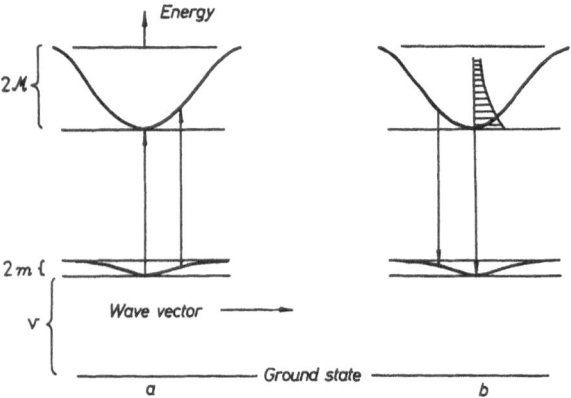

Fig.5.1a,b. Band-to-band transitions: a) light absorption; b) luminescence; the hatching denotes the exciton energy distribution

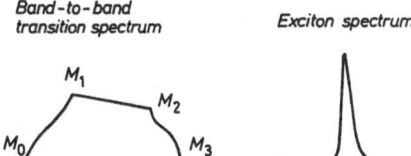 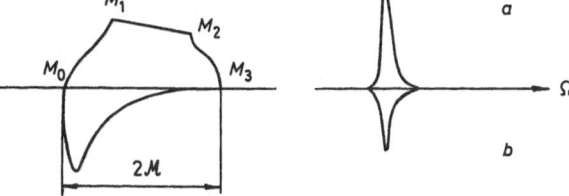

Fig.5.2a,b. Band-to-band transition spectra: a) absorption spectrum; b) luminescence spectrum

tion band, or when the band-to-band transitions corresponding to different phonons are superimposed on one another, the spectrum becomes very complicated and cannot be used to analyze the structure of the exciton band.

The vibronic spectrum of the intrinsic luminescence of a molecular crystal is always a band-to-band transition spectrum. An outline of such transitions involving one phonon is given in Fig.5.1b. In contrast to the absorption spectrum, the initial state of an emission spectrum is not the equilibrium one. In general, one cannot say that the exciton distribution in a band is Gibbsian. An analysis of the experimental data shows that the band excitons manage to thermalize [5.5] during the lifetime of the excited state ($\gtrsim 10^{-8}$ s). This means that their distribution is described by the usual $\exp\{-\varepsilon(\mathbf{k})/T\}$ law where T is the temperature of the lattice. However, since T is usually less than $2\mathcal{M}$, the different exciton states are not filled uniformly. As a result, the luminescence from a region (width $\approx$T) close to the bottom of the exciton band dominates. The emission spectrum is sketched in Fig.5.2b. It is seen that the band-to-band absorption and emission spectra differ substantially. This difference diminishes with increasing temperature.

Since the filling of the initial states (the phonon states for absorption and the exciton states for emission) is determined by the temperature, the term "*hot band spectroscopy*" (side by side with the term band-to-band transition) is used with reference to these spectra.

Under exciton thermalization conditions it is possible to establish a general relationship between the band-to-band absorption and emission spectra based on the generalization of the Einstein equations to a weakly absorbing dispersive medium. The spectral intensity of the luminescence flux in the s-direction with the electric vector polarized in the e-direction is given by

$$I_{\mathbf{e}}(\Omega,\mathbf{s}) \propto A_{\mathbf{e}}(\Omega,\mathbf{s})(\mathbf{e}\sigma(\Omega)\mathbf{e})\exp\left(-\frac{\Omega - E_{min} + \nu}{T}\right) \quad , \qquad (5.1)$$

where E_{min} is the energy of the bottom of the exciton band. The coefficient

A_e can be expressed in terms of the dielectric permeability tensor $\kappa(\Omega)$, i.e., in terms of the refractive indices of the crystal. The frequency dependence of $A_e(\Omega)$ is usually weak compared to the frequency dependence of $\exp\{-(\Omega - E_{min} + \nu)/T\}$ and, hence, can be neglected. In this case, the relationship between the emission and absorption spectra is simplified considerably.

The existence of a one-to-one correspondence between the band-to-band absorption and emission spectra implies that both spectra should, in principle, yield equivalent information on the structure of the exciton band. It is expedient to operate at the lowest temperature possible, so that the band broadening due to interactions with the external phonons is kept to a minimum. Since $2\mathcal{M}$ is usually less than ν, emission spectra are preferred.

However, the key question is: What specific information on the structure of the exciton band can be derived from the band-to-band transition spectra? To find the answer, we can write down the usual formula for a conductivity tensor σ [see (2.49)] with reference to band-to-band transitions. Of course, one must average over the Planck phonon distribution in the initial state. The corresponding equation has the following form

$$\sigma_{ij}(\Omega K) = \frac{\pi}{\Omega} \sum_{\lambda\mu k} n\left(\nu_\lambda(k)\right) \left\langle \lambda k \,|\, J_i^+(K) \,|\, \mu k + K \right\rangle$$
$$\cdot \left\langle \mu k + K \,|\, J_j(K) \,|\, \lambda k \right\rangle \delta(\Omega - E_\mu(k + K) + \nu_\lambda(k)) \quad , \qquad (5.2)$$

where μ and λ denote the exciton and phonon bands, respectively. The most comprehensive information on the structure of a band can be obtained from band-to-band absorption experiments if the following conditions are fulfilled simultaneously:

1) The Heitler-London approximation is applicable. This implies that either the configurational mixing of the electronic states in the given band is insignificant or that the admixture of higher exciton states is constant over the entire exciton band, i.e., is k-independent.

2. The phonon band is narrow compared to the temperature, i.e., $2m \ll T$. In this case, $n(\nu_\lambda(k)) \approx n(\nu)$ is practically constant.

3. The width of the internal phonon band is small compared to that of the exciton band. Hence, one can set $\nu_\lambda(k) \approx \nu = $ const. in the δ function of (5.2).

4. All of the sublattices are symmetrically dependent.

Assuming that the first condition is fulfilled and using (2.40) and (2.48), we obtain

$$\left\langle \mu k + K \,|\, J(K) \,|\, \lambda k \right\rangle = \frac{1}{\sqrt{\nu N}} \sum_\alpha \overset{*}{j}{}_\alpha^{(\nu)} B_\alpha(\mu, k + K) \overset{\smile}{B}{}_\alpha^*(\lambda k) \quad , \qquad (5.3)$$

where the coefficients $\overset{\smile}{B}_\alpha$ refer to phonon wave functions and are the analogues of the electronic coefficients B_α. The vectors $j_\alpha^{(\nu)}$ are matrix elements of the

166

current operator for the intramolecular vibronic transition. The internal phonons are unique in that the matrix element given by (5.3) only depends on the momentum $\mathbf{k}$ because of the coefficients B and $\tilde{B}$. Later on it will be shown that they do not appear in the final equations.

Taking into account the first and second conditions and the completeness condition for the coefficients $\tilde{B}_\alpha$, which is similar to (2.14), we obtain the following equation from (5.2,3) (at $\mathbf{K} \approx 0$):

$$\sigma(\Omega) = \frac{\pi}{\Omega} \bar{n}\,(\nu)\,\frac{1}{\sqrt{N}} \sum_{\alpha\mu\mathbf{k}} |B_\alpha(\mu\mathbf{k})|^2 \widehat{\mathbf{j}_\alpha^{(\nu)} \mathbf{j}_\alpha^{(\nu)}}^* \delta(\Omega - E_\mu(\mathbf{k}) + \nu) \qquad (5.4)$$

It is obvious that the value of E_μ is the same for all of the momenta obtained from the initial momentum $\mathbf{k}$ by applying to it the operations of the point symmetry group of the crystal (i.e., for all of the vectors belonging to the star of the vector $\mathbf{k}$) [5.6]. Since (5.4) contains a summation over the entire $\mathbf{k}$-space, the factors $|B_\alpha(\mu\mathbf{k})|^2$ only appear in the form of their sum over the star. Such a sum cannot depend on α because all of the sublattices are equivalent. Therefore, it is possible to average over α. This shows, with due regard for the normalization condition given in (2.13), that it is possible to replace $|B_\alpha(\mu\mathbf{k})|^2$ by $1/Z$, where Z is the number of molecules in the unit cell. The remaining sum over $\mu\mathbf{k}$ is expressed in terms of the density of states in the exciton band, ρ. It is convenient to write a final equation in terms of the frequency $\omega = \Omega - (E_\rho - \nu)$ and choose the mean frequency of the band-to-band transition, $E_\rho - \nu$, as an origin (here E_ρ is the electronic term in the crystal). Finally

$$\sigma(\omega) \approx \frac{\pi}{E_\rho \nu} \rho\,(\omega) \sum_\alpha \widehat{\mathbf{j}_\alpha^{(\nu)} \mathbf{j}_\alpha^{(\nu)}}^* \qquad (5.5)$$

As usual, here the frequency ω in $\rho(\omega)$ is counted from the term E_ρ.

If one can neglect the frequency dependence of A_e in (5.1), then it follows from (5.1,5) that the spectral distribution of luminescence is given by

$$I(\omega) \propto \rho(\omega) \exp\,[-(\omega - \varepsilon_{min})/T] \qquad (5.6)$$

In the last two equations it is seen that the band-to-band absorption spectrum exactly maps the density of states in the exciton band, and the luminescence spectrum is directly related to it. This led RASHBA to propose that the band-to-band transition spectrum could be used to quantitatively reconstruct $\rho(\varepsilon)$ [5.7]. This work was preceded by the observation of the specific temperature broadening of the vibronic emission spectrum of naphthalene by BROUDE et al. [5.8,9] (Sect.5.2.2).

The last factor in (5.5) shows that band-to-band absorption is weakly polarized. The frequency dependence of the absorption is the same in all of the components of the spectrum, while the intensity ratio is determined by the orientation of the dipole moment of the intramolecular vibronic transition relative to the crystallographic axes.

If the above assumptions are not valid, the intensity distribution in the spectrum is more complicated and $\rho(\varepsilon)$ cannot be directly determined from it. For example, if the third condition breaks down, the joint densities of state corresponding to the "dispersion laws", $E_{\mu\lambda}(\mathbf{k}) = E_{\mu}(\mathbf{k}) - \nu_{\lambda}(\mathbf{k})$, must be manifested in some way.

If the first condition is violated, the dependence of the matrix element of the transition on $\mathbf{k}$ leads to an absorption spectrum that is different from $\rho(\omega)$. This dependence is smooth with a continuous derivative, whereas $\rho(\omega)$ necessarily shows van Hove singularities [5.10].

The cause for the appearance of these singularities is very general and consists in the following. Any smooth, periodic function of $\mathbf{k}$ must have a definite number of critical points (i.e., zero-slope points) in the $\mathbf{k}$-space. This is also true for the function $\varepsilon(\mathbf{k})$, which determines the exciton dispersion law. The critical points are defined by the condition $\nabla\varepsilon(\mathbf{k}) = 0$. If the spectrum is not degenerate near the critical point, the expansion of $\varepsilon(\mathbf{k})$ in its vicinity has the following form:

$$\varepsilon(\mathbf{k}) = \sum_i a_i k_i^2 \quad . \tag{5.7}$$

Critical points are usually denoted by M_s. The value of s equals to the number of negative a_i's in (5.7). Obviously, M_0 corresponds to the minimum and M_1 and M_2 correspond to saddle points on the surface of $\varepsilon(\mathbf{k})$. A simple calculation shows that the behaviour of the density of states near the critical points ε_{cr} is non-analytical: root behaviour $\rho(\varepsilon) \propto |\varepsilon - \varepsilon_{cr}|^{1/2}$ must be observed on one side of each critical point. This has already been pointed out in Sect.3.2 with reference to the points M_0 and M_3. The variation in the density of states for four critical points is outlined in Fig.5.2a. Two thresholds (M_0 and M_3) and two edges (M_1 and M_2) are seen. The band degeneracy, the non-analytical behaviour of $\varepsilon(\mathbf{k})$ in the small-momentum region, and the quasi-one-dimensional and quasi-two-dimensional nature of the spectrum change the details of the behaviour near the individual critical points without affecting the general picture.

It is evident that the smooth dependence of the matrix elements will not change the singular behaviour of the absorption near the critical points. Thus, the band-to-band transitions can be used to determine the positions of a number of critical points in the exciton band. To do this, it is advisable to use the techniques of modulation spectroscopy [5.11]. The first steps have been taken in the thermomodulation spectroscopy of molecular crystals [5.12-15].

5.2.2 Experimental Investigation of Band-to-Band Spectra

The band-to-band transition method has been widely applied in the spectroscopy
of molecular crystals. The stimulus for using the band-to-band transition
spectrum to investigate the structure of the exciton spectrum was the dis-
covery of different temperature behaviour of the two groups of bands in the
fluorescence spectrum of naphthalene (Fig.5.3). The widths of bands that correspond to vibronic transitions in impurity centres are only slightly affected
by the temperature (curve 1, Fig.5.3), whereas the widths of the intrinsic
vibronic luminescence bands of naphthalene are strongly dependent upon the
temperature (curve 2, Fig.5.3). Their half-widths are nearly proportional to
T, which is in full agreement with the band-to-band transition scheme discussed at the beginning of the preceding section. The work described in [5.9]
resulted in the establishment of an effective experimental criterion for identification of the intrinsic part of the luminescence. Finally, in [5.16] the
possibility of using the vibronic luminescence spectrum to obtain the general
contours of the density-of-states distribution in the exciton band (the position of the band edges, the gaps in the spectrum, etc.) was formulated.

The first quantitative comparison of the band-to-band absorption and emission spectra was carried out for naphthalene crystals. The solid curves in
Fig.5.4 denote the experimental data. The width of the phonon band did not

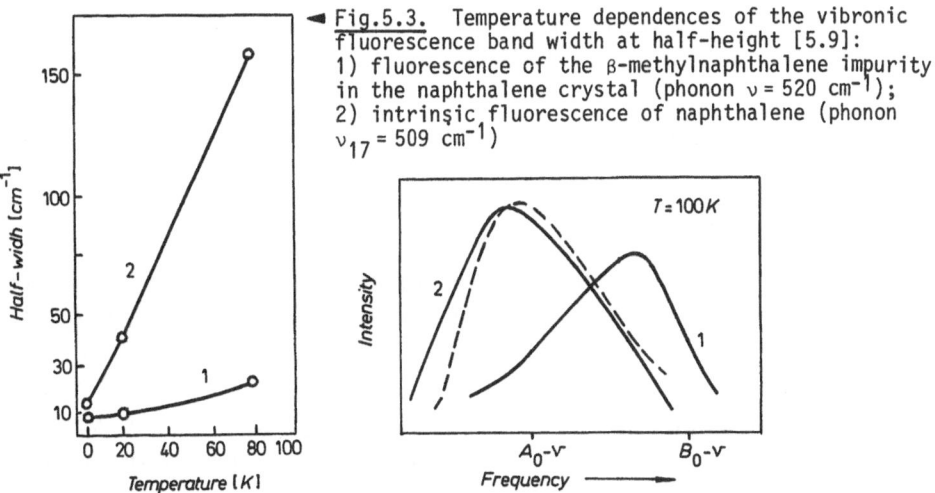

◄ Fig.5.3. Temperature dependences of the vibronic
fluorescence band width at half-height [5.9]:
1) fluorescence of the β-methylnaphthalene impurity
in the naphthalene crystal (phonon $\nu = 520$ cm^{-1});
2) intrinsic fluorescence of naphthalene (phonon
$\nu_{17} = 509$ cm^{-1})

Fig.5.4. Shape of the band-to-band transition bands in the spectrum of the
naphthalene crystal [5.15]: 1) absorption; 2) fluorescence (phonon $\nu_{17} =$
509 cm^{-1}); $A_0 - \nu$ and $B_0 - \nu$ denote transitions from the $\mathbf{k} = 0$ states of both
exciton bands. See text for explanation of dashed and solid lines

exceed 1-3 cm^{-1}.[3] The dashed line is the result of multiplying the absorption band by the exponent $\exp[(\omega - \varepsilon_{min})/T]$ (5.1,6). The central part of this curve closely coincides with the fluorescence band. This confirms both the quasi-equilibrium in the exciton band and the one-to-one correspondence between the bands of the band-to-band absorption and emission spectra (Sect.5.2.1).

These experimental results can be used to investigate the structure of the exciton band, provided that the conditions that were formulated in the preceding section are satisfied. The last three conditions are fulfilled in a naphthalene crystal. As for the first condition, it must be stated that the absorption spectrum resulting from the first electronic transition is due to configuration mixing (Sect.2.5.2). However, it can be assumed that the mixing with the upper states is nearly constant throughout the exciton band since the distance to the nearest electronic level exceeds the width of the band, $2\mathcal{M}$, by an order of magnitude. An unambiguous answer to this question would be a direct application of the band-to-band absorption curve (curve 1, Fig.5.4), which must, according to (5.5), represent $\rho(\varepsilon)$, for quantitative comparison with other experimental data, for instance, IDC spectra. However, the situation is complicated by an interaction with the external phonons which leads to a broadening of the bands and a smoothing of their structure. That this effect is strong follows at least from the fact that under the weak coupling conditions the vibronic absorption band would have to intersect the Ω-axis at a frequency equal to $A_0 - \nu$ because the A_0 band corresponds to a transition to the bottom of the exciton energy band. At the same time the experimental absorption curve (Fig.5.4) has an intensive low-frequency tail. The external phonons undoubtedly distort the band as a whole. Thus, the effect of the external phonons on the shape of the bands must be taken into account. To do this, COLSON et al. [5.16] investigated the changes in the shape of the absorption and fluorescence bands with temperature.

Figure 5.5 illustrates the temperature dependence of the band-to-band transition fluorescence spectrum of the naphthalene crystal. The absorption and fluorescence bands at 77 K are compared in Fig.5.6. Combining the data on low-temperature fluorescence and absorption, COLSON et al. obtained a "composite" function $\rho(\Omega)$ shown in Fig.5.7. The use of this function in the treatment of the spectra of the IDC of d-naphthalenes has shown that it accurately yields both the total width of the exciton spectrum and the position of the maximum of the density $\rho(\varepsilon)$. However, to obtain the correct positions of the IDC im-

---

[3] The recent calculation of the dispersion of the naphthalene internal phonon modes has proved this estimate [5.17].

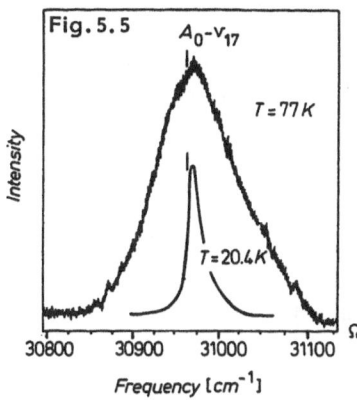

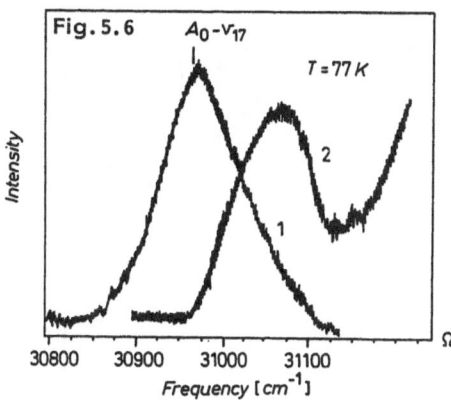

Fig.5.5. Temperature dependence of the shape of the band-to-band fluorescence band of the naphthalene crystal [5.18] (phonon ν_{17} = 509 cm^{-1})

Fig.5.6. Shapes of band-to-band transitions of the naphthalene crystal in (1) fluorescence and (2) absorption [5.18] (phonon ν_{17} = 509 cm^{-1})

purity levels, this function must be changed appreciably at the edges. Figure 3.19 compares the functions $\rho(\varepsilon)$ obtained by the band-to-band transition method and that obtained from IDC spectra. The external phonons lead to an inevitable difference in the functions $\rho(\varepsilon)$ determined by various methods [5.19]. The differences are not great when the coupling with the external phonons is weak.

A similar situation is observed for the benzene crystal [5.18,20]. As in naphthalene, the absorption and fluorescence spectra corresponding to the first electronic transition are due to a configurational mixing with the upper electronic states. The experimental investigations into the bands of band-to-band transitions are carried out at comparatively high temperatures, at which the effects related to the interaction with the external phonons become significant. The temperature broadening of the bands is illustrated in Fig.5.8. The high frequency of the phonon involved in the most intensive vibronic transition is the reason for carrying out the investigation of the band-to-band

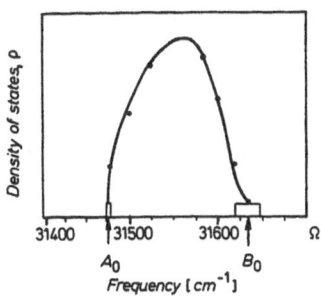

Fig.5.7. Composite density of states $\rho(\Omega)$ in the energy spectrum of the naphthalene crystal determined in the high-frequency region by the band-to-band absorption and in the low-frequency region by the band-to-band fluorescence [5.18]. A_0 and B_0 are the positions of the exciton band maxima

171

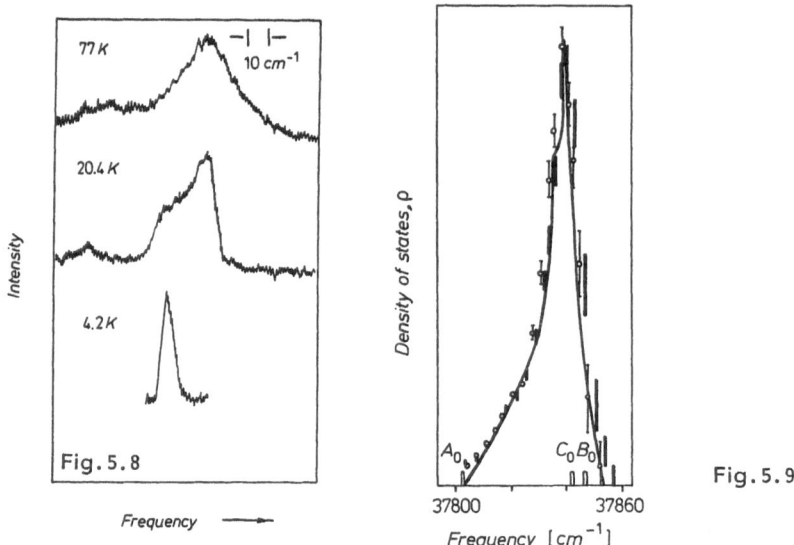

Fig.5.8. Temperature dependence of the shape of the band-to-band fluorescence band of the benzene crystal [5.18] (phonon $\nu_{18} = 606$ cm^{-1})

Fig.5.9. Density of states $\rho(\Omega)$ in the exciton energy spectrum in the benzene crystal determined from the band-to-band fluorescence at 20.4 K (vertical bars) and at 27 K (open circles) [5.18]; A_0, B_0, and C_0 are the positions of the exciton multiplet bands; solid curve) calculation

absorption at higher temperatures (T > 200 K). Of course, this reduces the amount of information obtained. Therefore, Fig.5.8 only contains fluorescence spectra. The density of states $\rho(\varepsilon)$ obtained from these spectra is shown in Fig.5.9. The exciton band of the benzene crystal is comparatively narrow ($2\mathscr{M} \approx 60$ cm^{-1}), so that spectra measured at temperatures not exceeding 27 K have been used.

This function $\rho(\varepsilon)$ was used to calculate the spectra of IDC of d-benzenes (Sect.3.3). The results are in excellent agreement with the experimental data. It should, however, be noted that the depth of the levels of the monomeric isotopic centres in IDC of d-benzenes is much greater than in IDC of d-naphthalenes. Therefore, their spectra are less sensitive to the structural details of the function $\rho(\varepsilon)$.

All of the necessary conditions as listed in Sect.5.2.1 are fulfilled for the anthracene crystal. Nevertheless, the experimental situation is much more complicated than for benzene or naphthalene. The energy of the phonon which most strongly interacts with the exciton is equal to 1400 cm^{-1}. This is too high for a productive investigation of the vibronic absorption. The vibronic fluorescence also has to be measured at high temperatures so as to ensure an

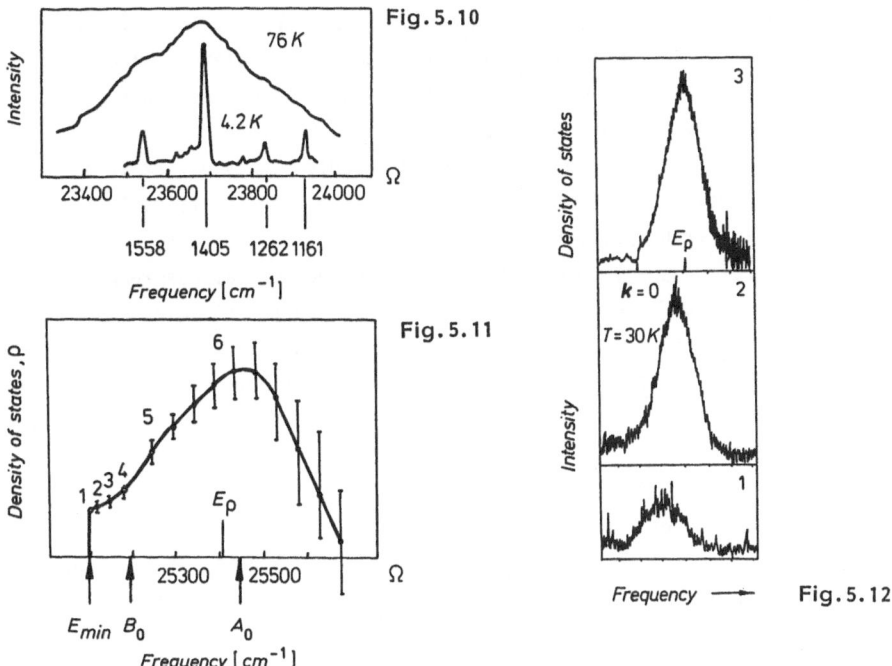

Fig.5.10. Part of the vibronic fluorescence spectrum of the anthracene crystal [5.21]

Fig.5.11. Density of states $\rho(\Omega)$ in the exciton energy spectrum of the anthracene crystal, obtained from band-to-band fluorescence [5.21]: the numbers denote the points at which the curves, measured at various temperatures in the range from 4.2 up to 250 K, are "joined together". E_{min}, B_0, and A_0 are the positions of the exciton band bottom and of the centres of gravity of the B_0 and A_0 bands of the exciton doublet. E_ρ is the density-of-states centre of gravity

Fig.5.12. Determining the exciton density of states from the vibronic fluorescence of the hmd$_0$ crystal (phonon $\nu = 454$ cm^{-1} [5.20]: 1) fluorescence spectrum recorded once by a photon counter; 2) sum of eight spectra of type 1 with an improved signal-noise ratio; 3) density of states obtained from spectrum 2

appreciable thermal population of the upper levels. Otherwise, the great width of the exciton band hinders an appreciable population at lower temperatures.

Figure 5.10 illustrates the fluorescence spectra of an anthracene crystal in the band-to-band transition range with the participation of the 1400 cm^{-1} phonon [5.21]. The density of states $\rho(\Omega)$ found by treatment of the curves of Fig.5.10 is shown in Fig.5.11. The accuracy is not high because one has to divide the fluorescence band into components which correspond to overlapping vibronic transitions and to subtract the contribution from the external phonons. Nevertheless, the function $\rho(\Omega)$ shown in Fig.5.11 may help to better familiarize oneself with the spectrum of the anthracene crystal, for which no reliable data are available at present. The width of the exciton band which follows from Fig.5.11 ($2\mathcal{M} \approx 600$ cm^{-1}) is rather close to the calculated value ($2\mathcal{M} \approx 700$ cm^{-1}) [5.22] and does not contradict the data on the absorption spectra of mixed crystals of d-anthracenes (Sect.4.5.6).

The band-to-band transition method is undoubtedly highly effective in determining the density of the exciton states in molecular crystals. It has now become one of the principal methods of studying exciton energy spectra. For example, the exciton spectrum of a hexamethylbenzene crystal was recently investigated by WOODRUFF and KOPELMAN using this method [5.20]. Figure 5.12 illustrates the technical achievements that have been made in determining the density of exciton states from band-to-band vibronic fluorescence spectra. The dispersion law in hexamethylbenzene was reconstructed from the resulting $\rho(\varepsilon)$ (Fig.2.7) [5.20].

Despite recent technical achievements no specific features of the function $\rho(\varepsilon)$ within the exciton band could be related to the van Hove singularities for any crystal. All of the functions obtained so far are smooth. No doubt, this is a result of the interaction between excitons and external phonons, which causes the smoothing of absorption and emission bands and therefore hinders the singling out of edges and thresholds.

Recently, time-resolved spectra of the transient band-to-band luminescence have been applied to measuring the exciton scattering times. These spectra are excited by a laser flash, preparing the exciton sub-system in a $k = 0$ initial state. It was found that thermalization of singlet excitons in naphthalene ($T = 1.5$ K) occurs within tens of nanoseconds [5.23]. The relaxation times of triplet excitons in 1,2,4,5-tetrachlorobenzene has been shown to have a microsecond scale ($T \approx 2$ K) [5.24]. These latter times were measured using a special type of band-to-band transitions, only existing for triplet excitons [5.25].

5.3 Band-to-Band Spectra of Mixed Crystals

5.3.1 Calculating the Band-to-Band Spectra of Mixed Crystals

The preceding section considered the possibilities for investigating the structure of exciton bands of perfect crystals by the band-to-band transition method. The method also opens up intriguing prospects for the study of mixed crystals. This question has actually already been briefly discussed in Sect.3.3.5 with reference to dilute solutions of isotopic impurities. It was shown that the vibronic fluorescence spectrum of an isolated large-radius monomeric impurity centre consists of two bands because of the difference in the vibrational frequencies of the guest and host molecules. The intensity of the bands directly depends on the distribution of the electronic excitation between the guest and adjacent host molecules. Thus, the vibronic fluorescence directly yields the distribution of the electronic excitation over the different isotopic components of the doped crystal.

This method can also be applied to crystals with a high concentration of isotopic impurity, which exhibit broadened local exciton levels and one or several regions of a continuous spectrum. As has already been noted, vibronic transitions take place between the exciton continuum and the continuum of the internal phonons. In analogy to the preceding section, we shall again call them band-to-band transitions and try to find out what information is contained in the corresponding spectra.

The intensity of the vibronic luminescence due to transitions to the vibrational levels of guest and host molecules is directly related to the squares of the excitation amplitudes of these molecules (Sect.3.3.6). In deriving (3.40), which describes this relationship, it was important that isotopic substitution could be assumed to be ideal and that a mixing of the internal vibrations of the host and guest molecules was non-existent. The last condition is fulfilled if the phonon band width is appreciably less than the difference between the vibrational frequencies of the isotopic components.

In considering concentrated mixed crystals, these two conditions and all four conditions formulated in Sect.5.2.1 are assumed to be fulfilled. The assumption that the phonon bands are narrow suggests that the phonons are localized at the sites. Therefore, the excitation of the vibrations of a definite isotopic component will occur exclusively at the sites occupied by this component. Considering the results of the preceding section, we can readily conclude that the intensity of the corresponding transitions is proportional to the contribution to the density of exciton states $\rho(E)$ of a mixed crystal from this isotopic component. This contribution to the total density will be called the *partial density of states* and denoted by $C_r\rho_r(E)$, where C_r is the concentration of the rth component and ρ_r is the partial density per molecule of the rth component.

A simple equation for $\rho_r(E)$ can be obtained using the CPA (Sect.4.4) [5.26]. To determine $\rho_r(E)$, the one-site CPA considers a single molecule of the r-component immersed in a medium with a potential U. Then, according to (2.34), we must find the value of the exciton's Green's function with coinciding arguments using (3.6). The density $\rho_r(E)$ is the contribution to the sum in (2.34) from those sites in which r-type molecules reside. After simple rearrangements we obtain

$$\rho_r(\Omega) = -\frac{1}{\pi} \text{Im} \left\{ \frac{G_0}{1 - (E_r - U)G_0} \right\} , \qquad (5.8)$$

where all of the symbols are the same as in Sect.4.4 and G_0 is determined by (4.35).

For a two-component mixture (4.36,37) yield

$$\frac{C_1}{1 - (E_1 - U)G_0} + \frac{C_2}{1 - (E_2 - U)G_0} = 1 . \qquad (5.9)$$

From (5.8,9) it follows that

$$C_1\rho_1(\Omega) + C_2\rho_2(\Omega) = -\frac{1}{\pi} \text{Im} \{G_0\} = \rho_{CP}(\Omega) . \qquad (5.10)$$

The last equality follows from (4.39). Thus, added together the partial densities yield the total density of state $\rho_{CP}(\Omega)$.

By way of example, Fig.5.13 illustrates the partial densities ρ_1 and ρ_2 of the two isotopic components of a mixed crystal of bd_0 and bd_1 calculated using the CPA. The absorption spectra of this mixed crystal are described in Sect.4.2. The densities shown in Fig.5.13 will be required below to describe the fluorescence spectra. According to the magnitude of the parameter $\Delta/\mathcal{M} \approx 1$, the energy spectrum of this crystal is two mode with a pseudogap. It is seen that although the shift Δ is comparatively small (33 cm^{-1}), i.e., less than the width of the spectrum of the one-component crystal ($2\mathcal{M} \approx 60$ cm^{-1}), the mixing of the excitations of different isotopic components is slight. A pseudogap separates the regions in either of which the excitation of one of the isotopic constituents dominates.

Let us see which band-to-band vibronic-fluorescence spectrum corresponds to such an energy spectrum. Two limiting situations are possible depending on the

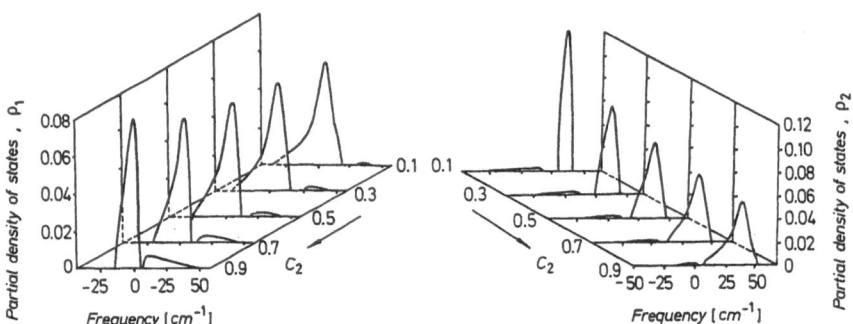

Fig.5.13. Partial densities of states ρ_1 of (bd_0) and ρ_2 of (bd_1) in the spectrum of bd_0/bd_1; calculation according to [5.27]

ratio between $\mathcal{M}$ and the isotopic shift in the vibrational frequencies, $\nu_1 - \nu_2$. The two situations are shown in Fig.5.14, where it is assumed that the low-frequency electronic spectrum corresponds to the lighter component, whose vibrational frequency $\nu_1 > \nu_2$. Under these conditions, if $\nu_1 - \nu_2$ exceeds the width of the exciton spectrum of a mixed crystal, the densities ρ_1 and ρ_2 manifest themselves independently in the band-to-band fluorescence spectrum. Conversely, if $\nu_1 - \nu_2$ is much less than the width of the exciton spectrum, the total density of states ρ manifests itself in the fluorescence spectrum. The picture is more complicated in the intermediate situation.

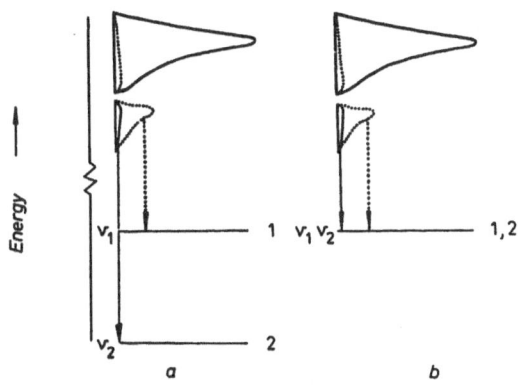

<u>Fig.5.14a,b.</u> Scheme of transitions in the band-to-band fluorescence spectrum of a two-component mixed crystal: dashed and solid curves denote the partial densities $C_1\rho_1$ and $C_2\rho_2$, respectively; ν_1 and ν_2 are phonon frequencies of the first and second components; a) $\nu_1 - \nu_2 > \mathcal{M}$; b) $\nu_1 - \nu_2 = 0$ [5.27]

Figures 5.15,16 show the band-to-band vibronic fluorescence spectra of a mixed crystal of bd_0 and bd_1 calculated as a function of the composition (Fig.5.15) and the temperature (Fig.5.16). Although the partial densities ρ_1 and ρ_2 overlap only slightly (as can be seen from Fig.5.13), a doublet structure must be observed in the low-temperature fluorescence spectrum. The high-frequency maximum corresponding to the heavy isotopic component is clearly seen on the wing of the main curve over a wide concentration range. For dilute solutions $(C_1 \rightarrow 0)$, the pattern agrees with the calculated results given at the end of Sect.3.3.6. The magnitude of the splitting is determined by the difference between the vibrational frequencies of the components bd_0 and bd_1 $(\nu_1 - \nu_2 \approx 10 \text{ cm}^{-1})$.

The pattern in Fig.5.16 shows how the structure of the fluorescence spectrum increases in complexity with rising temperatures. The difference of

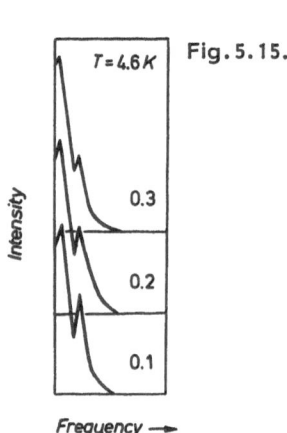

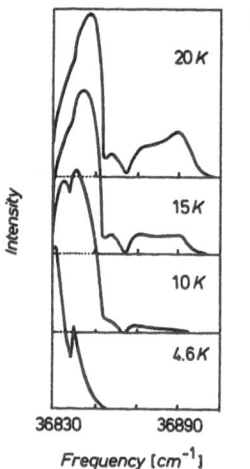

Fig.5.15.

Fig.5.16

T=4.6K

20 K

15K

10 K

4.6K

0.3

0.2

0.1

Intensity

Intensity

Frequency →

36830   36890

Frequency [cm⁻¹]

Fig.5.15. Band-to-band fluorescence spectrum of bd_0/bd_1 for three concentrations of bd_0; calculation according to the CPA (phonon ν_1 = 990 cm^{-1}; $\nu_1 - \nu_2$ = 10 cm^{-1}) [5.28]

Fig.5.16. Band-to-band fluorescence spectrum of bd_0/bd_1 (C_1 = 0.2, $C_2 = 0.8$) at different temperatures; calculation according to the CPA (phonon ν_1 = 990 cm^{-1}; $\nu_1 - \nu_2$ = 10 cm^{-1}) [5.28]

the phonon frequencies is less than the width of the exciton spectrum ($\nu_1 - \nu_2 \approx 10$ cm^{-1}, $2\mathcal{M} \approx 60$ cm^{-1}). Still it is relatively large and, therefore, the fluorescence spectrum (in contrary to that of a pure crystal) is not a product of the total density of states and the exponential energy distribution of the excitons. This example shows that the reconstruction of the exciton density of states from the band-to-band fluorescence spectrum is a much more difficult task for a mixed crystal than for perfect crystals. The reason is that actually both partial densities have to be reconstructed.

5.3.2 Experimental Investigation of the Band-to-Band Spectra of Mixed Crystals

Band-to-band transitions were observed in the fluorescence spectra of binary mixed crystals of nd_0 and nd_8 [5.29-31], bd_0 and bd_1 [5.28], and ad_0 and ad_{10} [5.32]. The temperature dependence was only measured for a mixed crystal of nd_0 and nd_8 [5.30,31]. The data have not yet allowed to restore the density of states of mixed crystals because of the complexity of the spectra (see the end of Sect.5.3.1).

The main feature of the low-temperature fluorescence spectra of all three systems is the doublet structure of the bands. The magnitude of the splitting is practically independent of the composition of the mixed crystal. A change

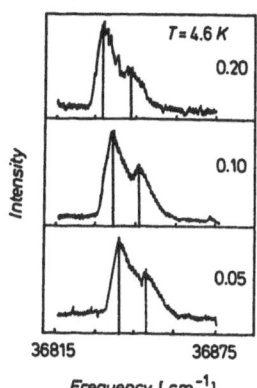

Fig.5.17. Band-to-band fluorescence of bd_0/bd_1 for three concentrations of bd_0 (phonon $\nu_1 = 990$ cm^{-1}; $\nu_1 - \nu_2 = 10$ cm^{-1}) [5.28]

in composition only affects the ratio of the intensities of the bands that make up the doublet. The doublet as a whole shifts to lower frequencies if the concentration of the lighter component increases. The foregoing is demonstrated in Fig.5.17, which illustrates the concentration dependence of the fluorescence spectra of a mixed crystal of bd_0 and bd_1.

The experimentally observed pattern is in full agreement with the theoretical predictions. The magnitude of the splitting of the fluorescence bands is approximately equal to 10 cm^{-1}, i.e., equal to the difference of the frequencies of the internal phonons of bd_0 and of bd_1, $\nu_1 - \nu_2$. The intensity of each component of the doublet is determined by the product of the relevant partial density of states $C_r \rho_r$ (Fig.5.13) and by the Gibbs distribution function of the excitons near the pseudobottom of the spectrum. The width of the Gibbs distribution is of the order of T. The doublet structure shown in Fig.5.17 is well resolved because $T < \nu_1 - \nu_2$ at T = 4.3 K. The calculated values of the band intensities at the maxima are shown in Fig.5.17 as vertical segments. The high-frequency maximum corresponds to the band of the heavier component superimposed on the wing of the main band. The low-frequency shift of the spectrum with an increase in concentration of the lighter component bd_0 is a reflection of the concentration shift of the pseudobottom (Sect.4.5.6).

It was noted in Sect.5.3.1 that the mixing of the states of different isotopic components in a crystal of bd_0 and bd_1 is slight. The pseudogap separates the regions in which the excitation of bd_0 or bd_1 molecules is dominant (Fig.5.13). Indeed, the fluorescence spectra shown in Fig.5.17 reflect the substantial prevalence of the excitation of the lighter component even for relatively low concentrations.

Quite a different situation is observed in the vibronic-fluorescence bands of a mixed crystal of ad_0 and ad_{10}. The energy spectrum of this crystal is

179

close to the limit of amalgamated bands (Sect.4.5.6). As a result, the excitations near the pseudobottom of the energy spectrum are the collective modes in which both isotopic constituents are strongly involved over the entire concentration range. The last circumstance, in turn, results in an enormous variation in the intensities of the bands of the band-to-band doublet associated with the excitation of the phonons ν_1 (of ad_0) and ν_2 (of ad_{10}). Figure 5.18 shows the concentration dependence of the ratio of the intensities, $I_1(\nu_1)$ and $I_2(\nu_2)$. The total change in this ratio is almost 400-fold when the concentration of the lighter component ad_0 is increased from $C_1 = 0.025$ to $C_1 = 0.95$ [5.32]. The intensity I_2 of the band corresponding to the excitation of the heavier component phonon dominates at low concentrations of ad_0. At $C_1 \approx 0.2$, the intensities I_1 and I_2 are equal. The intensity I_1 rapidly increases with further increases in C_1.

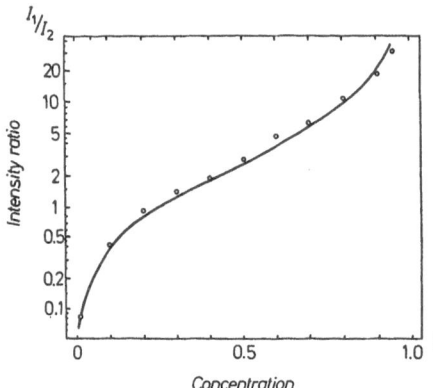

Fig.5.18. Dependence of the intensity ratio of the bands constituting the band-to-band fluorescence doublet in ad_0/ad_{10} on the ad_0 concentration (ad_0 phonon $\nu_1 = 394$ cm^{-1}, $\nu_1 - \nu_2 = 14$ cm^{-1}); open circles) experiment [5.32]; solid line) calculation

In contrast to the calculations for mixed crystals of benzene the calculations of I_1/I_2 using the CPA are not possible now because reliable data on the density of states ρ_0 of one-component anthracene crystals are not available. However, to estimate the I_1/I_2 ratio one can use the excitation amplitudes of molecules of different species which describe the integral intensities of the absorption bands in the average-amplitude approximation (AAA, Sect.4.2).

In this approximation, the intensity of the vibronic fluorescence associated with the optical transitions that occur in r-type molecules is proportional to the density of the excitation at these molecules (in this case, near the pseudobottom of the energy spectrum):

$$I_r \propto C_r \rho_r \propto C_r a_r^2 \quad ,$$

where the a_r are excitation amplitudes [see (4.11)]. The ratio of the excitation amplitudes for different isotopic components is naturally different for the different quantum states of a mixed crystal. If we neglect this difference and adopt the same ratio of amplitudes for the pseudobottom that is obtained in the AAA for the low-frequency absorption band of a mixed crystal polarized parallel to **b**, we find from (4.12) that

$$\frac{I_1}{I_2} = \frac{C_1}{C_2} \left(\frac{\mathscr{E}_b - E_2}{\mathscr{E}_b - E_1}\right)^2 \quad , \tag{5.11}$$

where $\mathscr{E}_b$ is the position of the absorption band of a mixed crystal and E_1 and E_2 are the electronic terms of the one-component crystals. In the amalgamated bands approximation, the dependence of $\mathscr{E}_b$ on C_1 is determined by the linear law

$$\mathscr{E}_b = E_1 + \varepsilon_b + (1 - C_1)\Delta \quad . \tag{5.12}$$

The solid line in Fig.5.18 represents the intensity ratio I_1/I_2 calculated using (5.11) for the values of E_1 and E_2 given in Sect.4.5.4. The calculation agrees surprisingly well with the experimental data. Although such high accuracy is certainly accidental, the agreement between the curves evidently speaks in favour of the correctness of the approach to the interpretation of the experimental data on mixed crystals of d-anthracenes and, at the same time, to the interpretation of the spectrum of the pure anthracene crystal.

6. Vibronic Spectra of Molecular Crystals

The shape of the exciton absorption bands is determined by the exciton-phonon interaction. A general theory to explain the shape of the bands cannot be developed because the type of final states in a system strongly depends on a number of parameters. In addition, the exciton-phonon interaction Hamiltonian is practically unknown. However, the situation corresponding to vibronic absorption, i.e., when the phonon that is produced simultaneously with the exciton is an internal one, permits a general approach. A detailed theory can be developed if $\mathcal{M} \ll \nu$, i.e., if the half-width of the exciton band is much less than the phonon frequency. Within the framework of this theory the behaviour of the system is purely dynamical. The exciton and the phonon (or several phonons) that are produced upon the absorption of light may be treated as stable interacting particles. The uniqueness of the situation also lies in the fact that this final-state interaction does not contain any new unknown parameters and is expressed via the exciton dispersion law. Therefore, this dynamical theory allows one to quantitatively describe the energy spectrum of the vibronic states and the vibronic absorption spectrum. Naturally, it can also be applied to doped and mixed crystals. The present chapter introduces this dynamical theory, which is also used as a basis for an analysis of the vibronic spectra of a number of crystals. As a result of such analyses it was possible to classify the bands and to ascertain the type of quantum states relevant to the exciton + phonon system (bound, dissociated, impurity vibron, etc.). The success that was in the quantitative description of the vibronic spectrum of the naphthalene crystal is particularly encouraging.

In conclusion the summarized data, which give an indication of the efficiency of the methods developed to describe molecular excitons, are discussed.

6.1 Exciton-Phonon Interaction

This section introduces the basic concepts related to the vibronic interaction, i.e., the interaction between excitons and internal phonons. To this end we first of all obtain the vibronic-interaction Hamiltonian, discuss its basic features which distinguish the vibronic interaction from other exciton-phonon interactions, and, on this basis, briefly consider the classification of the exciton states in the presence of the exciton-phonon interaction. This will enable us to develop a quantitative theory describing the vibronic absorption spectra for high-frequency internal phonons.

6.1.1 Vibronic Interaction Hamiltonian

It is quite natural do develop a theory describing the vibronic interaction on the same general approach which lies at the basis of the whole concept of molecular excitons, namely the predominance of intramolecular interactions over intermolecular ones. It should also be assumed in the spirit of the concept that the intramolecular vibronic interaction dominates and that the direct interaction between electronic and vibrational excitons at different sites can be neglected.

The vibronic interaction in a molecule (Sect.1.2 and Appendix A) can be written in the form of (A.18), i.e., in the secondary-quantization representation for both excitons and phonons. Consequently, the vibronic-interaction Hamiltonian in a crystal is equal to

$$H_v^0 = \gamma v \sum_{n\alpha} a_{n\alpha}^+ a_{n\alpha}(b_{n\alpha}^+ + b_{n\alpha}) + \frac{\delta}{2} \sum_{n\alpha} a_{n\alpha}^+ a_{n\alpha}(b_{n\alpha}^+ + b_{n\alpha})^2 \quad , \qquad (6.1)$$

i.e., the sum of the vibronic Hamiltonians of the individual molecules, where the $a_{n\alpha}$ are exciton operators and the $b_{n\alpha}$ are phonon operators. The constant γ represents the linear interaction, which leads to a shift in the equilibrium position of the nuclei in the electronically excited molecule and a Franck-Condon lowering of the energy level of the molecule in the electronically excited state ($E_{FC} = \gamma^2 v$). With reference to non-degenerate electronic levels, the linear interaction only exists for totally symmetric (TS) intramolecular vibrations and the resulting TS phonons. The second term in (6.1) corresponds to a quadratic interaction which causes a shift in vibrational frequency when the molecule is excited electronically. The frequency in the excited state $v^* = v\sqrt{1 + 2\delta/v}$ (A.27). If $\delta \ll v$, the frequency shift $\Delta_v = v^* - v \approx \delta$. For the sake of simplicity, the Hamiltonian in (6.1) is written for the interaction with one intramolecular vibration. Actually, an interaction with all of the vibrations takes place. As a result, H_v^0 contains a sum over all of the vibrations. Although, in this case, the coefficient δ is transformed to a matrix with off-diagonal elements resulting in a coupling between the vibrational modes, these off-diagonal elements can be neglected because they are usually small. Their inclusion is beyond the accuracy currently obtainable in the analysis of vibronic spectra. Practically speaking, H_v^0 can be reduced to a single term. As a result, the complete Hamiltonian of the crystal is given by

$$H = H_e^0 + H_{ex}^0 + H_{ph} + H_v^0 \quad , \qquad \text{where} \tag{6.2}$$

$$H_e^0 = E_e \sum_{n\alpha} a_{n\alpha}^+ a_{n\alpha} \quad ,$$

$$H_{ex}^0 = \sum_{n\alpha \neq m\beta} M_{n\alpha m\beta}^0 a_{n\alpha}^+ a_{m\beta} \quad , \qquad H_{ph} = \nu \sum_{n\alpha} b_{n\alpha}^+ b_{n\alpha} \quad . \tag{6.3}$$

The dispersion of the phonon branches can be neglected in H_{ph} because it is assumed to be much smaller than the exciton dispersion. Actually, its inclusion does not complicate the situation. The matrix shift D is already taken into account in the energy E_e.

An important feature of the vibronic-interaction Hamiltonian in (6.1) is that it does not introduce any new interaction constants other than γ and δ, which appear in the Hamiltonian of isolated molecules and, at least in principle, can be determined from molecular spectra. This considerably simplifies the problem of vibronic interactions in a crystal (Sect.5.1).

It would be useful to transform the Hamiltonian given in (6.1-3) from the site to the momentum representation by transforming the operators in accordance with (2.7). For a crystal with a single molecule in the unit cell

$$a_n = \frac{1}{\sqrt{N}} \sum_k \exp(ikn) a_k \quad ; \qquad b_n = \frac{1}{\sqrt{N}} \sum_q \exp(iqn) b_q \quad , \tag{6.4}$$

and the Hamiltonian is

$$H = \sum_k (E_\rho + \varepsilon(k)) a_k^+ a_k + \nu \sum_q b_q^+ b_q + \frac{\gamma \nu}{\sqrt{N}} \sum_{kq} a_k^+ a_{k+q} (b_q^+ + b_{-q})$$

$$+ \frac{\delta}{2N} \sum_{k_1 k_2 q} a_{k_1}^+ a_{k_2} (b_q^+ b_{k_2-k_1-q}^+ + b_q b_{k_1-k_2-q}$$

$$+ b_q^+ b_{k_1-k_2+q} + b_q b_{k_2-k_1+q}) \quad . \tag{6.5}$$

It is significant that the electron-phonon interaction constants are independent of k and q, i.e., the momenta of the exciton and the phonon, respectively. With several molecules in the unit cell this property is lost because the expression under the summation sign becomes a function of the coefficients $B_\alpha(\mu k)$. However, since Frenkel's restricted model (Sect.2.2) only takes into account interactions between nearest neighbours, the Hamiltonian is usually again transformed to the simple form given in (6.5).

Despite the fact that the Hamiltonian H_v^0 does not contain any unknown parameters, the complete solution to the problem remains inaccessible. Indeed, the inclusion of only the first term in H_v^0, which is linear with respect to the phonon amplitudes b^+ and b, corresponds to the problem of polaron theory with all the diverse situations arising in it. The second term in H_v^0 complicates the picture considerably. Therefore, only definite models can be investigated.

For weak vibronic interaction one can use standard perturbation theory. Such an approach was developed by McRAY [6.1] for $\Delta_v = 0$. It is valid for $\gamma^2 \ll 1$. Non-interacting excitons and internal phonons are zero-approximation states. Transitions, whereby an exciton and an internal phonon are created simultaneously, are possible if an interaction between these states is assumed ($\gamma \neq 0$).

The picture is similar to the one for the external phonon-assisted exciton absorption of light, which has widely been discussed in the literature. The spectrum of vibronic absorption consists of broad absorption bands whose band width is of the order of $\mathscr{M}$. With such an approach the specifics of vibronic absorption could be revealed only in that the vibronic-absorption spectrum can be expressed in terms of the density of states in the exciton band $\rho(\varepsilon)$, due to the independence of the coupling constant on the momentum (6.5) and the absence of phonon dispersion. However, such a method of reconstructing $\rho(\varepsilon)$ has apparently never been applied to electronic excitons[1]. The change in the energy spectrum of an exciton under weak coupling conditions is only moderate. Qualitative changes only occur when the exciton dispersion law is one-dimensional [6.3-5].

An alternative model is described in DAVYDOV's earlier works [6.6] and corresponds to a description of vibronic excitations similar to Frenkel electronic excitons, i.e., as an intramolecular vibronic excitation propagating through a crystal. Such a formation shall be termed a *molecular vibron*. Davydov's model was developed in a number of subsequent works [6.7-9] and was used to analyze experimental data.

Despite a number of advantages, these models do not create a unified picture for the vibronic spectrum, and do not use explicitly the parameter which is generally the distinguishing feature of vibronic spectra, i.e., the fact that the phonon frequency ν is large. The analysis of vibronic spectra in this chapter will be based on a model which is valid for $\nu \gg \mathscr{M}$, i.e., for a phonon frequency that is large compared to the width of the exciton band. As a result, no restrictions will be imposed on the strength of the exciton-

---

[1] Recently it was successfully applied to vibrational excitons in some molecular solids where the dispersion law of internal phonons was reconstructed [6.2].

phonon coupling. In this case, the characteristic features of the energy and optical spectra of vibronic excitations are clearly visible. Although the theory is quantitatively correct only when the condition $\nu \gg \mathcal{M}$ is fulfilled, qualitative conclusions can also be drawn if $\nu \gtrsim \mathcal{M}$. The opposite limiting case, $\nu \ll \mathcal{M}$, is much less interesting.

6.1.2 Effect of the Exciton-Phonon Interaction on the Energy Spectrum of Excitons

Let us consider the change in the energy spectrum of molecular excitons due to the exciton-phonon interaction. We restrict ourselves to the exciton spectrum, i.e., to states in which one exciton is excited but no real phonons are present. An exciton naturally causes the surrounding lattice to become deformed, i.e., it gives rise to virtual phonons which renormalize the energy spectrum of the exciton.

The basic principles of the exciton-phonon interaction theory were developed in the early works of PEIERLS [6.10] and FRENKEL [6.11]. They were discussed by DAVYDOV [6.6] with reference to molecular excitons. The further development of the theory was influenced considerably by LANDAU's approach [6.12] to the problem of self-trapping, the theory of large [6.13] and small [6.14] polarons, and the theory of the shape of impurity absorption bands [6.15-17].

We shall now consider a classification of the possible situations. Three competing energy parameters are involved: $\mathcal{M}$, E_{FC}, and the characteristic phonon frequency ν. In [6.18,19] the determining role is attributed to the ratio between the parameters $\mathcal{M}$ and E_{FC}. The former describes the decrease in kinetic energy (as compared to E_e) due to the free motion of the exciton through the crystal, while the latter describes the decrease in the potential energy of the system due to the interaction of the exciton with the field of a surrounding deformation. Operating with only these two quantities ($\mathcal{M}$ and E_{FC}) one concludes that the effect of the phonons is insignificant if $\mathcal{M} \gg E_{FC}$ and the exciton motion is practically completely suppressed if $\mathcal{M} \ll E_{FC}^2$. In [6.10,11] attention is focussed on the relation between the parameters $\mathcal{M}$ and ν. If $\mathcal{M} \gg \nu$, the deformation does not manage to follow the exciton. Hence, it moves practically freely. On the contrary, if $\mathcal{M} \ll \nu$, the deformation follows the exciton. As a result, the velocity of the exciton may be reduced considerably[3]. In actuality, the physical picture is much more complicated and interesting.

---

[2] In the literature these two limiting cases are sometimes called strong- and weak-resonance coupling.

[3] Some authors use the terms "free" and "localized" exciton.

It is convenient to begin a classification with reference to the parameter $\mathcal{M}/\nu$. Excitons with $\mathcal{M} \ll \nu$ would naturally be called *heavy*, while those with $\mathcal{M} \gg \nu$ would be called *light* [6.20,21].

We shall first of all consider heavy excitons. For singlet excitons in aromatic crystals this criterion is best satisfied for the internal phonons. Therefore, it is instructive to consider the Hamiltonian given in (6.2).

In the lower order of the perturbation theory, the exciton term H_{ex}^0 may be neglected in the Hamiltonian (6.2). Then H breaks up into a sum of vibronic Hamiltonians of individual molecules. Each Hamiltonian can be diagonalized by a canonical transformation (A.20). The same transformation must be performed with H_{ex}^0. Each of the exciton operators is transformed according to (A.29). In transforming H_{ex}^0 we restrict ourselves to the leading term. Hence, we put $\xi = 0$, i.e., we neglect the correction for the change in vibrational frequencies. Taking into account (A.30) a transformed Hamiltonian (with $\tau = -\gamma$) is obtained:

$$H_{ex}(\tau) = \sum_{n\alpha \neq m\beta} M_{n\alpha m\beta} a_{n\alpha}^+ a_{m\beta} \exp[\gamma(b_{n\alpha}^+ - b_{m\beta}^+)]\exp[-\gamma(b_{n\alpha} - b_{m\beta})] \quad ,$$

$$M_{n\alpha m\beta} = M_{n\alpha m\beta}^0 \exp(-\gamma^2) \quad . \tag{6.6}$$

Subsequent difficulties are caused by the presence of the term $H_{ex}(\tau)$ since both the exciton transfer and the interaction with the phonons are incorporated in it. Until now, all of the transformations of the Hamiltonian have been exact. The aim was to take the advantage of the fact that the perturbation theory is applicable to the Hamiltonian given in (6.6), even when the intramolecular exciton-phonon interaction is not small (for example, $\gamma \sim 1$). Indeed, the coefficients of the exponential function are the values of $M_{n\alpha m\beta}$, which are small if $\mathcal{M}/\nu \ll 1$. For the time being, we shall restrict ourselves to the behaviour of heavy excitons in the absence of real phonons. It can be shown (Sect.6.2) that under these conditions it is sufficient to isolate the part of the Hamiltonian that operates on the subspace containing zero-phonon states, i.e., to take the matrix element of the Hamiltonian between the zero-phonon states. If we use (6.2,3,6) and (A.25,26), we finally obtain

$$H = H_e + H_{ex} \quad , \quad H_e = E_\rho \sum_{n\alpha} a_{n\alpha}^+ a_{n\alpha} \quad , \quad H_{ex} = \sum_{n\alpha \neq m\beta} M_{n\alpha m\beta} a_{n\alpha}^+ a_{m\beta} \quad . \tag{6.7}$$

The renormalized electronic term E_ρ differs from E_e in the Franck-Condon energy and the correction associated with the change in the energy of zeropoint vibrations.

Thus, the initial exciton band described by the matrix elements $M_{n\alpha m\beta}^0$ is transformed into a new band of renormalized exciton states which exhibits the same structure, but is compressed by a factor $e^{-\gamma^2}$. If the finite value of the parameter $\mathcal{M}/\nu$ is taken into account, the compression of the band will naturally be asymmetrical. The asymmetry is small in this parameter [6.22]. Attention should also be drawn to the fact that the degree of compression is determined by the factor $\gamma^2 \equiv E_{FC}/\nu$. Therefore, the band can retain a considerable width even if $\nu \gg \mathcal{M}$, provided that γ^2 is not too large. The above calculations were made for the interaction with a single intramolecular vibration. Actually, the formula for $M_{n\alpha m\beta}$ (6.6) must contain a sum of γ^2 for all of the TS phonons.

As a result of the transformation, the Hamiltonian given in (6.7) was obtained instead of the exciton Hamiltonian H_{ex}^0 given in (6.3). The former retains the same structure but differs from the latter in the numerical values of the coefficients. The states within the exciton band remain undamped. The interaction with the phonons only affects the exciton dispersion law. States which correspond to the simultaneous existence of interacting excitons and phonons are higher in energy by about ν from the exciton band. The next section is devoted to these states.

In the preceding chapters, the Hamiltonian given in (6.7) was introduced as an "exciton" Hamiltonian and served as a basis for the consideration of the spectrum of a mixed crystal, band-to-band transitions, etc. without an explicit allowance for a vibronic interaction. The question arises as to whether these results are still valid if the renormalizations at $\gamma^2 \sim 1$ are taken into account. An analysis [6.23] has shown that the results do indeed remain valid if the renormalized exciton spectrum is given by $\varepsilon(\mathbf{k})$ and the intensities are corrected for the factor $e^{-\gamma^2}$. However, corrections must be made in powers of the parameter $\mathcal{M}/\nu$ for all the quantities under consideration, including the polarization ratio P for the exciton multiplet of the perfect crystal. Sometimes their effect decreases because the magnitude of the numerical coefficients is small in first-order perturbation theory.

The situation is quite different for light excitons ($\mathcal{M} \gg \nu$). Of course, physically we are still dealing with the same excitons, but while we previously considered their interaction with high-frequency internal phonons, we now focus our attention on the external phonons; for them, usually, $\nu \lesssim \mathcal{M}$. First of all, the exciton-phonon interaction results in the emission of real phonons by an exciton. At $T \neq 0$ the phonon absorption processes are added. These processes lead to a damping of the exciton states $\sim \hbar/\tau_{fr}$, where τ_{fr} is the mean free time of the exciton. When $\mathcal{M} \lesssim \hbar/\tau_{fr}$, such notions as the exciton band and the exciton dispersion law lose their meaning altogether.

A strong interaction between the light excitons ($\mathcal{M} \gg \nu$) and the phonons ($E_{FC} \gg \mathcal{M}$) leads to an overall rearrangement of the exciton spectrum. Self-

trapped exciton states are formed below the exciton band. Although some damping
also occurs in the exciton band itself, it may be relatively small [6.24-27],
e.g., in the crystals of solid rare gases [6.28]. Most of the molecular crystals
of aromatic compounds do not exhibit any experimental evidence in favour of the
formation of self-trapped excitons. On the contrary, the experimental data suggest a weak interaction with external phonons. The existence of excimer fluorescence strongly shifted to the low-frequency side of the exciton absorption in
some crystals is an exception [6.29-31]. The displacement of large organic molecules in a crystal due to the formation of excimers may be very large. Thus, the
existence of excimer states could hardly serve as a basis for conclusions as
to the magnitude of interaction with external phonons near the configuration
corresponding to an unexcited crystal. This problem has not yet been studied.

Henceforward, we shall assume that the coupling to external phonons is
weak and can be neglected. We shall mainly be concerned with a systematic description of the effects due to the interaction of an exciton with the internal phonons.

Earlier in this section we divided the phonons into high-frequency ($\nu \gg \mathcal{M}$)
and low-frequency ($\nu \ll \mathcal{M}$) phonons in order to make the physical picture as
clear as possible. Actually, the intermediate case, $\nu \sim \mathcal{M}$, may also prove to
be important. At present it can only be investigated within the framework of
variational methods or interpolation schemes. The dynamical coherent-potential
approximation [6.32,33] is apparently the most effecitve scheme of this kind.

6.2 Dynamical Theory of Vibronic Spectra

Vibronic absorption is accompanied by the simultaneous creation of an exciton
and one or several internal phonons. The aim of this section is to show
that for heavy excitons ($\mathcal{M} < \nu$) the task of giving general solutions of
the problem of vibronic spectra can be reduced to the problem of stable particles (e.g., an exciton + a phonon) which exhibit a very unusual interaction
law. With this formulation of the problem one does not have to impose restrictions on the strength of the vibronic interaction. In the following subsection,
we basically present the results of [6.34,35].

6.2.1 Dynamical Theory Hamiltonian

Let us refer to (6.6) for the Hamiltonian of an exciton interacting with internal phonons. This equation may be simplified considerably if the smallness
of the parameter $\mathcal{M}/\nu$ is to be taken into account.

The Hamiltonian $H_{ex}(\tau)$ can be divided into two parts: the diagonal H_d and the off-diagonal H_{nd}. The former includes the $H_{ex}(\tau)$ averaged over the zero-phonon state, the matrix elements of $H_{ex}(\tau)$ between one-phonon states, etc. It conserves the total number of phonons. The latter includes the matrix elements of $H_{ex}(\tau)$ between all the states with non equal number of phonons. Naturally, we are concerned with the number of real phonons. The virtual phonons that form the "dressed" exciton are eliminated by the transformation given in (A.30). Both parts have the same order of magnitude $\mathcal{M}$. However, their effect on the spectrum is completely different for $\mathcal{M} \ll \nu$. To see this more clearly, we can carry out an approximate diagonalization of the Hamiltonian $H_0 + H_1$ (with $H_1 \ll H_0$) using the canonical transformation e^T:

$$ e^T(H_0 + H_1)e^{-T} \simeq (1+T)(H_0 + H_1)(1-T) \simeq H_0 + \{H_1 + [T, H_0]\} + \ldots \qquad . \qquad (6.8) $$

To cancel the off-diagonal matrix elements that are first order in H_1 we require that

$$ T_{nm} \approx (H_1)_{nm} \big/ (E_n - E_m) \quad , \quad n \neq m \quad , \qquad\qquad (6.9) $$

where the E_n are the eigenvalues of H_0. In this case, the roles of H_1 and H_0 are played by H_{nd} and $H_{ph} + H_0$, respectively. Hence, the differences $|E_n - E_m|$ are approximately equal to $l\nu$, where l is an integer. Therefore, the matrix T is small in the parameter $\mathcal{M}/\nu$. The formal procedure for expanding this matrix in the parameter $\mathcal{M}/\nu$ consists in isolating H_d, performing the transformation e^T, and calculating the corrections to H_d (which will be of the order of $\mathcal{M}H_{nd}/\nu$), etc. Let us consider the first step of this procedure by singling out the diagonal part of the operator $H_{ex}(\tau)$. This was how the zero-phonon states Hamiltonian given in (6.7) was obtained.

To determine the one-phonon state's Hamiltonian the terms must be kept bilinear in the operators b^+ and b in the expansion of the exponents in (6.6). In addition, similar terms must be retained in the second summand of (6.1). As a result, the following expression is obtained for the total Hamiltonian of the one-phonon states:

$$ H = H_e + H_{ex} + H_{ph} + H_v \quad , \qquad\qquad (6.10) $$

where the H_e, H_{ex}, and H_{ph} are determined by (6.3,7), and the vibronic interaction Hamiltonian H_v is equal to [6.35]

$$ H_v = \gamma^2 \sum_{n\alpha \neq m\beta} M_{n\alpha m\beta} a^+_{n\alpha} a_{m\beta} (b^+_{n\alpha} b_{m\beta} + b^+_{m\beta} b_{n\alpha} - b^+_{n\alpha} b_{n\alpha} - b^+_{m\beta} b_{m\beta}) $$
$$ + \Delta_\nu \sum_{n\alpha} a^+_{n\alpha} a_{n\alpha} b^+_{n\alpha} b_{n\alpha} \quad . \qquad\qquad (6.11) $$

Thus, H_v is completely determined by the renormalized dispersion law (i.e., by a set of transfer integrals $M_{n\alpha m\beta}$) and the linear and quadratic electronic-vibrational interaction constants γ and $\Delta_v \approx \delta$. In this respect the vibronic interaction is unique. Usually, the exciton-phonon interaction Hamiltonian contains a number of unknown constants. Of course, a simple form of H_v can be obtained by only taking into account the intramolecular vibronic interaction. The intermolecular vibronic interaction can easily be taken into account by introducing new matrix elements into (6.11). However, a multitude of new and unknown coefficients appear. Hence, the use of such a Hamiltonian at this stage is not very helpful. At the same time H_v, written in the form shown in (6.11), describes the experimental data quite well (Sect.6.4). This speaks in favour of the smallness of the intermolecular vibronic interaction which is in complete agreement with the basic interaction schemes valid for molecular crystals.

A very important difference between the Hamiltonians H_v and H_v^0 is that the former conserves the number of particles, whereas the latter does not. There-fore, the problem with the Hamiltonian H_v^0 (6.1) is a field-theory problem for which a general solution cannot be found. The problem with the Hamiltonian H_v is, on the contrary, a purely dynamical one, i.e., a quantum-mechanical problem involving the motion of two stable interacting particles - an exciton and a phonon. In some cases the problem can be solved exactly; in others, the solutions possessing a sufficient degree of accuracy may be found. The theory that is based on a Hamiltonian of the type shown in (6.11) and that describes the interaction of stable particles, will be called the *dynamical theory of vibronic spectra*. This theory can be easily modified to take into account a larger number of phonons. Only terms of the type b^+b^+bb must be retained in the expansion of the same exponents. It is not difficult to write out the cor-responding expressions. However, they become increasingly bulky as the number of phonons increases.

The reduction of the field problem to that of the motion of stable inter-acting particles, regardless of whether or not an exact solution can be found, enables one to give a simple classification of the quantum states. A con-tinuous spectrum - *dissociated states* of the exciton + phonon pair - always exists. The width of the continuous spectrum is equal to that of the exciton band $2\mathcal{M}$ [4]. Furthermore, if the interaction is strong enough, one or several *bound states* of this pair will arise. The states belonging to a continuous

---

[4] It is assumed that the phonon dispersion can be neglected.

spectrum will be called *two-particle states*, while the bound states will be called *one-particle states*, or *vibrons*.

There exists a class of entities that are neither molecules nor crystals. They are clusters consisting of several weakly bound parts: dimers, trimers, etc. A vibronic interaction in them is rather similar to that for a crystal. An analysis of such entities, which was begun by McCLURE [6.36], was advanced considerably in a number of subsequent papers [6.37-40]. The limiting cases were studied analytically while the intermediate case was investigated numerically. However, the methods that have been proven to be effective for these clusters cannot be applied to crystals. Besides, on passing to a crystal, i.e., to an infinite system, the emphasis is shifted considerably and the very formulation of the problem changes. In particular, the concepts of one- and two-particle states are meaningless when dealing with finite systems.

The particle interaction law corresponding to the Hamiltonian H_v is quite unusual. Only the last term has a simple structure. This term represents the contact interaction of an exciton and a phonon due to the shift in vibrational frequency Δ_v (Sect.1.2). It is universal in the sense that it exists for both the TS and the NTS phonons. The other summands only exist, in the approximation concerned, for the TS phonons (for the NTS phonons $\gamma = 0$). Let us now consider the physical meaning of each of these summands.

The first term represents the joint transfer of an exciton and a phonon from one site to another. The second term describes the exchange transfer, i.e., when the exciton and the phonon exchange sites. The next two terms describe the change in the integral of exciton transfer between two sites due to the presence of a real phonon at one of these sites. The separate terms of the Hamiltonian are presented graphically in Fig.6.1. It is important that dispersionless TS phonons became mobile due to their interaction with an exciton.

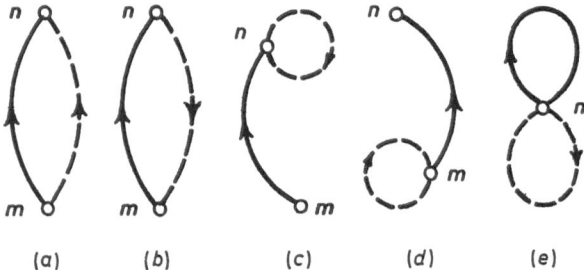

(a)   (b)   (c)   (d)   (e)

Fig.6.1a-e. Graphical representation of the separate terms in the Hamiltonian H_v: a) joint transfer; b) exchange transfer; c) and d) effect of the phonon on the exciton transfer integral; e) interaction via vibrational frequency change

It will be instructive to compare the relative role of the linear and quadratic electronic-phonon interactions in different effects. The contribution of the linear interaction to the vibronic absorption intensity is of the order of γ^2, while that of the quadratic interaction is of the order of $(\Delta_\nu/\nu)^2$ (see the end of Appendix A). On the other hand, their contribution to the Hamiltonian H_ν of the dynamical theory is $\gamma^2 \mathcal{M}$ and Δ_ν, respectively. It is obvious that in this case the role of the quadratic interaction increases considerably. In particular, it is possible that $|\Delta_\nu| \gtrsim \gamma^2 \mathcal{M}$ even if $(\Delta_\nu/\nu)^2 \ll \gamma^2$, 1. We shall show below that the competition between $|\Delta_\nu|$ and $\mathcal{M}$ is substantial in the formation of the energy spectrum of the system. Since $\mathcal{M}$ is always much less than ν for systems to which the dynamical theory can be applied, the inequality $|\Delta_\nu| \gtrsim \mathcal{M}$ can be fulfilled for $|\Delta_\nu| \ll \nu$, i.e., when the corrections to the transition intensities due to the frequency shift are small.

The dynamical theory of vibronic spectra was first introduced by RASHBA [6.34,35]. Different versions of the theory were subsequently developed [6.41-47]. The role of the vibrational-frequency shift Δ_ν as a mechanism of exciton-phonon interaction was noted in [6.48]. For NTS phonons H_ν is reduced to the last term. As a result in this special case the dynamical theory becomes mathematically equivalent to the theory describing the phonon-phonon interaction by mechanism of fourth-order anharmonicity [6.49-51]. The approach developed in the dynamical theory was also used in the theories describing the diffusion of molecular excitons and electrical conductivity in molecular crystals [6.52].

6.2.2 Conductivity Tensor

We can now write down the expression for the intensities in the vibronic absorption spectrum at $T = 0$. If the vibronic-transition intensities are due to an electronic-vibrational interaction, it is best to begin with the general expression for the conductivity (2.43) and transform the operators in Green's functions using (A.30). As a result, the exciton Green's function of the crystal,

$$G_{n\alpha m\beta}(t-t') = -i\left\langle Ta_{n\alpha}(t)a^+_{m\beta}(t')\right\rangle \quad ,$$

is transformed to

$$-ie^{-\gamma^2}\left\langle Ta_{n\alpha}(t)a^+_{m\beta}(t') \exp\left[\gamma(b^+_{m\beta}(t') - b^+_{n\alpha}(t))\right] \exp\left[-\gamma(b_{m\beta}(t') - b_{n\alpha}(t))\right]\right\rangle \quad .$$

After isolating the part responsible for the one-phonon vibronic transitions, the function is reduced to

$$-i\gamma^2 \exp(-\gamma^2) \left\langle Ta_{\mathbf{n}\alpha}(t) b_{\mathbf{n}\alpha}(t) b_{\mathbf{m}\beta}^+(t') a_{\mathbf{m}\beta}^+(t') \right\rangle \equiv \gamma^2 \exp(-\gamma^2) F_{\mathbf{n}\alpha\mathbf{m}\beta}(t-t') \quad .$$

$$(6.12)$$

Thus, the exciton-phonon Green's function F is obtained instead of the exciton Green's function G. The factor $\gamma^2 \exp(-\gamma^2)$ standing in (6.12) renormalizes the matrix element of the current. The matrix elements of the electronic transition in the absence of the electronic-phonon coupling $\mathbf{j}_{\mathbf{n}\alpha}^0$ yield the matrix elements of the vibronic transitions $\mathbf{j}_{\mathbf{n}\alpha}^{(v)}$:

$$\mathbf{j}_{\mathbf{n}\alpha}^{(v)} = \gamma \exp(-\gamma^2/2) \, \mathbf{j}_{\mathbf{n}\alpha}^0 \quad . \tag{6.13}$$

This relation between matrix elements agrees with the intensity distribution for vibronic transitions that take place under the Condon approximation in an isolated molecule (A.16). The same matrix elements $\mathbf{j}_{\mathbf{n}\alpha}^{(v)}$ appear in the band-to-band transition theory (5.3). As a result, the conductivity tensor can be written as follows:

$$\sigma(\Omega\mathbf{K}) = -\frac{1}{E_\rho} \frac{1}{vN} \mathrm{Im}\left\{ \sum_{\mathbf{n}\alpha\mathbf{m}\beta} F_{\mathbf{n}\alpha\mathbf{m}\beta}(\Omega) \exp[-i\mathbf{K}(\mathbf{n}_\alpha - \mathbf{m}_\beta)] \widehat{\mathbf{j}_\alpha^{(v)} \mathbf{j}_\beta^{(v)*}} \right\}$$

$$= -\frac{1}{E_\rho v} \mathrm{Im}\left\{ \sum_{\alpha\beta} F_{\alpha\beta}(\Omega\mathbf{K}) \widehat{\mathbf{j}_\alpha^{(v)} \mathbf{j}_\beta^{(v)*}} \right\} \quad , \tag{6.14}$$

where $F_{\alpha\beta}(\Omega\mathbf{K})$ is the Fourier component of the exciton-phonon Green's function.

If we consider the vibronic transitions involving NTS phonons, (6.14) is still valid but the matrix elements that arise as a result of an intramolecular configurational mixing (Sect.1.2) appear in it as $\mathbf{j}_\alpha^{(v)}$.

6.2.3 Energy Spectrum. Limiting Cases

The detailed structure of the energy spectrum in the vibronic region depends on the exciton dispersion law and on the values of the parameters γ and Δ_v. To get an idea of the possible types of spectrum we shall consider a number of limiting cases.

a) NTS Phonon

In this case, $\gamma = 0$ and H_v is reduced to the last term in (6.11):

$$H_v = \Delta_v \sum_{\mathbf{n}\alpha} a_{\mathbf{n}\alpha}^+ a_{\mathbf{n}\alpha} b_{\mathbf{n}\alpha}^+ b_{\mathbf{n}\alpha} \quad . \tag{6.15}$$

Since the phonon dispersion is neglected [see (6.3)] and (6.15) is diagonal in the site representation, the phonon remains fixed. Hence, the problem with the interaction Hamiltonian in (6.15) is reduced to that of a local exciton in the vicinity of an isotopic impurity. The only difference is that the role of the isotopic shift Δ_{ex} is played by the vibrational-frequency shift Δ_ν. Accordingly, all of the conclusions drawn in Sect.3.2 can be transferred automatically; in particular, if $|\Delta_\nu| > |\Delta_{cr}|$ a single discrete level corresponding to the bound state of the exciton and phonon must appear. Of course, strictly speaking, this will not be a level but a narrow band of one-particle states (vibronic band), whose width is determined by mechanisms which have not yet been considered (dispersion in the phonon band, dipole-dipole interactions for vibronic trans-itions[5], etc.). The dissociated states of the exciton + NTS phonon pair, i.e., the spectrum of two-particle states, correspond to the continuous spectrum of the local exciton.

It is clear that even in one-particle states the exciton and phonon can either occupy the same site, *joint configuration*, or different sites, *separated configurations*. The relative contribution of either configuration is deter-mined by the wave function of the internal motion in the vibron (i.e., of the relative motion of the exciton and phonon), which coincides with the wave func-tion of the local exciton (3.17). The vibron energy ε^V is determined by (3.8) in which Δ is replaced by Δ_ν. On the absolute energy scale, the energy of the vibron $E^V = E_\rho + \nu + \varepsilon^V$. The quantity $E_\rho + \nu$ will be called the *vibronic term of the separated configuration*. Thus, for an NTS phonon the problem of vibronic states has an exact solution.

b) TS Phonon: General Properties

Unfortunately, the problem cannot be solved exactly when the constant γ in H_V has a finite value. It is easy to see where the principal difficulty arises if one, first of all, transforms the Hamiltonian. Let us pass on to the **k** representation in order to be able to separate the motion of the centre of gravity. For a crystal with a single molecule in the unit cell,

$$H_V = \frac{\gamma^2}{N} \sum_{k_1 k_2 k} \{(\Delta_\nu + \varepsilon(k)) + \varepsilon(k_1 + k_2 - k)$$

$$- \varepsilon(k_2) - \varepsilon(k_1)\} a^+_{k_1} b^+_{k-k_1} b_{k-k_2} a_{k_2} , \qquad (6.16)$$

where **k** is the total momentum. For a fixed value of **k**, there are only two sum-mation indices: k_1 and k_2. Furthermore, since the phonon dispersion is still equal to zero, the problem at hand is completely equivalent to the zero-phonon

---

[5] Since the oscillator strength of intramolecular vibronic transitions in-volving NTS phonons is usually small ($f \gtrsim 0.001$), the band width produced by dipole-dipole mechanism will also be small.

problem with the Hamiltonian $H_{eff}(\mathbf{k})$:

$$H_{eff}(\mathbf{k}) = E_\rho + \nu + H_{ex} + H_v^{eff}(\mathbf{k}) \quad , \tag{6.17}$$

where H_{ex} is defined by

$$H_{ex} = \sum_{\mathbf{k}_1} \varepsilon(\mathbf{k}_1) a_{\mathbf{k}_1}^+ a_{\mathbf{k}_1} \quad , \tag{6.18}$$

and H_v^{eff} is equal to

$$H_v^{eff}(\mathbf{k}) = \frac{\gamma^2}{N} \sum_{\mathbf{k}_1 \mathbf{k}_2} \{(\Delta_\nu + \varepsilon(\mathbf{k})) + \varepsilon(\mathbf{k}_1 + \mathbf{k}_2 - \mathbf{k})$$

$$- \varepsilon(\mathbf{k}_2) - \varepsilon(\mathbf{k}_1)\} a_{\mathbf{k}_1}^+ a_{\mathbf{k}_2} \quad , \tag{6.19}$$

and $\mathbf{k}$ is the parameter of a one-exciton Hamiltonian. The complete equivalence of the two problems is seen in the fact that all of the matrix elements of the two Hamiltonians coincide.

It would be useful to change from the momentum to the site representation by means of the transformation given in (6.4). Having done so H_{ex} transforms to the form given in (6.7) and H_v^{eff} becomes

$$H_v^{eff}(\mathbf{k}) = \gamma^2 \{(\Delta_\nu + \varepsilon(\mathbf{k})) a_0^+ a_0 + \sum_{\mathbf{l} \neq 0} M_\mathbf{l} \exp(-i\mathbf{k}\mathbf{l}) a_{-\mathbf{l}}^+ a_\mathbf{l}$$

$$- \sum_{\mathbf{l} \neq 0} M_\mathbf{l} (a_0^+ a_\mathbf{l} + a_\mathbf{l}^+ a_0)\} \quad . \tag{6.20}$$

Having separated the centre-of-gravity motion, (6.19,20) become expressions for the Hamiltonians of the internal motion of the exciton + phonon pair in which the total momentum $\mathbf{k}$ enters a c-number.

The Green's function $F(\Omega \mathbf{k} \mathbf{k}_1 \mathbf{k}_2)$ as well as the effective exciton Hamiltonian given in (6.17) are matrices in the momenta $\mathbf{k}_1$ and $\mathbf{k}_2$, depending on $\mathbf{k}$. In the equation $F = F^0 + F^0 H_v F$, the operator H_v is also a matrix whose indices are $\mathbf{k}_1$ and $\mathbf{k}_2$ and whose matrix elements are defined by the curly brackets in (6.16, 19). The Green's function for a free particle is given by $F^0(\Omega \mathbf{k} \mathbf{k}_1 \mathbf{k}_2) = (\Omega - E_\rho - \varepsilon(\mathbf{k}_1) - \nu)^{-1} \delta_{\mathbf{k}_1 \mathbf{k}_2}$. It is shown in Appendix B that the degenerate-perturbation theory yields exact solutions for matrices of the type given by (B.1) - a sum of the products of two factors, each factor being dependent on one of the indices only (in this case, $\mathbf{k}_1$ and $\mathbf{k}_2$). Only the second term in (6.16,19) describing the exchange jumps does not belong to this type. There-fore, it is precisely this term that does not allow the problem to be solved exactly. Nevertheless, it can be shown that one can effectively solve the auxiliary equation in which only this term is retained in H_v [6.35]. As a result only the interaction between a small number of nearest neighbours must be taken into account. The effect of the remaining three terms is then taken into account by degenerate-perturbation theory. The solution for naphthalene (Sect.6.3.2) was obtained this way. Similar conclusions readily follow from the site-representation form of the effective Hamiltonian (6.20), which also conveniently leads to a numerical solution.

Since even dispersionless phonons become mobile as a result of the interactions with excitons (Sect.6.2.1), the dispersion of the one-particle branch at $\gamma^2 \sim 1$ is of the same order of magnitude as the dispersion in the exciton band. Since H_v^{eff} is a function of $\mathbf{k}$ (6.19), it may turn out that the one-particle branch only exists in a restricted region of the $\mathbf{k}$ space. Generally, the number of one-particle branches may differ from the number of molecules in the unit cell (and, hence, from the number of exciton bands). A typical spectrum is shown in Fig.6.2.

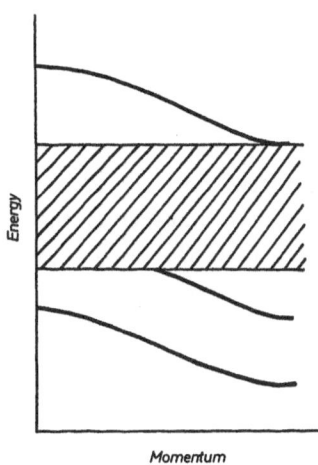

Fig.6.2. Scheme of a vibronic spectrum involving a TS-phonon; hatched area) region of the two-particle spectrum; solid curves) branches of the one-particle spectrum

c) Weak Vibronic Coupling

A vibronic interaction is weak and may be treated by perturbation theory if the H_v that is determined by (6.11) is small compared to the width of the exciton band. The above condition is fulfilled if

$$\gamma^2 \ll 1 \quad , \quad |\Delta_v| \ll \mathcal{M} \quad . \tag{6.21}$$

In the case of a three-dimensional exciton energy spectrum there are no one-particle states and the entire vibronic spectrum consists of two-particle states. This corresponds to the approximation considered by McRAY [6.1] (Sect.6.1).

In low-dimensional systems (layer compounds, polymer chains) one-particle states may arise even under weak vibronic-interaction conditions. However, their binding energy will be small to the extent of the smallness of the vibronic-interaction constants.

d) Strong Vibronic Coupling

We restrict ourselves to a strong vibronic coupling at $\Delta_\nu = 0$. Then the vibronic interaction Hamiltonian H_ν (6.11) exceeds the exciton Hamiltonian H_{ex} (6.7) by the factor γ^2. For this reason, in the main order in parameter γ^2 the eigenfunctions of H_ν will be the eigenfunctions of the total Hamiltonain $H_{eff}(\mathbf{k})$ (6.17).

Considering the effective Hamiltonian H_ν^{eff} in (6.20), one can easily understand that the number of eigenfunctions is equal to the number of involved sites: the central site ($\mathbf{l} = 0$) and also those sites ($\mathbf{l} \neq 0$) direct interaction of which with the central site is included into the Hamiltonian. The overall energy scale of the spectrum is of the order of $\gamma^2 \mathcal{M}$. The joint and separated configurations contribute approximately equally to the individual states. If the term H_{ex} is taken into account, a two-particle spectrum region with a width of $2\mathcal{M}$ arises. Those solutions of the zero-approximation problem which fall into this region are included in the two-particle spectrum; the others form one-particle branches. Near the edge of the two-particle spectrum, the one-particle branches will be strongly perturbed. Those branches which intersect the two-particle spectrum edge will terminate at this edge.

e) Molecular Limit

Under what sort of conditions can the molecular vibrons (Sect.6.1.1) appear in the vibronic spectrum? From the above-considered cases it is clear that this is only possible if the large value of Δ_ν is responsible for a strong vibronic coupling ($|\Delta_\nu| >> \mathcal{M}$ and the value of γ is not too large, $\gamma^2 \lesssim 1$). Actually, this limit must be attained rather soon, already at $|\Delta_\nu| \gtrsim \mathcal{M}$, as was the transition to $a^2 \approx 1$ in the theory of isotopic impurity centres (Sect.3.2.5). The approximate wave function of the molecular vibron is

$$|\mu\mathbf{k}\rangle \approx \frac{1}{\sqrt{N}} \sum_{\mathbf{n}\alpha} B_\alpha(\mu\mathbf{k}) \exp{(i\mathbf{k}\mathbf{n}_\alpha)} a_{\mathbf{n}\alpha}^+ b_{\mathbf{n}\alpha}^+ |0\rangle \quad , \tag{6.22}$$

where $|0\rangle$ represents the ground state of the crystal and the $B_\alpha(\mu\mathbf{k})$ are the coefficients that determine the wave function of the exciton (2.8). The smallness of the contribution from the separated configurations is due to the fact that $|\Delta_\nu| >> \mathcal{M}$. As a result, the predominant contribution to the energy is made by the first and the last terms in the Hamiltonian of (6.11), which are responsible for joint jumps and the frequency shift, respectively. Hence,

$$E_{\mu}^{v}(\mathbf{k}) = E_{\rho} + v + \varepsilon_{\mu}^{v}(\mathbf{k}) \quad , \qquad \varepsilon_{\mu}^{v}(\mathbf{k}) \approx \Delta_{v} + \gamma^2 \varepsilon_{\mu}(\mathbf{k}) \quad , \tag{6.23}$$

where $\varepsilon_{\mu}(\mathbf{k})$ represents the exciton dispersion law. The expression
$E_{\rho} + v + \Delta_{v} = E_{\rho} + v^*$ determines the *vibronic term of the joint configuration*.
With this approach, the structure of the vibronic band is identical to that
of the exciton band. The only difference is that its width is changed by a
factor of γ^2. The Davydov splittings of the vibronic and exciton bands differ
by the same factor. Therefore, the Davydov splitting for TS vibrons ($\gamma^2 \neq 0$)
greatly exceeds the splitting for NTS vibrons ($\gamma^2 = 0$). This regularity is
a well-known experimental fact; it was explained by CRAIG and WALMSLEY [6.9].

Of course, even in the molecular limit the vibronic spectrum does not only
consist of molecular vibrons. A spectrum of two-particle states always exists.
Furthermore, at $\gamma^2 \sim 1$, additional one-particle branches may exist. Indeed,
since $|\Delta_{v}| \gg \mathcal{M}$, joint configurations enter with a very small weight into any
state with the exception of the molecular vibron state. Therefore, of the first
four terms of the vibronic Hamiltonian in (6.11) the second term, corresponding
to exchange jumps, proves to be the most important one. It ensures an effective
exciton-phonon interaction ($\sim \gamma^2 \mathcal{M}$), that can compete with the exciton-band
width ($\sim \mathcal{M}$) at $\gamma^2 \sim 1$ and, thus, causes the formation of new vibronic branches.

f) Quasi-One-Particle States

In conclusion we wish to mention a remarkable fact: Exact solutions exist at
$\gamma^2 = 1$ and at arbitrary values of Δ_{v}. These solutions are the functions given in
(6.22). Indeed, applying the Hamiltonian (6.10) to $|\mu\mathbf{k}\rangle$, we obtain

$$H|\mu\mathbf{k}\rangle = (E_{\rho} + v^* + \varepsilon_{\mu}(\mathbf{k}))|\mu\mathbf{k}\rangle \quad , \qquad v^* = v + \Delta_{v} \quad . \tag{6.24}$$

This results from the fact that the terms that are obtained from the action
of H_{ex} and the fourth term in (6.11) on the function $|\mu\mathbf{k}\rangle$ cancel one another,
while the second and third terms in (6.11) when acting on $|\mu\mathbf{k}\rangle$ yield zero.
If the value of $\Delta_{v} = v^* - v$ is such that the energy obtained from (3.8) (see
also Sect.6.2.3a) lies outside of the two-particle spectrum, then $|\mu\mathbf{k}\rangle$ de-
scribes a molecular vibron - a stationary one-particle state. If, however,
the energy falls inside of the region of two-particle states, the resulting
state is a quasi-one particle state. It is unstable and decomposes through
extraneous mechanisms into two-particle states (for example, interaction with
external phonons, a small difference between γ^2 and unity).

6.2.4 Optical Absorption Spectrum

We must now consider the polarization of the absorption bands and the intensity distribution in the spectrum. In describing vibronic spectra it is convenient to count the frequency $\omega = \Omega - (E_\rho + \nu)$ from the vibronic term of the separated configuration $E_\rho + \nu$. Accordingly, the energy ε, counted from E_ρ, will be considered as exciton energy and the frequency ν must be omitted. The vibron energy ε^V is counted from $E_\rho + \nu$.

Let us suppose that a one-particle branch of vibronic states with an energy equal to $\varepsilon_\lambda^V(\mathbf{k})$ is formed. Isolating the contribution to F from this branch, substituting into (6.14) and setting $\mathbf{k} = 0$, we obtain [with $\varepsilon_\lambda^V \equiv \varepsilon_\lambda^V \ (\mathbf{k} = 0)$]

$$\sigma(\omega) = \frac{\pi}{E_\rho \nu N} \sum_{\mathbf{n}\alpha \mathbf{m}\beta} \left\langle 0 | a_{\mathbf{n}\alpha} b_{\mathbf{n}\alpha} | \lambda 0 \right\rangle \left\langle \lambda 0 | b_{\mathbf{m}\beta}^+ a_{\mathbf{m}\beta}^+ | 0 \right\rangle$$

$$\cdot \widehat{j_\alpha^{(v)} j_\beta^{(v)}}^* \delta(\omega - \varepsilon_\lambda^V) \quad . \tag{6.25}$$

If the wave function of a vibron with $\mathbf{k} = 0$ is written in the form

$$| \lambda, \mathbf{k} = 0 \rangle = \frac{1}{\sqrt{N}} \sum_{\mathbf{n} \mathbf{l} \alpha \beta} \psi_{\alpha\beta}^{(\lambda)}(\mathbf{l}) a_{\mathbf{n}\alpha}^+ b_{\mathbf{n}+\mathbf{l},\beta}^+ | 0 \rangle \tag{6.26}$$

then $\psi_{\alpha\beta}^{(\lambda)}(\mathbf{l})$ is the internal-motion function. Inserting it into (6.25), we obtain

$$\sigma_\lambda(\omega) = \frac{\pi}{E_\rho \nu} \widehat{\mathbf{j}_\lambda^{(v)} \mathbf{j}_\lambda^{(v)}}^* \delta(\omega - \varepsilon_\lambda^V) \quad ,$$

$$\mathbf{j}_\lambda^{(v)} = \sum_\alpha \psi_\alpha^{(\lambda)} \mathbf{j}_\alpha^{(v)} \quad , \qquad \psi_\alpha^{(\lambda)} \equiv \psi_{\alpha\alpha}^{(\lambda)} \ (\mathbf{l} = 0) \quad . \tag{6.27}$$

It is seen that the conductivity is related to the ψ_α^λ's (the amplitudes of the residence of the exciton and phonon at the same site), i.e., to the contribution of the joint configurations to the wave function of the vibron. The results are particularly simplified for NTS phonons, in which case the bands with different values of λ are degenerate and the summation can be carried out over them. As a result, it can be shown that the total conductivity is equal to

$$\sigma(\omega) = \frac{\pi}{E_\rho \nu} a^2 \widehat{\mathbf{j}_\alpha^{(v)} \mathbf{j}_\alpha^{(v)}}^* \delta(\omega - \varepsilon_\lambda) \quad , \qquad a^2 = \sum_\lambda | \psi_\alpha^{(\lambda)} |^2 \quad , \tag{6.28}$$

where a^2 is the probability that the exciton and phonon are at the same site. The absorption defined by (6.28) is weakly polarized. The polarization ratio depends on the orientation of the molecular dipole.

The presence of strongly polarized bands is usually considered as a strong indication of the exciton absorption. Therefore, the weakly polarized absorption corresponding to the one-particle branches associated with NTS phonons is usually called *molecular* (Sect.1.3) while the corresponding absorption bands are called M *bands*.

For TS phonons, the transition frequencies ε_λ^v corresponding to different values of λ do not coincide. The absorption $\sigma_\lambda(\omega)$ described by (6.27) is polarized along the crystallographic directions and consists of a multiplet of sharply polarized bands. They are usually called *crystal bands* (Sect.1.3) or K_1 *bands* by analogy with the K_0 bands of the exciton multiplet. The subscript 1 indicates that the vibronic transition involves a single TS phonon. The number of one-particle branches near $\mathbf{k} = 0$ is a function of the values of γ and Δ_v. It can either be smaller or larger than Z, the number of molecules in the unit cell. Therefore, a K_1 multiplet can differ substantially from a K_0 multiplet, in which the total number of components (allowed and forbidden) is always equal to Z. In this sense, K_0 is always complete. In addition, the polarization ratio of the K_1 bands may differ from that of the K_0 bands even for a complete K_1 multiplet. The reason is that the joint configurations contribute differently to the one-particle states corresponding to the individual components of the K_1 multiplet. The polarization ratios of the K_0 and K_1 multiplets only coincide in the limit of the molecular vibron.

Along with the narrow-band absorption in the M and K_1 bands there must always exist an absorption corresponding to two-particle states, i.e., to the dissociated exciton + phonon pairs. The relevant absorption bands are called D *bands*. These are usually located on the high-frequency side of the corresponding M and K_1 bands. The intensity of the D bands is due to the contribution of the joint configurations of the exciton and phonon to the wave functions of the two-particle states. D bands arise but are very weak, even when one-particle states are molecular vibrons. The contours of the D bands are complicated and depend strongly on the parameters of the problem (Fig.6.12). However, in each case the magnitude of the band width is given by $2\mathscr{M}$, the width of the exciton band (Fig.1.10). As a result, in thick crystals they have the shape of "pillars" with an almost universal width. This regularity has been observed experimentally by PRIKHOTJKO [6.53,54].

Thus, narrow-band vibronic absorption (K and M bands) is due to the formation of one-particle states[6]. These, in turn, are the result of a sufficiently strong vibronic coupling, for instance, a large value of Δ_v. The possibility of considering the one-particle and two-particle spectra on the same basis is one of the advantages of the dynamical theory. The clear-cut formulation of

---

[6] For $\gamma^2 = 1$ quasi-one-particle states.

the simultaneous manifestation of the two classes of states in vibronic spectra arose within the framework of this theory.

The absorption in the D band can only be calculated analytically at $\gamma = 0$. In this case the phonon is immobile and the Green's function $F_{n\alpha m\beta}$, defined by (6.12), is diagonal. Also the Hamiltonian H_v (6.15) is equivalent for an exciton to the isotopic defect potential. Hence

$$F_{n\alpha m\beta}(\Omega) = G_{n\alpha n\alpha}(\omega)\delta_{n\alpha m\beta} = \delta_{n\alpha m\beta}\Big/(G_0^{-1} - \Delta) \quad , \tag{6.29}$$

where $G_{n\alpha n\alpha}$ is the Green's function of the exciton in a doped crystal in which the defect is at the same site $n\alpha$. This function is determined by (3.6). The last expression in (6.29) is based on it. Inserting (6.29) into (6.14) and keeping in mind that $G_0 = G_0' - \pi i\rho$ (2.30) and $\mathbf{K} = 0$, we obtain

$$\sigma(\omega) = \frac{\pi}{E_\rho v}\sigma_0(\omega) \sum_\alpha \widehat{j_\alpha^{(v)} j_\alpha^{(v)*}} \quad , \qquad \sigma_0(\omega) = \frac{\rho(\omega)}{[1 - \Delta_v G_0'(\omega)]^2 + [\pi\Delta_v\rho(\omega)]^2} \quad . \tag{6.30}$$

If the first expression in the denominator has an isolated zero outside of the continuum, i.e., if a one-particle state arises, the intensity of the one-particle peak is given by (6.28). Otherwise, (6.30) describes the intensity distribution in the two-particle continuum. The polarization ratio is constant within the absorption band. It is worth noting that the factor $\sigma_0(\omega)$ also appears in (3.29) for absorption induced in the exciton-continuum by the impurity.

The conductivity $\sigma_0(\omega)$ is normalized, i.e., $\int \sigma_0(\omega)\, d\omega = 1$. The fraction of the intensity allotted to the one- and two-particle states is equal to a^2 and $1 - a^2$, respectively (6.28). The inversion formula can be derived using the dispersion relations [6.55]. It enables one to reconstruct the density of states from the vibronic absorption in which an NTS phonon is involved:

$$\rho(\omega) = \frac{\dot{\sigma}_0(\omega)}{\left(1 + \Delta_v \dashint \frac{\sigma_0(\omega')}{\omega - \omega'}\, d\omega'\right)^2 + (\pi\Delta_v\sigma_0(\omega))^2} \quad . \tag{6.31}$$

The integration in (6.31) is performed over the one- and two-particle spectrum. The cross-bar indicates that the integral was taken in the sense of a Cauchy main value.

Since it is usually impossible to obtain an analytical expression for $\sigma(\omega)$, it would be helpful to give the exact integral relations for the first two moments of the spectrum:

$$\int \sigma(\omega)d\omega = \frac{\pi}{E_\rho v} \sum_\mu \widehat{j_\mu^{(v)} j_\mu} (v)^* \quad , \tag{6.32}$$

$$\int \omega\sigma(\omega)d\omega = \frac{\pi}{E_\rho v} \sum_\mu (\Delta_v + \gamma^2 \varepsilon_\mu) \widehat{j_\mu^{(v)} j_\mu} (v)^* \quad . \tag{6.33}$$

The matrix elements $j_\mu^{(v)}$ and $j_\alpha^{(v)}$ are related to one another by the same expressions (2.44) as the matrix elements of the exciton transitions. In both
(6.32,33), the integration is extended over the entire vibronic spectrum. The
first equation reflects the conservation of the total oscillator strength of
the vibronic transition. The second yields a simple relation for the general-
ized Davydov splitting of a vibronic transition $\gamma^2(\varepsilon_{\mu_1} - \varepsilon_{\mu_2})$. It is defined as
the difference between the centres of gravity of the respective components of
the spectrum.

Up until now the spectrum intensity distribution was discussed exclusively
with respect to two-particle absorption, while the one-particle bands were
assumed to be δ-functions. Their width is determined by extraneous mechanisms
such as impurity scattering, external phonons, etc. The theory that describes
the shape of these bands must be developed by analogy with the theory describing the shape of the exciton absorption bands.

A different type of optical transition is associated with the one-particle
states. It becomes possible at $T \neq 0$ and consists of the formation of a vibron
from a thermal internal phonon and an optically created exciton [6.56].For an
NTS phonon, this absorption at $\Delta_v < 0$ is observed on the immediate low-frequency
side of the exciton band. It exhibits all of the features of the Rashba effect
with isotopic impurities (Sects.3.2,3). The absorption must be structural and,
therefore, the low-frequency wing of the exciton absorption band is expected
to exhibit a non-Urbach behaviour.

6.3 Vibronic Absorption Spectra of the Solid Aromatic Compounds

6.3.1 General Structure of Vibronic Spectra

In considering the vibronic absorption spectra of solid aromatic compounds
we shall proceed from the main premise that a set of bound and dissociated
states of interacting quasi-particles (an exciton and one or several internal
phonons) is associated with each vibronic state of a free molecule in a molecular crystal. The larger the number of quasi-particles, the more complicated
is the energy spectrum of such a system.

The absorption spectrum of a crystal is naturally a reflection of the complicated structure of its energy spectrum. To interpret the absorption spec-

trum it is necessary to establish at first the type of vibronic excitation the bands correspond to. We shall mostly restrict ourselves to one-phonon vibronic transitions. They are located in the low-frequency part of the absorption spectra of molecular crystals. A quantitative theory (Sect.6.2) which enables one to analyze the energy and optical spectra has been developed.

This restriction, nevertheless, does not imply that the principal features of the vibronic spectra of crystals and the differences with respect to the spectra of free molecules shall not be considered. The one-phonon transitions are selected since precisely in them all these features manifest themselves most prominently. Despite the obvious increased complexity of the energy spectrum of multi-phonon vibronic excitations, the absorption spectrum in the region of composite tones and overtones (in molecular terminology) may actually be simpler than the one-phonon absorption spectrum. As the number of phonons involved in a vibronic transition increases, so does the energy of the quadratic exciton-phonon interaction due to the summation of the phonon frequency shifts. As a result, the relative role of the joint vibronic configurations increases. Thus, the fraction of dissociated (many-particle) states in the total vibronic absorption decreases and the intensity of the one-particle bands increases. These bands approach the molecular-vibron limit. If NTS phonons are involved, the shape of the absorption spectrum of the crystal in the region of multi-phonon vibronic transitions approaches that of an oriented molecular gas. It acquires a "molecular character", which is often mentioned in the literature. The only difference is that the intensity distribution in the vibronic series changes due to configurational mixing (Sect.1.3). These properties of multi-phonon vibronic states shall be illustrated for the spectrum of a naphthalene crystal in Sect.6.4.4.

The energy spectrum of one-phonon vibronic excitations is determined by the motion of two quasi-particles. It consists of a continuous two-particle spectrum and a spectrum of one-particle states. The optical transitions from the ground state to these excited states determine the one-phonon vibronic absorption spectrum at low temperatures. An absorption band can only be classified as a one- or two-particle band by comparing the energy and optical spectra of the crystal since there is no crucial difference in the shape of one- and two-particle absorption bands which allows to distinguish them unambiguously. Indeed, the participation of external phonons in a real absorption spectrum can lead to great deviations in the expected shape of a band.

If the positions and widths of the exciton bands and the values of the phonon frequencies ν in the crystal ground state are known, the edges of the energy spectrum of the two-particle states can be constructed by simply shifting the position of the edges of the exciton band by the value of the phonon energy ν. Comparing this spectrum with the absorption spectrum and bearing in mind that bound states can only exist outside of the continuous spectrum, one can divide the observed absorption into one- and two-particle absorption. This work was performed for a number of aromatic compounds [6.57,58].

6.3.2 Benzene

a) One-Photon Absorption

The general appearance of the absorption spectrum of the benzene crystal in the region of the first two vibronic transitions is shown in Fig.6.3b. The first vibronic transition corresponds to the molecular transition $B_{2u} \cdot e_{2g} \leftarrow A_{1g}$, which involves an NTS vibration $\nu_{18}{}^7$ ($\nu = 606$ and $\nu^* = 520$ cm^{-1} in the ground and excited state of the molecule, respectively). The vibrational frequency shift Δ_ν is equal to -86 cm^{-1}.

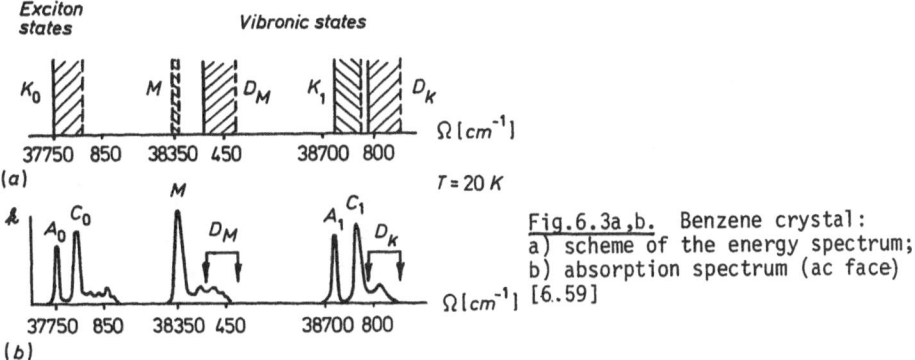

Fig.6.3a,b. Benzene crystal: a) scheme of the energy spectrum; b) absorption spectrum (ac face) [6.59]

The second vibronic transition corresponds to the intramolecular transition $B_{2u} \cdot a_{1g} \leftarrow A_{1g}$, which involves a TS vibration ν_2 ($\nu = 992$ and $\nu^* = 923$ cm^{-1}). The vibrational frequency shift for this transition is equal to -69 cm^{-1}. The square of the linear electronic-vibrational interaction constant, γ^2, is approximately equal to 1 (Table 1.4).

The width $2\mathcal{M}$ of the energy spectrum of the excitons is about 60 cm^{-1} (Sect.2.5). The low frequency edge of the spectrum is marked by the position of the A_0 band of the exciton triplet. The energy spectra of the two-particle states D_M and D_K (Fig.6.3a) are $2\mathcal{M}$ wide and are shifted relative to the exciton spectrum by ν_{18} and ν_2, respectively.

---

7 Further on, we denote the different phonons in accordance with their numbering in the molecule, e.g., the phonon ν_{18} has the same number "18" as the corresponding vibration in the free molecule, where the system of numeration is well established.

Figure 6.3b illustrates the general shape of the absorption spectrum of the crystal in the region under consideration. The vibronic absorption bands M, A_1, and C_1 lie outside of the energy spectrum of dissociated states, and, therefore, must be assigned to one-particle states. The corresponding energy bands of the one-particle vibronic states are shown in Fig.6.3a to the left of the spectrum of the two-particle states. The band width is shown rather conventionally. The vertical arrows in Fig.6.3b define the regions in which two-particle absorption is possible. The intensity of the absorption in these regions is only a small fraction of the intensity of one-particle absorption for both vibronic transitions. The predominance of one-particle absorption is due to the fact that $|\Delta_\nu| > \mathscr{M}$. The difference between $|\Delta_\nu|$ and $\mathscr{M}$ is still greater in two-phonon and multi-phonon transitions. Therefore, the entire vibronic one-photon absorption spectrum of the benzene crystal is mostly of a one-particle nature that approaches the molecular limit (Sect.6.2.3e). We note, however, that additional bands, whose nature is not clear, are observed in the region of the K_1 transition [6.60]. Some of these bands may be related to the additional branches of the one-particle states for the ν_2 phonon due to the large value of the constant γ ($\gamma^2 \approx 1.1$, Sect.6.2.3).

b) Two-Photon Absorption Spectrum

Since the first electronic transition of the benzene molecule is of B_{2u} symmetry, the one-photon vibronic absorption spectrum only involves g-type vibrations (g-spectrum). On the other hand, the two-photon vibronic spectrum is a u-spectrum, since only NTS vibrations of u-type are involved. For them Δ_ν completely determines the exciton-phonon interaction. Since the values of Δ_ν are usually large for benzene molecule, the one-particle absorption bands are dominant in the two-photon vibronic absorption. Two-photon absorption by the benzene crystal has been investigated by HOCHSTRASSER et al. [6.61]. A part of the fluorescence excitation spectrum is shown in Fig.6.4. On the whole, the spectrum corresponds to the molecular limit, i.e., narrow one-particle bands predominate. The two regions (ν_4 and ν_{15} phonons) marked with asterisks are exceptions. The relevant bands are wider and of a two-particle nature. The Δ_ν values for the ν_4 and ν_{15} phonons are equal to -22 and -6 cm^{-1}, respectively, i.e., one-particle states do not arise because $|\Delta_\nu| < |\Delta_{cr}| \approx 30$ cm^{-1} (Sect.3.3.1).

6.3.3 Naphthalene

a) Stress-Free Crystal

The one-photon absorption spectrum of the naphthalene crystal in the region of the first two vibronic transitions is illustrated in Fig.6.5b. The first

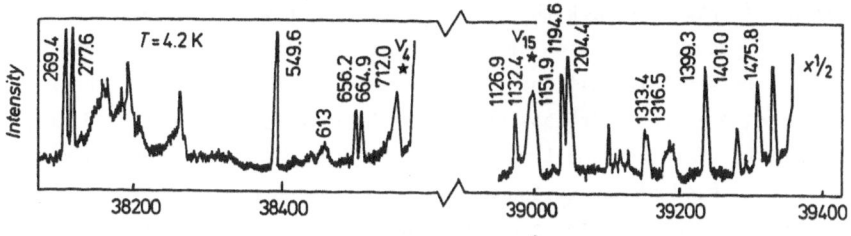

Fig.6.4. Two-photon fluorescence-excitation spectrum of the benzene crystal (bc face). Light is polarized parallel to $\underline{c}$ [6.61]; frequencies of phonons ν_4 and ν_5 are $\nu/\nu^* = 712/690$ and $1152/1146$ cm^{-1}, respectively

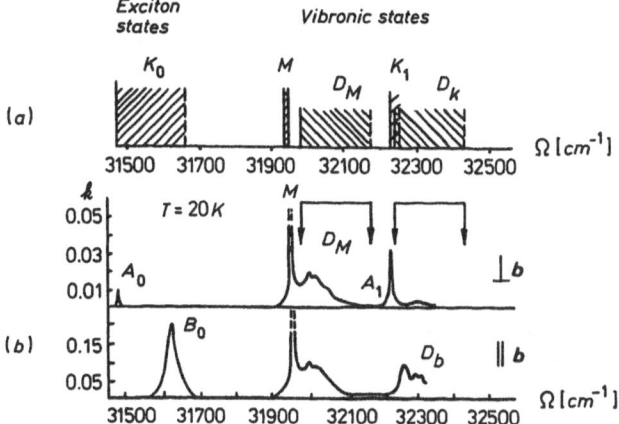

Fig.6.5a,b. Naphthalene crystal: a) scheme of the energy spectrum; b) absorption spectrum (in polarized light) [6.62]

transition corresponds to the intramolecular transition $B_{3u} \cdot b_{1g} \leftarrow A_g$ which involves an NTS vibration ν_{17} ($\nu = 509$ cm^{-1}, $\nu^* = 438$ cm^{-1}). The magnitude of the electronic-vibrational interaction is determined by the vibrational frequency shift, $\Delta_\nu = -71$ cm^{-1}. The second vibronic transition $B_{3u} \cdot a_g \leftarrow A_g$ involves a TS vibration ν_8 ($\nu = 758$ cm^{-1}, $\nu^* = 702$ cm^{-1}). The magnitude of the electronic-vibrational interaction is determined by the value of Δ_ν (-56 cm^{-1}) and by the linear interaction constant γ ($\gamma^2 \approx 0.2$; Table 1.4).

The width of the exciton energy spectrum is ≈ 180 cm^{-1}. Its low-frequency edge coincides with the A_0 band of the exciton doublet. The regions of the two-particle excitation of the first and second vibronic transitions are denoted by D_M and D_K in Fig.6.5a. They are spaced from the low-frequency edge of the exciton spectrum by frequencies ν_{17} and ν_8, respectively. The widths of the two spectra are each practically equal to 180 cm^{-1}, since the widths of the phonon bands do not exceed several wave numbers.

The arrows in Fig.6.5b mark the spectral regions in which two-particle absorption may occur. Absorption of perceptible intensity occurs in these regions (for the first transition - the side band of the M band; for the second transition - the entire absorption in the b component, this band will be denoted by D_b). At the same time the intense M band of the first transition and the A_1 band of the second are located outside of the two-particle continuum and correspond to the excitation of the one-particle states. These absorption bands mark the position of the narrow M energy band and the low-frequency edge of the K_1 energy band of the one-particle vibronic states. They are shown to the left of the spectra of two-particle states in Fig.6.5a. The K_1 multiplet is incomplete, i.e., it only includes the A_1 band. Thus, the vibronic absorption of the naphthalene crystal is of a mixed nature. The one- and two-particle absorptions contribute equally to the complete spectrum because $|\Delta_\nu| < \mathcal{M}$.

b) Stressed Naphthalene Crystal

An extremely thin naphthalene crystal in optical contact with a quartz substrate is in a state of uniform elastic stretching at low temperatures. As a result, the distance between molecules increases and the transfer integrals become smaller. This leads to a narrowing of the exciton band ($2\mathcal{M} \approx 100$ cm^{-1}). Δ_ν and γ^2 remain practically unchanged and the system enters the molecular vibron regime $|\Delta_\nu|/\mathcal{M} \gtrsim 1$ (Sect.6.2.3e).

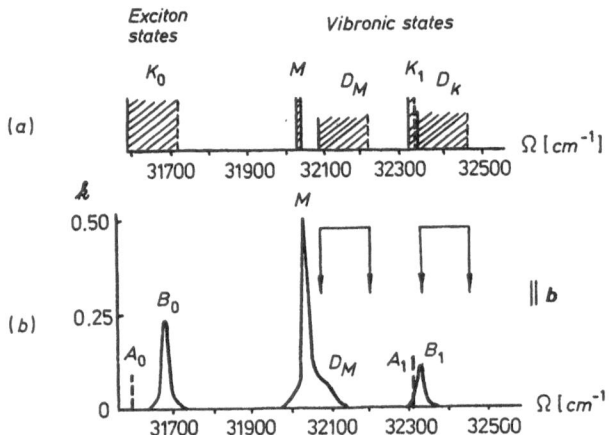

Fig.6.6a,b. Deformed naphthalene crystal: a) scheme of the energy spectrum; b) absorption spectrum in a light polarized parallel to **b**, crystal thickness d = 0.1 μm [6.63]. The dashed lines in b) denote the positions of the A_0 and A_1 bands

On the whole, the absorption spectrum of a stretched crystal (Fig.6.6b) is appreciably narrower. The energy spectrum is shown in Fig.6.6a. The greater part of the absorption for both vibronic transitions is outside of the region of two-particle states (restricted by the arrows). The intensity of the D_M band adjacent to the M band is considerably smaller. This can naturally be attributed to a reduction in the fraction of two-particle absorption due to an increase in the ratio $|\Delta_\nu|/\mathcal{M}$. The greatest changes occur in the region of the second transition: a one-particle B_1 band can be observed. As a result, the K_1 multiplet becomes complete. The two-particle absorption following the B_1 band is relatively weak. The positions of the one-particle energy bands are shown to the left of the two-particle continua in Fig.6.6a. Thus, the absorption spectrum of a stretched crystal is predominantly of a one-particle type.

c) Two-Photon Absorption Spectrum

The two-photon vibronic absorption spectrum of the naphthalene crystal is a u-spectrum. Calculations have shown that the frequency shifts for u-vibrations are often as large as $100 - 200$ cm^{-1}. They greatly exceed the usual values of $|\Delta_\nu|$ for g-type vibrations. The value of the parameter $|\Delta_\nu|/\mathcal{M}$ determines the shape of the spectrum in this case. For most transitions $|\Delta_\nu|/\mathcal{M} > 1$, so that two-photon vibronic absorption must mainly produce one-particle states. The one- and two-photon absorption spectra are compared in Fig.6.7. The asterisks mark the two-particle absorption bands. There are several such regions in the g-spectrum. The u-spectrum is dominated by narrow bands which have practically no high-frequency side bands. The only two-particle absorption region corresponds to the excitation of the vibration ν_{38} with $\nu = 1128$ and $\nu^* = 1107$ cm^{-1} ($\Delta_\nu = - 14$ cm^{-1}). Since $|\Delta_\nu| < |\Delta_{cr}|$, no one-particle states arise.

6.3.4 Anthracene

The vibronic absorption spectrum of the anthracene crystal, as well as the absorption spectrum of the molecule, can be constructed using only TS vibrations (Fig.6.8b). The bands are wide and the spectrum is relatively simple. Of the dozen vibrations visible in the fluorescence spectrum, only the most intensive vibronic transitions involving the vibrations $\nu_{12} = 400$ and $\nu_6 = 1400$ cm^{-1} have been reliably identified in the absorption spectrum. The upper part of Fig.6.8 exhibits the fluorescence spectrum of the crystal. In comparing these two spectra, which do not exhibit the mirror symmetry inherent in the spectrum of the free molecule, one must explain: i) why the absorption bands belonging to the exciton doublet are wide, and ii) why the vibronic absorption bands are wide and the absorption spectrum does not exhibit numerous fine details that are seen in the fluorescence spectrum.

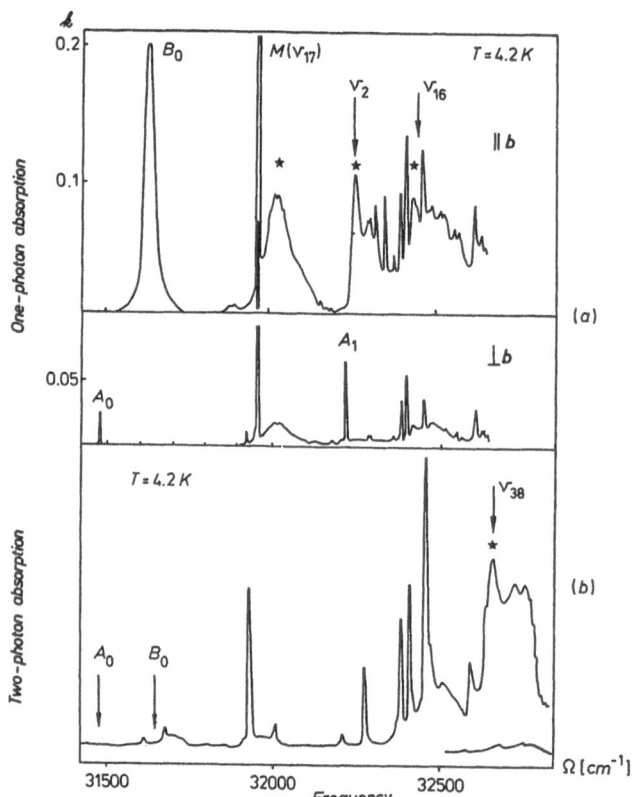

Fig.6.7a,b. Absorption spectrum of the naphthalene crystal: a) one-photon absorption [6.64]; b) two-photon absorption (non-polarized light) as measured by the fluorescence-excitation spectrum [6.65]

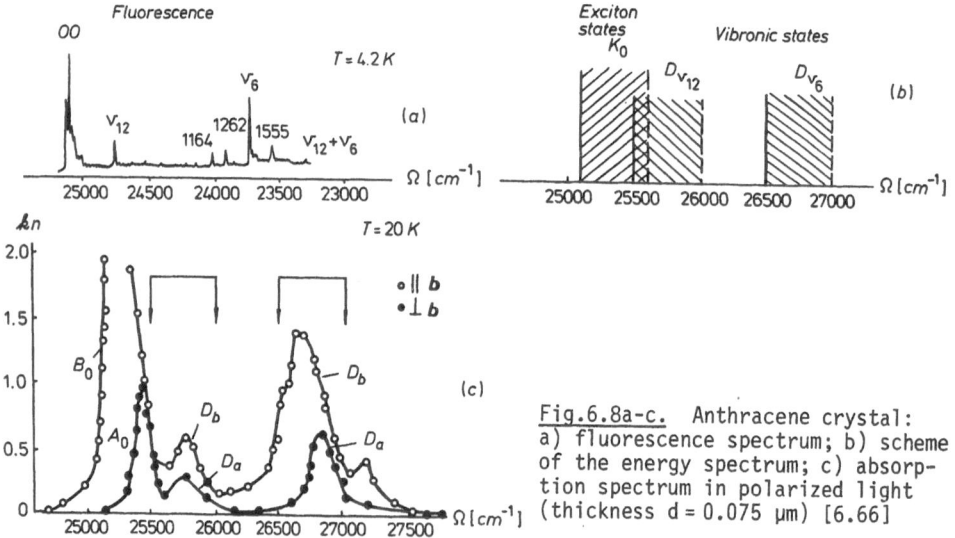

Fig.6.8a-c. Anthracene crystal: a) fluorescence spectrum; b) scheme of the energy spectrum; c) absorption spectrum in polarized light (thickness d = 0.075 μm) [6.66]

The narrow bands in the fluorescence spectrum suggest that the interaction between excitons and external phonons must be weak and that it can be neglected in analyzing the structure of the energy spectrum of vibronic states and the shape of the two-particle absorption bands. At present it is not clear why the exciton absorption bands are so wide. It is possible that this is typical of an energy spectrum with a large contribution from long-range interactions, or it may be that the bottom of the lower band is located at $\mathbf{k} \neq 0$. It is clear, however, that the width of these absorption bands is not a direct measure of the strength of the coupling of excitons to external phonons.

The energy spectrum of excitons in the anthracene crystal is much less understood than the corresponding spectra for benzene and naphthalene. However, a qualitative analysis of the vibronic absorption spectrum only requires a knowledge of the position of the bottom of the exciton spectrum and the extent of the spectrum. The position of the bottom is known. It is determined by the headline of the low-temperature fluorescence spectrum of the crystal (E_{min} = 25098 cm^{-1}) [6.66]. A reasonable estimate of the width $2\mathcal{M}$ of the band is 500 cm^{-1}) (Sects.4.5 and 5.2). Both the energy spectrum of the anthracene crystal in the exciton region and the region of the two vibronic transitions (involving the phonons ν_{12} = 395 cm^{-1} and ν_6 = 1403 cm^{-1}) shown in Fig.6.8b are based on this estimate. The dispersion in the phonon bands was neglected.

For both transitions the vibronic absorption takes place within the energy ranges of two-particle absorption (restricted by vertical arrows in Fig.6.8c). Thus, there are no one-particle vibronic states in the energy spectrum of the anthracene crystal. The absorption spectrum consists of wide D bands of two-particle absorption ($D_{\nu_{12}}$ and D_{ν_6}). The distances between the centres of gravity of the components of these bands (for different polarizations D_a and D_b) agree with the values of the Davydov splittings obtained using (6.33), the value of the exciton splitting $\Delta_{Dav} \approx 240$ cm^{-1} and the coupling constants $\gamma_{12}^2 \approx 0.15$ and $\gamma_6^2 \approx 0.9$ found from the integral intensities of the bands (6.13, 32). The anthracene crystal illustrates how difficult it sometimes is to establish the nature of a spectrum from its external appearance. Thus, the doublet of the D_{ν_6} bands is externally so similar to the exciton doublet of the $B_0 - A_0$ bands that its interpretation as a K_1 doublet was not questioned for a long time.

In the first excited state of the anthracene molecule, $\Delta_\nu \approx 0$ (to within 3 - 5 cm^{-1}) for all vibrations. The vibronic interaction is linear. For most of the TS vibrations of the anthracene molecule $\gamma^2 \ll 1$. Hence, for them the criterion of weak vibronic coupling holds (Sect.6.2.3c) and the vibronic states are dissociated. Only for the vibronic transition involving the phonon ν_6 is

$\gamma^2 \approx 0.9$. The result is the appearance of quasi-one-particle (quasi-stationary) states (Sect.6.2.3f). The relative "simplicity" of the absorption spectrum compared to the fluorescence spectrum is due to the fact that it is made up of the broad superimposed absorption bands. These bands are a result of the excitation of two-particle states involving the vibrations at 627, 784, 1006, 1164, 1262 cm^{-1} and others. Because the frequencies of these vibrations are close and the width of the exciton band is large, the bands overlap to form a continuous background. Only the bands of the strongest vibronic transitions (with ν_{12} and ν_6) stand out against this background.

PHILPOTT and TOURLET [6.67] obtained much better spectra of anthracene. Several edge anomalies are seen in them. These anomalies were assigned to the edges of two-particle spectra, in which various internal phonons are involved.

The vibronic spectrum corresponding to the phonon ν_{12} was investigated the most carefully by TOKURA et al. [6.68]. They measured the thermoreflectance spectra from which they reconstructed the absorption spectrum shown in Fig.6.9. Since the dynamical-theory criterion $\mathscr{M} \ll \nu$ is strongly violated for this phonon, the absorption spectrum was calculated using the dynamical CPA [6.69]. A few trial functions $\rho(\varepsilon)$ were used in the calculations and a bandwidth of magni-

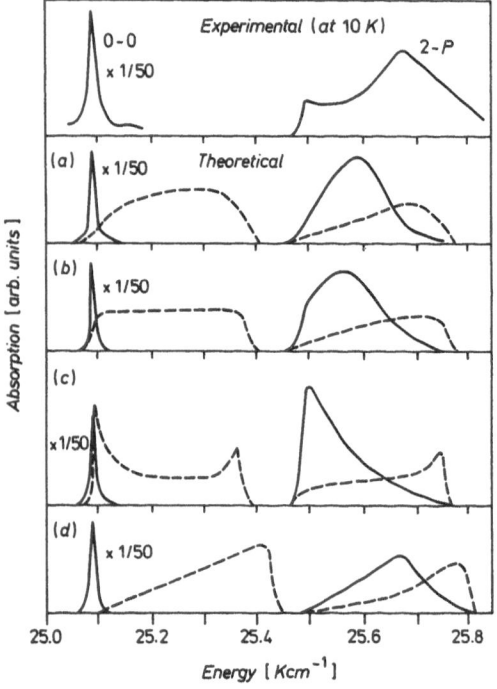

Fig.6.9. Exciton and vibronic absorption spectra of anthracene, b-component [6.68]. Solid curves show absorption spectra and dashed curves show density of states. The top spectrum is the experimental result. The theoretical results (a-d) are obtained assuming the densities of states are as shown

tude $2\mathcal{M}\approx300$ cm^{-1} was used [6.70]. The vibronic absorption is entirely
two-particle and authors stress the existence of the two characteristic features in the broad absorption band: a step-like rise of the vibronic absorption near its low-frequency edge and a broad absorption peak. Despite the fact
that neither of the trial functions $\rho(\varepsilon)$ gave a satisfactory result, the authors are in favour of the curves (c) and (d) in Fig.6.9. They suppose that
the real function $\rho(\varepsilon)$ possesses some specific features of both of the trial
functions: an abrupt rise at the left edge and the maximum shifted to the
right edge. It should be noted that the curve shown in Fig.5.11 has a roughly
similar shape. However, its width is about twice as large as that in Fig.6.9.

It is usually impossible to analyze the vibronic absorption bands deep into
the vibronic spectrum of a crystal because of the strong absorption. "Relaxation spectra" or secondary-emission spectra (Raman scattering, hot fluorescence)
may be of considerable help. The secondary-emission spectrum of the anthracene
molecule [6.71] in a matrix of n-hexane or neon at 4.2 K includes both the Raman
scattering and hot fluorescence. The latter is seen in the vibronic excitation
region ≈2000 cm^{-1} wide. This observation enables one to estimate the relaxation times of the vibronic states in the isolated molecule (1 - 10 ps). In the
secondary emission spectrum of a crystal, no hot fluorescence is observed in
a region 3000 cm^{-1} wide with an identical time resolution [6.72,73]. From this
follow subpicosecond relaxation times from vibronic states of the crystal. A
decrease in the relaxation times in the crystal relative to those for an isolated molecule in a solid matrix by more than one order of magnitude clearly
suggests a change in the nature of the vibronic states themselves. This change
is understandable if one-particle states with relaxation times of the same
order of magnitude as in the molecule are lacking among the vibronic states of
the crystal. On the other hand, the relaxation time from two-particle states
only depends on the time of spatial separation between exciton and phonon and
is approximately equal to $\hbar/\mathcal{M}$, which yields values that lie in the range
10^{-14} - 10^{-13} s.

6.3.5 Alkylbenzenes

These compounds are of interest because it is possible to trace the effect of
polymorphic transformations on the formation of exciton bands, and, hence, on
the structure of the vibronic absorption spectra. The absorption spectra of
the high-temperature (HTM) and low-temperature (LTM) structural modifications
of a crystal of 4-(1'-ethylpropyl)toluene are shown in Fig.6.10. They differ
so much that they can readily be taken for spectra of different compounds.

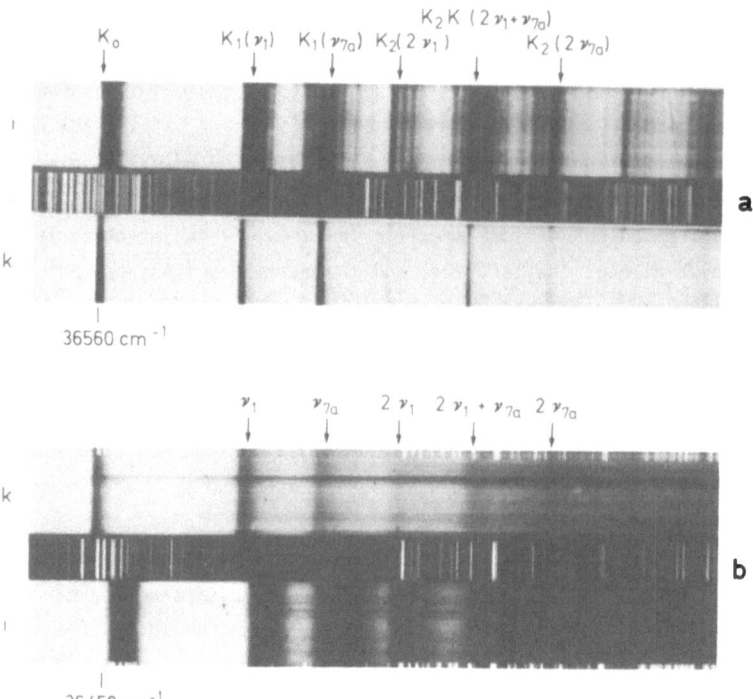

Fig.6.10a,b. Absorption spectra of a 4.1'-ethylpropyltoluene crystal in polarized light, T = 20 K [6.74]: a) high-temperature modification; b) low-temperature modification; frequencies of phonons ν_1 and ν_{7a} are 775 and 1185 cm^{-1}, respectively; ik) unidentified crystallographic face

The energy spectra of the excitons in both modifications have not yet been investigated. On the basis of the qualitative features observed in the spectra of solid aromatic compounds one can speculate on the reasons for the observed rearrangement of the vibronic spectrum at a phase transition.

The intensity of the electronic transition in the molecule is rather high, and, therefore, only TS phonons are visible in the vibronic spectrum. The magnitude of the Davydov splitting in the exciton absorption of the HTM crystal is ≈ 12 cm^{-1}, which is characteristic of a narrow exciton energy band. On the other hand, the vibrational frequency shifts in alkylbenzenes, including those of the TS vibrations, are of the order of tens of wave numbers. It is thus natural to conclude that $|\Delta_\nu| > \mathcal{M}$ in an HTM crystal and that the vibronic spectrum is predominantly of a one-particle nature (close to the molecular limit). This is in agreement with the form of the spectrum, which consists of numerous narrow, low-intensity (hence, $\gamma^2 \ll 1$) bands that are the result of different K transitions. Only two vibrations, ν_1 and ν_{7a} with $\gamma^2 \sim 1$, can be singled out.

The magnitude of the Davydov splitting of the exciton doublet in LTM crystals increases to 165 cm^{-1}, so that the width of the exciton band may be of the order of 200 cm^{-1}. As a result, $|\Delta_\nu| < \mathcal{M}$ and the situation resembles the one in the anthracene crystal. The smoothening of the structure of the vibronic spectrum of the LTM is naturally due to the fact that the entire spectrum of weak transitions involving TS phonons is of a two-particle nature. They set up a continuous background against which only the bands of the intensive vibronic transitions with ν_1 and ν_{7a} stand out. These bands cannot be classified because the positions of the edges of the exciton bands are not known.

Similar analyses of the vibronic spectra of other compounds have been made [6.75].

6.4 A Quantitative Interpretation of the Vibronic Spectra of Crystals Based on the Dynamical Theory

6.4.1 Calculation of the Vibronic Absorption Spectrum

The dynamical theory of the vibronic spectra of molecular crystals presented in Sect.6.2 is based on the assumption that the parameter $\mathcal{M}/\nu$ is small. The inequality $\mathcal{M}/\nu \ll 1$ is valid for the vast majority of vibronic transitions in molecular crystals, and therefore, the dynamical theory can serve as a basis for a quantitative interpretation of the spectra.

In the exciton-phonon Hamiltonian of the dynamical theory (6.10), the exciton-phonon interaction H_ν is given by two separate terms whose values are determined by the intramolecular constants Δ_ν and γ (6.11). If these values are known (for example, from the gas-phase spectra or from the spectra of impurities in a solid matrix), the exciton-phonon interaction Hamiltonian is completely determined by the integrals of the excitation transfer $M_{n\alpha,m\beta}$.

In calculating the vibronic spectrum of a crystal, it should be remembered that the constants Δ_ν and γ^2 may undergo a slight change on passing from the free molecule to the crystal. The molecular values of these quantities are a good zero-order approximation. A good agreement between the experimental and calculated data is obtained by varying the values of these constants just slightly.

There is one more peculiarity of spectra specific to crystals with weak exciton transitions, e.g., for benzene and naphthalene.

The integral intensity of the first exciton transitions in these crystals is determined by a configurational mixing with higher-lying electronic states. The differences in the energies of the electronic states to be mixed are usually somewhat smaller in a crystal than in the free molecule and in some cases (naphthalene) become comparable with the phonon frequency. Under such conditions the different vibronic states belonging to a certain electronic level mix independently with the next higher-lying electronic state. As a result, not only is the Condon approximation violated, but also a much more general adiabatic approximation, strictly speaking, is on the verge of applicability. For vibronic transitions involving NTS phonons this does not cause any special additional complications because the change in intensity can be taken into account by introducing corrections into the matrix elements of the current in (6.28). This circumstance is much more important for vibronic transitions involving TS phonons since the vibronic Hamiltonian (6.1) is written in the Condon approximation, i.e., the dependence of the electronic matrix elements on nuclear displacements is neglected. This is also true for the transfer integrals $M_{n\alpha m\beta}$ and for the matrix elements of the current in (6.28). The equation for the intensity distribution in the K_n-transition series (where n is the number of phonons),

$$I(n) = I(0)\gamma^{2n}/n! \quad , \tag{6.34}$$

can only be obtained if the matrix elements of the current are constant. The universal constant γ^2 does not appear in the Hamiltonian of (6.1) if a Condon intensity distribution is not observed.

However, the problem is simplified and soluble when the ratio of the oscillator strengths of the mixing transitions is small even after the mixing is taken into account, i.e., $f_1/f \ll 1$. As a result, one can assume the interlevel mixing to be relatively small (the admixture of the function Ψ to Ψ_1 is not greater than $\sqrt{f_1/f}$). If the contribution of dipole-dipole interactions to the formation of the exciton band is moderate (this is usually the case for crystals with weak electronic transitions) and f_1/f is small, the dynamics of the quasi-particles will be described by the Hamiltonian given in (6.11) as before. This Hamiltonian will also correctly describe the intensity distribution within each vibronic transition, including the one- and two-particle spectra. However, the distribution of the integral intensities in the K_n series will no longer be described by (6.34). Due to configurational mixing, it must depend on the constants of the exciton-phonon interaction for both exciton states.

Hence, the energy and absorption spectra are determined by the magnitude of Δ_ν and γ^2, by the transfer integrals $M_{n\alpha m\beta}$, and by the density of states $\rho(\varepsilon)$ determined by them. The results of a calculation [6.76] using the improved plane-network approximations (Sect.3.5.2) with $M_{12} = 22.5$ cm^{-1} and the $\rho(\varepsilon)$ of Fig.3.5 shall be given and later on applied to an analysis of the naphthalene spectrum. They make it possible to illustrate the gross features of the spectrum.

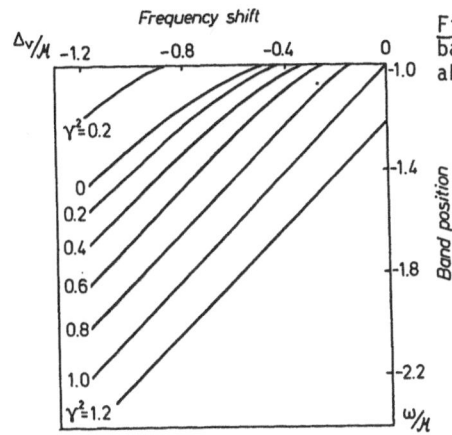

Frequency shift

$\Delta_\nu/\varkappa$ -1.2   -0.8   -0.4   0

-1.0

$\gamma^2 \approx 0.2$

0

0.2

0.4

0.6

0.8

1.0

$\gamma^2 \approx 1.2$

-1.4

-1.8

-2.2

Band position

$\omega/\varkappa$

Fig.6.11. Dependence of the δ-shaped-band position of one-particle vibronic absorption on Δ_ν and γ^2 [6.76]

The positions of the δ-shaped, one-particle absorption bands are determined according to (6.14), by the isolated poles of the exciton Green's function $F_{n\alpha m\beta}$. The frequencies of these poles as a function of Δ_ν are shown in Fig.6.11. We shall only consider the negative Δ_ν's since they correspond to the vast majority of situations that are interesting from an experimental point of view. At $\Delta_\nu < 0$, the one-particle bands are mainly located below the low-frequency edge of the two-particle absorption. The parameter of the curves in Fig.6.11 is γ^2. If $\gamma^2 = 0$, which corresponds to the interaction of an exciton with an NTS phonon, the equation for finding the poles ω_M of the function $F_{n\alpha m\beta}$ is identical to the one for an isotopic impurity (Sect.3.2):

$$\Delta_\nu \int \frac{\rho(\omega')}{\omega_M - \omega'} d\omega' = 1 \quad . \tag{6.35}$$

The integration is performed over the entire energy range of the two-particle states. The frequency ω is counted from the vibronic term $E_\rho + \nu$.

If $\gamma^2 \neq 0$, two different frequencies are obtained for the poles of $F_{n\alpha m\beta}$. They give the frequencies of bands polarized perpendicular and parallel to **b**. The curves located above and below the curve $\gamma^2 = 0$ correspond to a parallel and perpendicular polarization of the incident light, respectively. One-particle states only arise if $|\Delta_\nu| > |\Delta_{cr}|$, $|\Delta_{cr}|$ being dependent on γ^2. The absolute values of Δ_{cr} for the component perpendicular to **b** are appreciably smaller than those for the component parallel to **b**. Therefore, the one-particle band only arises in the component perpendicular to **b** of the spectrum for a wide range of Δ_ν's, i.e., the vibronic doublet is incomplete.

The intensity distribution in the two-particle absorption calculated using (6.14) is shown in Fig.6.12 for various values of Δ_ν and γ^2. The narrowing of

217

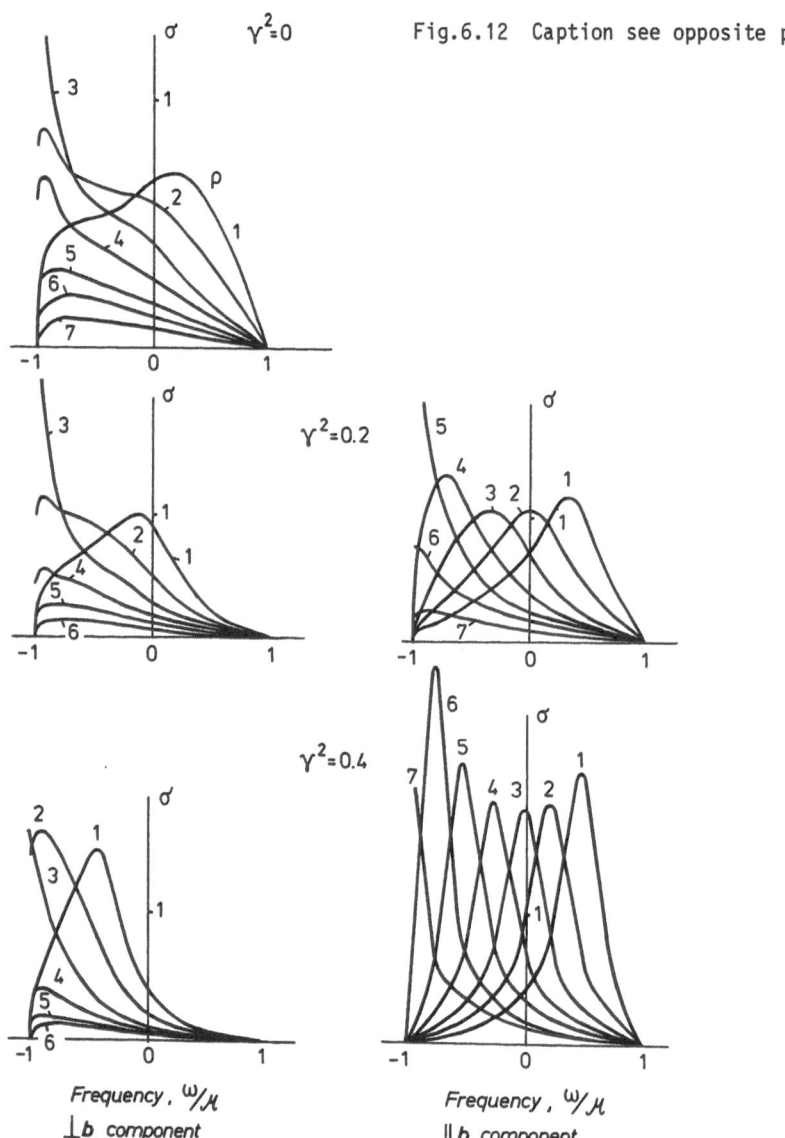

Fig.6.12 Caption see opposite page

Frequency, ω/μ
⊥ b component

Frequency, ω/μ
∥ b component

the absorption bands as γ^2 approaches unity is clearly visible. According to the dynamical theory, there arise quasi-one-particle states at $\gamma^2 = 1$ to which correspond the δ functions in the absorption spectrum (Sect.6.2.3f). Even for values of γ^2 that differ appreciably from unity, the two-particle bands have a clearly defined maximum that can be seen in Fig.6.12. The curves for small values of γ^2 (~0.1) and almost critical values of Δ_ν are exceptions. In these cases the two-particle spectrum consists of either a wide band with a gently

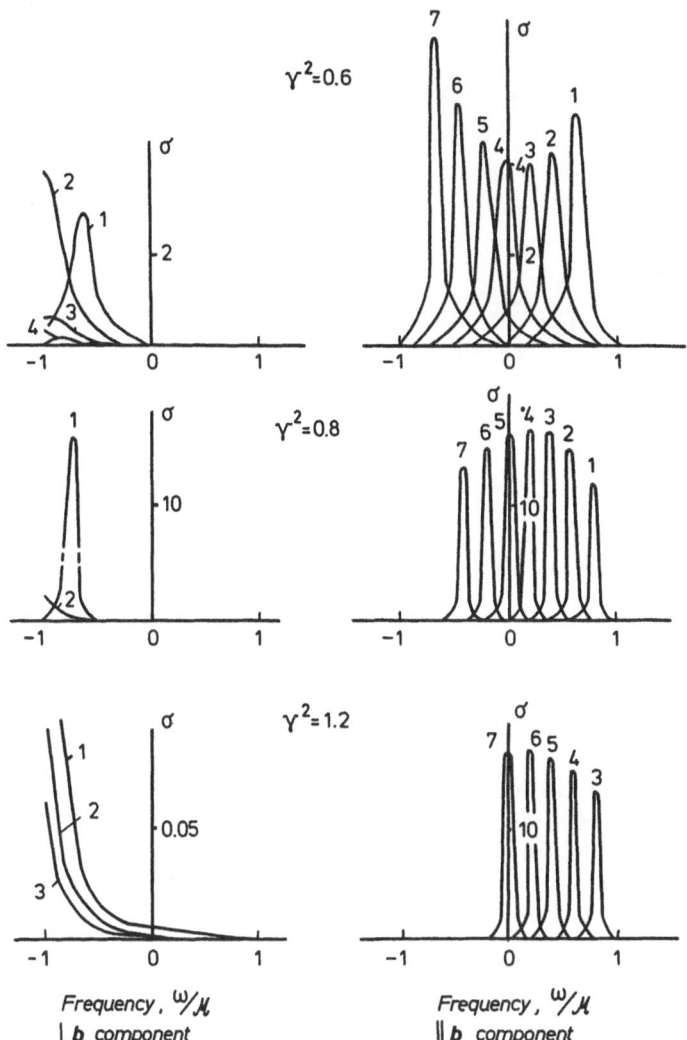

Fig.6.12. Intensity distribution of two-particle absorption in two polarizations [6.76]

sloping maximum or a band of a more intricate shape. At $\gamma^2 = 0$ the shape of the two-particle absorption band is described by (6.30).

The area under the calculated curves in Fig.6.12 (including the one-particle band that is not shown in this figure) is normalized to unity. According to (6.33), the first moment of the absorption spectrum determines the position of the centre of gravity relative to the term for the joint vibronic configuration, $E_\rho + \nu^*$. For the Davydov splitting of the K_1 transition $\Delta_{Dav}^1 = \gamma^2 \Delta_{Dav}$, where Δ_{Dav} is the Davydov splitting in the exciton spectrum.

6.4.2 Vibronic Spectra of Benzene and Naphthalene Crystals Involving NTS Phonons

We begin a consideration of the experimental vibronic spectra with the simplest case involving a single NTS phonon. In this case the vibronic-state problem is reduced to one with a local potential, i.e., the vibrating molecule can be represented by a potential well for the exciton that is Δ_ν deep. This is what creates a unified language for describing such vibronic states and levels of the local exciton in IDC (Chap.3).

The form of the vibronic spectrum is determined by the value of Δ_ν. The change in the spectral distribution and the total intensity of the two-particle absorption as a function of Δ_ν is illustrated in Fig.6.12 (fragment $\gamma^2 = 0$). The polarization ratio is independent of the frequency and can be determined using the oriented-gas model.

From the standpoint of dynamical theory a quantitative analysis of vibronic absorption involving an NTS phonon is simple if it is correlated with an analysis of the local-exciton spectra in IDC. However, the picture of vibronic absorption may be greatly complicated by the contribution made by the external phonons. The determination of this contribution becomes the most difficult task in passing from a qualitative to a quantitative analysis of the spectrum.

a) Benzene Crystal

The absorption spectrum of a single crystal of benzene in the region of the M transition (involving the NTS phonon ν_{18}) is shown in Fig.6.13. The region

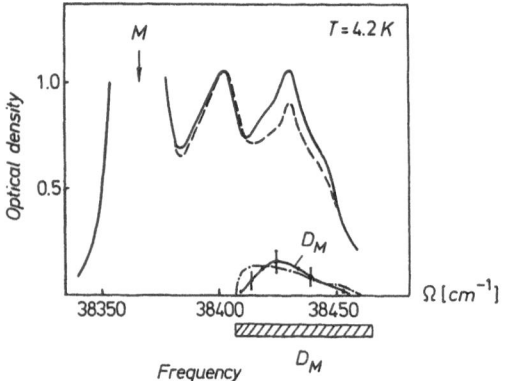

Fig.6.13. Absorption spectra of a benzene crystal (solid curve) and IDC of $\overline{b}d_6$ in $\overline{b}d_0$ (dashed curve) in the range of the vibronic M transition [6.77]. Shown at the bottom are the experimental (solid curve) and calculated (dot-and-dash curve) D_M bands. Vertical segments are experimental errors

of two-particle states is indicated under the frequency scale. The distance from the maximum of the M band to the edge of the D_M spectrum is $e_M = -49$ cm^{-1}. Since $\Delta_\nu = -86$ cm^{-1} and $2\mathcal{M} \approx 60$ cm^{-1}, the ratio $\mathcal{M}/|\Delta_\nu|$ is $\approx 1/3$. Therefore, the value of a^2 in (6.28) must be close to unity and the one-particle M band must dominate. However, the integral intensity of the high-frequency side band that follows the one-particle M band is ≈ 0.4 of the integral transition intensity. Such a strong band about a half of which is outside of the range of the two-particle spectrum is indicative of a contribution from the external phonons. One of the ways of singling out this absorption is shown in Fig.6.13. The dashed curve represents the same side band for a local vibron (Sect.6.5) bound to the isotopic impurity bd_6 in a crystal of bd_0. The two-particle vibronic absorption for this deep impurity centre ($\Delta_{ex} = 200$ cm^{-1}) must be negligibly small. Hence, the excessive absorption in the region of the second peak can naturally be interpreted to be of a two-particle nature. The shape of this absorption is shown at the bottom of Fig.6.13. Its integral intensity is 5% of the intensity of the M band. The value of a^2, estimated as $I_M/(I_D + I_M)$, is 0.95.

According to (6.35) and (3.18) $e_M = -49$ cm^{-1} and $a^2 = 0.98$. The spectral distribution $\sigma(\omega)$ in the D_M band according to (6.30) is denoted by the dashed curve at the bottom of Fig.6.13. Considering that the experimental curve was obtained by subtracting two similar spectra, the agreement between the calculated and experimental curve may be considered to be satisfactory.

b) Naphthalene Crystal

The absorption spectrum of the naphthalene crystal in the region of the M transition is shown in Fig.6.14. The D_M region of the two-particle states spectrum is shown under the frequency scale. The M band is located 25 cm^{-1} from its edge. The relative intensity of the side band is 0.48. It has not been possible to determine the contribution of the external phonons to this side band from the absorption spectra of the isotopic impurity. An estimation of the above contribution from the mirror-symmetrical transitions in the fluorescence spectrum is a rather crude one because the Condon approximation is not valid for the contribution from external phonons into the weak transitions spectra [6.78].

The values of ω_M, a^2, and $\sigma(\omega)$ were determined from the exciton density of states given in Fig.3.5. The values of e_M and a^2 were found to be -25 cm^{-1} and 0.75 cm^{-1}, respectively. The spectral distribution of $\sigma(\omega)$ is represented by the dashed curve in Fig.6.14. The difference between the experimental and the theoretical curve describes the approximate shape of the external phonon

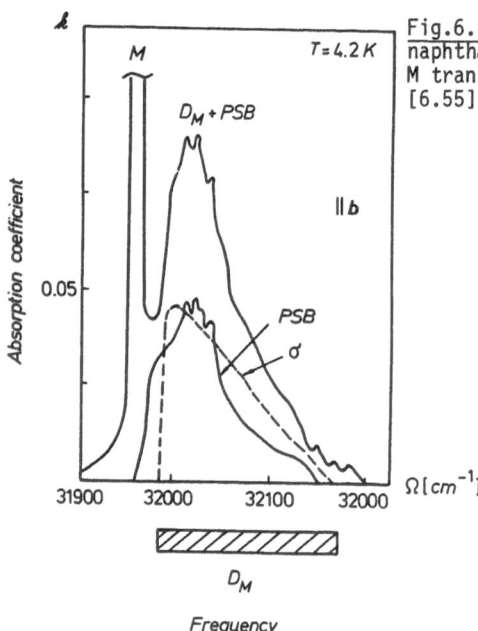

Fig.6.14. Absorption spectrum of the naphthalene crystal in the range of the M transition [6.64]: σ) calculated D_M band [6.55]; PSB) phonon side band of the M band

contribution into the side band of the M band. Its maximum is shifted by ≈ 50 cm^{-1} relative to the M band and can be correlated with the known maximum in the density of states of the external phonons [6.79] determined from the spectrum of inelastic neutron scattering.

Of all the vibronic transitions involving an NTS phonon investigated so far at the quantitative level, only for the transition under discussion is the intensity of the D_M band large (about 25 per cent of the total intensity of the M transition). In this context it is an appealing prospect to use this transition to directly reconstruct the exciton density of states in naphthalene using (6.31). However, the uncertainty associated with the determination of the contribution from the external phonons prevented such a reconstruction [6.55]. Later, this was successfully carried out for the two-phonon spectra of a number of inorganic crystals. Available data [6.80-84] indicate that combinations of internal phonons generally show broad features as a result of multiparticle transitions. Sometimes the broad band is also accompanied by sharp lines due to one-particle transitions. A combination of two internal modes yields density-of-states functions for one component provided that the other component has a very small dispersion. The procedure is based on (6.31). The density-of-states functions were obtained in this way for crystals of $NaNO_3$ [6.81] and UF_6 [6.83] and have recently been discussed for HCl [6.85].

6.4.3 Vibronic Absorption in Naphthalene Involving a Single TS Phonon

The absorption spectrum of an nd_0 crystal in the region of the K_1 transition is shown in Fig.6.15. The component perpendicular to **b** is dominated by a narrow A_1 band -11 cm^{-1} from the edge of the two-particle spectrum. In the component parallel to **b**, the absorption is represented by a broad D_b band. Superimposed on it are the bands I-IV which correspond to one-particle vibronic transitions involving other vibrations.

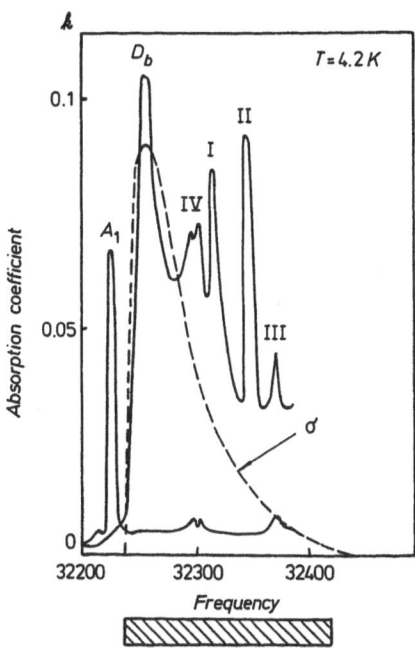

Fig.6.15. Absorption spectrum of the naphthalene crystal (nd_0) in the range of the K_1 transition [6.76]: solid curve) experiment; dashed curve) calculation. See text for explanation of I - IV

The comparison between theory and experiment is based on the data of Figs.6.11,12. The parameters Δ_ν and γ^2 were chosen so as to ensure that the positions of the A_1 band and the maximum of the D_b band are correct. The only pair of values satisfying these conditions is $\Delta_\nu = -57$ cm^{-1} and $\gamma^2 = 0.20$. In a free molecule, $\Delta_\nu = -56$ cm^{-1} and $\gamma^2 \approx 0.2$. With these parameters a one-particle absorption band only exists in the component perpendicular to **b** (see Fig.6.11). This is in full accordance with experiment. The calculated value of the fraction due to two-particle absorption is $\approx 40\%$ in this component. However, because of the low-absolute intensity, the broad-band two-particle absorption is at the level of the background. The calculated two-particle absorption spectrum $\sigma(\omega)$ in the component parallel to **b** is shown in Fig.6.15 as a dashed curve. Due to the strong configurational mixing it has been impossible to com-

223

pare the absolute intensities of the calculated and the experimental band. The scale for $\sigma(\omega)$ along the ordinate was chosen so that the difference between these bands is a minimum. The overall shapes of the curves coincide closely. The data for the K_1 transition in nd_8 are similar. They are shown in Fig.6.16. The optimum values of the parameters are $\Delta_\nu = -52$ cm^{-1} (free molecule -54 cm^{-1}) and $\gamma^2 = 0.16$. Since one-particle absorption is absent in the component parallel to **b** and the high-frequency side band of the A_1 band in the component perpendicular to **b** is so weak that it cannot be a real object of discussion, the effect of the external phonons turns out to be less important in the K_1 transition than in the M transition.

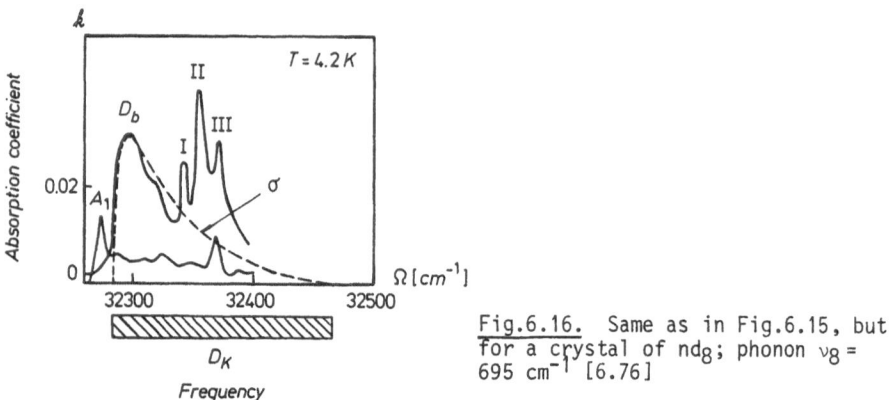

Fig.6.16. Same as in Fig.6.15, but for a crystal of nd_8; phonon $\nu_8 = 695$ cm^{-1} [6.76]

It is worth noting that the above values for γ^2, determined from the shape of the vibronic spectra, are close to the values estimated from the intensity distribution in the gas-phase spectra. They differ considerably, however, from the ratios of the absorption intensities for the K_1 and K_0 transitions in the crystals ($I_{K_1}/I_{K_0} = 0.86$ for nd_0 and 0.33 for nd_8 [6.86]), due to a strong configurational mixing.

6.4.4 Composite Transitions and Overtones in a Vibronic Spectrum

The dynamical theory can be generalized to include the participation of two or more phonons in a transition. If these two phonons are identical, we shall call the corresponding vibronic transitions *overtones*. If they are different, the corresponding transitions are called *composite transitions*.

The resulting three-particle problem is, in principle, much more complicated than the two-particle problem. In particular, in addition to a one-particle state (a vibron that corresponds to the bound state of all three particles) and a three-particle continuum D_3, partly dissociated states similar to those found in a two-particle continuum D_2 are possible. They include a vibron of the exciton + phonon type and a second phonon not bound to it. Nevertheless, due to a summation of the frequency shifts of the two phonons, the absorption spectrum is often simplified since one-particle absorption dominates the spectrum.

a) Composite Vibronic Transitions

The composite transitions that involve a TS and an NTS phonon (whose frequency shifts are equal to Δ_ν^{TS} and Δ_ν^{NTS}, respectively), are of great interest. In this case the quadratic interaction can be written as a sum of two terms, both similar to the last term in (6.11), for each of the phonons. The NTS phonon remains completely immobile.

Let us discuss the above situation with reference to the composite transition $B_{3u} \cdot b_{1g} a_g \leftarrow A_{1g}$ in the naphthalene crystal which involves a TS phonon $\nu_8 = 764$ cm^{-1} and an NTS phonon $\nu_{17} = 509$ cm^{-1}. This situation is best understood by using the scheme of Fig.6.17, which shows the configurations (i.e., the arrangements of the exciton and the phonons on the molecules) dominating the lower-energy states of a composite vibronic transition. It is seen from the energies[8] of each configuration that the ground state is usually represented by the first configuration in which the exciton and both phonons are located at the same site. Since the entire oscillator strength is associated with this configuration, the relevant optical transition will be strong [$(MK)_{m=0}$ band].

There also exists a set of states that usually adopt the second configuration, i.e., a vibron of the type ex + ph$_{NTS}$ to which the phonon ph$_{TS}$ is bound but weakly. The corresponding series of closely spaced bands converges towards the edge of the two-particle continuum. The contribution of the first con-

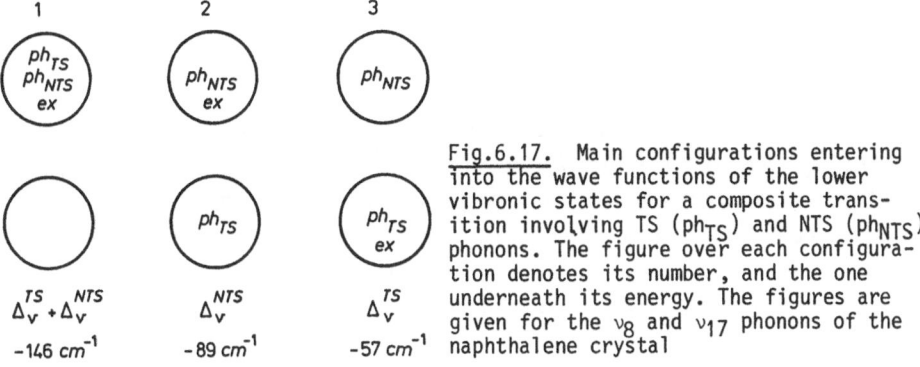

Fig.6.17. Main configurations entering into the wave functions of the lower vibronic states for a composite transition involving TS (ph$_{TS}$) and NTS (ph$_{NTS}$) phonons. The figure over each configuration denotes its number, and the one underneath its energy. The figures are given for the ν_8 and ν_{17} phonons of the naphthalene crystal

---

[8] The energies were calculated neglecting the excitation transfer.

figuration to the wave functions of these states is small. As a result, the intensity of the corresponding optical transitions is low. Calculations[9] have shown that it is equal to about 1% of the intensity of the first transition [6.76]. The results of the calculations are given in Table 6.1.

Table 6.1. Calculated values of the positions and intensities of the absorption bands of the naphthalene crystal corresponding to the quantum levels of MK-vibronic excitation $B_{3u} \cdot b_{1g} a_g$ [6.76]

Band	Energy[a] of dominating configuration [cm^{-1}]	Energy[a] [cm^{-1}] and quantum number m	Intensity[b]
$(MK)_{m=0}$	- 146	- 153.5 (m = 0)	0.91
$(MK)_m$	- 89	- 111.6 (m = 1)	0.009
		- 109.7 (m = 2)	0.0007
		- 109.5 (m = 3)	0.0004

[a] The energy is calculated from the term of the separated vibronic configuration ($E_\rho + \nu_8 + \nu_{17} = 31557 + 764 + 509 = 32830$ cm^{-1}).

[b] The sum of the intensities of all vibronic bands is taken as unity.

The spectra of crystals of nd_0 and nd_8 exhibit strong bands whose positions agree fairly well with the calculated positions of the $(MK)_{m=0}$ bands. A weak satellite whose position agrees with the calculated position of the dense series of $(MK)_m$ bands with m > 0 was observed in nd_0 [6.87]. The data are compared in Table 6.2.

b) Overtones of the Vibronic Spectrum

An excitation of the overtones of NTS phonons is not observed in aromatic molecules of the naphthalene type. Therefore, the overtone problem is of practical interest only in the K_2 series of TS phonons. The qualitative features of the vibronic absorption of naphthalene can be traced to the overtone 2ν for a TS phonon $\nu_8 = 764$ cm^{-1} (Table 6.2).

The main parameter of the overtone is the shift in energy for the joint configuration 1 of Fig.6.17. It is equal to $2\Delta_\nu = -114$ cm^{-1} for nd_0. As in the case of the composite transition, one-particle A_2 and B_2 bands can be observed because $2|\Delta_\nu| > \mathcal{M}$. They will dominate the overtone spectrum in dis-

---

[9] The calculations are similar to those made for local vibrons (Sect.6.5.3) and do not contain any new parameters other than those determined from the one-phonon spectra.

Table 6.2. Positions of the bands of composite vibronic transitions and overtones in the naphthalene crystal [6.87]

Band notation	Crystal	Phonons[a]	Term of joint configuration[b] $[\text{cm}^{-1}]$	Band position $[\text{cm}^{-1}]$ calcul.	experim.	Davydov splitting $[\text{cm}^{-1}]$ calcul.
a) Exciton + NTS phonon + TS phonon						
$(MK)_{m=0}$	nd_0	509 + 764				
		(423 + 707)	32687	32680	32674	0
$(MK)_m$	ditto	same		32720	32725	
$(MK)_{m=0}$	nd_8	494 + 695				
		(412 + 643)	32727	32717	32718	0
b) Exciton + 2 TS phonons						
(A_2B_2)	nd_0	2 · 764				
	·	(2 · 707)	32971	–	32956	3
(A_2B_2)	nd_8	2 · 695	32958	–	32954	1.6
		(2 · 643)				

[a] The parenthesis shows the frequencies of the phonons in the excited state, ν^*.

[b] The term of the joint configuration is defined as $\Omega = E_\rho + \nu^*_{TS} + \nu^*_{NTS}$ (or ν^*_{TS}).

tinction to the one-phonon K_1 transition, where the B_1 band is absent and the absorption is entirely of a two-particle nature. Another important characteristic of the K_n series of TS phonons is the rapid decrease in the magnitude of the Davydov splitting. The Davydov splittings in the overtone and exciton spectra are related by (6.34) so that $\Delta_{Dav}^{(2)} = (\gamma^4/2)\Delta_{Dav}$. At $\gamma^2 = 0.2$ and $\Delta_{Dav} = 150\ \text{cm}^{-1}$, the value of $\Delta_{Dav}^{(2)}$ ($\approx 3\ \text{cm}^{-1}$) is practically unmeasurable. It should be emphasized that the Davydov splittings in the K_n series of the naphthalene crystal rapidly decrease with increasing n due to the small value of γ^2. On the other hand, the intensities decrease slowly, or even increase, because the configurational mixing increases as the second electronic transition is approached.

6.5 Vibronic Spectra of Imperfect Crystals

The spectra of benzene and naphthalene doped with isotopic molecules, the spectra of mixed crystals of nd_0 and nd_8, spectra of amorphous films, and spectra of defect centres shall be discussed in this section. An analysis of the experimental data is preceded by a description of the formation of the vibronic spectra of impurity centres and of the calculation methods in terms of the dynamical theory. This section will show that the dynamical theory has much in common with the theory of electronic spectra of imperfect crystals (Chap.3). This enables one to discuss a wide range of problems in identical terms.

6.5.1 Terms Used in the Calculation of the Spectra of Local Vibrons

A local vibron is a vibronic excitation (exciton and phonon) that is bound to an impurity centre or a defect in a crystal. In some quantum states one or both of these particles are in the continuum spectrum. The relevant absorption is similar to induced absorption in the local-exciton theory (Sect.3.2.9). From now on the term *local vibron* shall be used in this extended sense.

To find the vibronic spectra of crystals doped with isotopic impurities[10], the Hamiltonian of (6.10,11) should be supplemented by the interaction Hamiltonian of the vibronic excitation and an impurity centre. It would be natural to adopt the isotopic substitution model used in Chaps.3 and 4. In this section, the isotopic shift of the electronic level will be denoted by Δ_{ex}[11], the isotopic shift of the vibrational frequency by $\Delta_{ph} = \nu' - \nu$ (ν' is the vibrational frequency of an unexcited guest molecule), and the isotopic change in the shift $\Delta_\nu = \nu^* - \nu$ of the vibrational frequency by Δ_{ex-ph}. Using these terms, the total isotopic shift of the vibronic level Δ_ν and the change in the vibrational frequency of the guest molecule at its electronic excitation Δ'_ν are equal to

$$\Delta_\nu = \Delta_{ex} + \Delta_{ph} + \Delta_{ex-ph} \quad , \quad \Delta'_\nu = (\nu')^* - \nu' = \Delta_\nu + \Delta_{ex-ph} \quad . \tag{6.36}$$

The contribution to the Hamiltonian due to the presence of a guest molecule at the site $\mathbf{n}_0 \alpha_0$ is equal to

$$H^\Delta_{imp} = \Delta_{ex} a^+_{\mathbf{n}_0\alpha_0} a_{\mathbf{n}_0\alpha_0} + \Delta_{ph} b^+_{\mathbf{n}_0\alpha_0} b_{\mathbf{n}_0\alpha_0} + \Delta_{ex-ph} a^+_{\mathbf{n}_0\alpha_0} a_{\mathbf{n}_0\alpha_0} b^+_{\mathbf{n}_0\alpha_0} b_{\mathbf{n}_0\alpha_0} \quad . \tag{6.37}$$

---

[10] A more complicated case is considered in Sect.6.5.4d.

[11] It is equivalent to the quantity used in Chaps.3 and 4. The subscript is used to distinguish the shift of the electronic level from other similar quantities (Sect.1.4).

Furthermore, H_{imp} must contain a term that takes into account the isotopic change in the linear coupling constant γ. If the value of this constant in the guest molecule is denoted by γ', the corresponding term in H_{imp} will be

$$H^{\gamma}_{imp} = \gamma(\gamma' - \gamma) \sum_{n\alpha \neq n_0\alpha_0} M_{n_0\alpha_0 n\alpha} a^+_{n_0\alpha_0} a_{n\alpha} (b^+_{n_0\alpha_0} b_{n\alpha} + b^+_{n\alpha} b_{n_0\alpha_0}$$

$$- b^+_{n_0\alpha_0} b_{n_0\alpha_0} - b^+_{n\alpha} b_{n\alpha}) + (n_0\alpha_0 \rightleftarrows n\alpha) \quad . \tag{6.38}$$

The second summand in (6.38) is obtained from the first by permuting the subscripts $n\alpha$ and $n_0\alpha_0$. The origin of the coefficient in H^{γ}_{imp} can readily be understood if one takes into account that the probability of excitation transfer between a guest and a host molecule is proportional to $\gamma\gamma'$. The total impurity Hamiltonian is then equal to

$$H^{V}_{imp} = H^{\Delta}_{imp} + H^{\gamma}_{imp} \quad . \tag{6.39}$$

With this interaction Hamiltonian the energies are counted from the vibronic term of the separated configuration for the host crystal $E_\rho + \nu$.

6.5.2 Spectrum of a Local Vibron with an NTS Phonon

The picture is simplest of all with NTS phonons. Since an NTS phonon is immobile, the energy spectrum of a local vibron is determined by the position of the sites $R_0 = n_0\alpha_0$ and $R_1 = n_1\alpha_1$ (at which a guest molecule and a phonon are located, respectively). Thus, R_1 plays the role of the quantum number of the local vibron.

Two distinct situations are possible. One of them is that $R_1 = R_0$, i.e., the phonon is located at the impurity site. In this case, the problem is equivalent to the one of finding the energy spectrum of a local exciton in the monomer centre with an effective "isotopic" shift equal to

$$\Delta' = \Delta_{ex} + \Delta_\nu + \Delta_{ex-ph} = \Delta_{ex} + \Delta'_\nu \quad . \tag{6.40}$$

The modulus of Δ' must exceed the critical value in order for a discrete level to exist. Therefore, it is clear that the relationship between the signs of Δ_{ex} and Δ'_ν can play a decisive role. If such a level does exist, the corresponding absorption band of the local vibron will be denoted by M'. The frequency of the transition is equal to $\omega_{M'} = \varepsilon'_i + \Delta_{ph}$, where ε'_i is the position of the local exciton level having undergone an isotopic shift Δ' (3.8). The intensity of the transition is reduced compared to that of the vibronic transition in the isolated molecule by the factor a^2, which is equal to the square of the excitation amplitude at the impurity site. In addition to the narrow

M' band, a broad absorption band corresponding to the transitions to states
in which the exciton is found in the continuum can be observed (while the
phonon is still at the impurity site). This continuum and the corresponding
absorption band will be denoted by $D_{M'}$. Its intensity is determined by the
factor $(1 - a^2)$. The M' level is split off from the $D_{M'}$ continuum by the
perturbation Δ'.

The second situation arises when $\mathbf{R}_1 \neq \mathbf{R}_0$. In this case, the exciton experi-
ences a perturbing potential Δ_{ex} and Δ_v at the sites $\mathbf{R}_0$ and $\mathbf{R}_1$, respectively.
This situation is equivalent to the one for an asymmetric-pair centre and is
similar to the problems considered in Sect.3.4. For a constant value of $\mathbf{R}_1$
the number of discrete levels does not exceed 2. They are determined from
the equation

$$(1 - \Delta_{ex}G_0(\varepsilon))(1 - \Delta_v G_0(\varepsilon)) = \Delta_{ex}\Delta_v |G_{\mathbf{R}}(\varepsilon)|^2 \quad ,$$

$$G_{\mathbf{R}}(\varepsilon) \equiv G_{\mathbf{n}\alpha\mathbf{n}_0\alpha_0}(\varepsilon) \quad . \tag{6.41}$$

For large values of $\mathbf{R} = \mathbf{R}_1 - \mathbf{R}_0$ the right-hand side of (6.41) is small and the
levels corresponding to different configurations are crowding and converge to
their limit position at $\mathbf{R} = \infty$. Only the bands corresponding to small values of
R, denoted by M", M''', ..., can be resolved. Sometimes the entire set of bands
shall be called M" bands. Their frequencies are equal to $\omega = \varepsilon_{\mathbf{R}}$, where $\varepsilon_{\mathbf{R}}$ are
the roots of (6.41) for different values of $\mathbf{R}$.

The detailed arrangement of the levels depends on the values of the parameters. However, the general trends can readily be understood by analogy with
the levels of the aggregate centres (Sect.3.5). Specific situations will be
discussed in the following sections.

Let us consider, by way of example, the particular case presented in
Fig.6.18. It corresponds to a situation in which $\Delta_{ex} < \Delta_v < 0$ and both Δ_{ex} and
Δ_v are large enough so that the discrete level K_0' of the local exciton and

Dissociated states

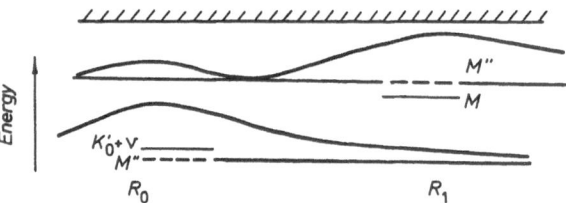

Fig.6.18. Scheme showing the origination of a pair of M" bands and the distribution of the exciton wave function squared in two M" states

230

the M band of one-particle states can be observed. Their energies are indicated by solid lines. As a result of a mixing of the $K_0' + \nu$ levels with the M band the absorption spectrum acquires two M" bands, whose positions are indicated by the dashed lines in Fig.6.18. These bands may be considered to be the "impurity" and "phonon" components of a pair of M" bands. The shifts with respect to $K_0' + \nu$ and M increase as R decreases. Since the system is asymmetric ($\Delta_{ex} \neq \Delta_\nu$), transitions to both levels are allowed. The intensities of the corresponding bands may, however, differ widely. Indeed, each of the intensities is determined by $\psi^2(\mathbf{R}_1)$, i.e., the square of the wave function at the site where the NTS phonon resides. If $|M_{\mathbf{R}_0\mathbf{R}_1}| << |\Delta_{ex} - \Delta_\nu|$, the wave functions centred at the sites $\mathbf{R}_0$ and $\mathbf{R}_1$ hardly overlap at all. The group of M" bands that is developed from the M band must be the more intensive group. It may, however, merge with the broad intrinsic M band to form a high-frequency wing. The weaker M" bands, whose shift relative to the K_0' band of the local exciton is almost equal to ν, may turn out to be more distinct.

When phonons are created near an impurity centre, the absorption distribution throughout the D_M band must change somewhat. This change can be described as the appearance of the $D_{M"}$ band, whose edges coincide with those of the D_M band. However, since the $D_{M'}$ and $D_{M"}$ bands overlap, they are difficult to separate. The total continuous absorption in these bands will be denoted by D_M^*. The outlined appearance of vibronic spectra for some relations between basic parameters is shown in Fig.6.19.

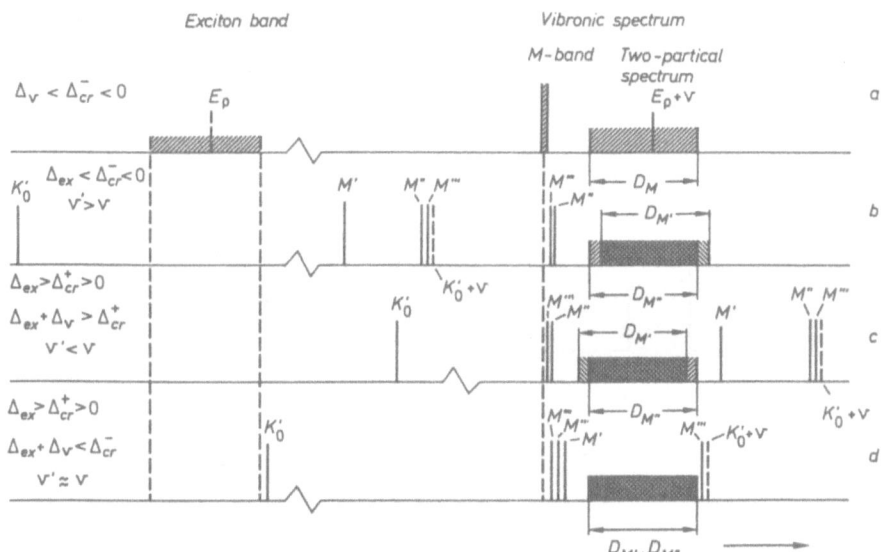

<u>Fig.6.19a-d.</u> Scheme of the vibronic spectrum with an NTS phonon: a) perfect crystal; b-d) additional spectrum of a doped crystal

6.5.3 Spectrum of a Local Vibron with a TS Phonon

The picture is much more complicated for TS phonons, i.e., for local vibronic
absorption in the region of the K_1 bands of the intrinsic absorption spectrum.
Due to the motion of the phonon, which is allowed by the vibronic Hamiltonian
H_v, R_1 is no longer a quantum number and the above classification of the levels
cannot be used. The problem takes into account the motion of two interacting
particles (an exciton and a phonon) in the field of an impurity centre and can
only be solved numerically. To understand the behaviour of the wave functions
of the individual levels, it is convenient to carry out a qualitative analysis
of the configurations contributing to the wave functions of the impurity vibron. These configurations are outlined in Fig.6.20. The calculated energies
of the configurations take into account all of the terms of the Hamiltonian
$H_v + H_{imp}^v$ except the transfer integrals. An essential difference from the spectrum involving an NTS phonon is that the configurations 1 and 3 (with the
phonon at the guest site) are mixed with the configurations 2, 4, and 5 (with
the phonon on the host molecules). The transitions to quantum states where the
joint configurations 1 or 4 dominate must be the strongest. An isolated level
is expected in a spectrum in which the configuration 1 dominates (K_1' by analogy
with M'). Also a converging infinite sequence of levels corresponding to a
progressively increasing average distance between the phonon and the impurity
site (K_1'', K_1''', ..., etc.) should be observed. The limit of the sequence corresponds to the removal of the phonon from the impurity exciton. The K_1',
K_1'', ... states are similar to the $(MK)_{m=0}$ and $(MK)_m$ states of the composite
vibronic transitions discussed in Sect.6.4.4.

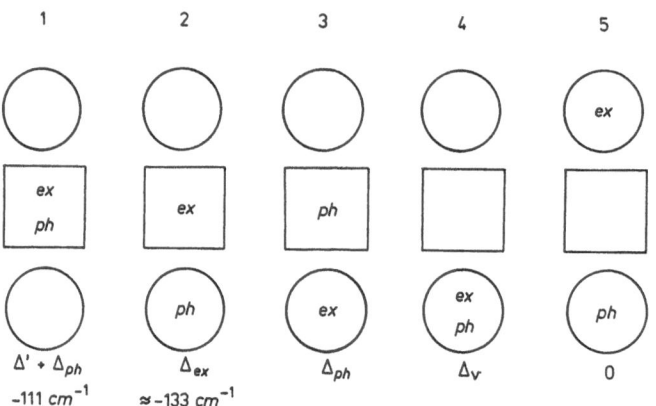

Fig.6.20. Configurations entering into the wave function of a local vibron
for the $\overline{K_1}$ transition in IDC of nd_0 in nd_8; open square) guest molecule;
open circles) host molecules; the figure over each configuration denotes its
number, the one underneath its energy

Of course, along with the $K_1^!$, $K_1^"$, ... states, a $D_{K_1}^*$ continuum and $D_{K_1}^*$ absorption must arise, which include the $D_{K_1}^!$ band (the exciton is removed and the phonon is localized on the guest molecule) and the $D_{K_1}^"$ band (change in the absorption in the region of the D_{K_1} band). The spectrum, on the whole, closely resembles Fig.6.19.

6.5.4 Vibronic Absorption in Doped and Mixed Crystals Involving an NTS Phonon

Vibronic absorption in doped crystals has been investigated in a comparatively small number of compounds; predominantly in isotopically doped crystals of d-benzenes and d-naphthalenes. Practically all of the possible situations have been observed because the value of the parameter $\Delta'/\mathcal{M}$ varies widely in these systems.

a) IDC of d-Benzenes

In the vibronic absorption spectra of the IDC of d-benzenes, the transition involving the NTS phonon ν_{18} has been studied comprehensively. The corresponding transition in a pure crystal was described in the preceding section. According to (6.40), the main parameter of the transition is $\Delta' = \Delta_{ex} + \Delta_\nu'$. The value of Δ_ν' for the phonon ν_{18} varies from -88 to -78 cm^{-1} for different isotopes. The average value of Δ_{ex} is 30 cm^{-1} per replaced atom. The sign of Δ_{ex} is determined by the mass ratio of the guest and host molecules.

If $\Delta_{ex} < 0$ (the impurity is lighter than the solvent), then $|\Delta'| > |\Delta_\nu|$. The vibronic absorption spectrum of a pure crystal approaches the molecular limit (Sect.6.4.2a). In all IDC with $\Delta_{ex} < 0$ the molecular-limit criterion $|\Delta'| \gg \mathcal{M}$ is fulfilled even better, so that the impurity absorption must be represented by a narrow M' band whose position coincides with the term of the joint vibronic configuration $E_\rho + (\nu')^*$. The observed spectrum corresponds to the diagram shown in Fig.6.19b. The intensities of the M", M"', ... bands and of the D_M^* band are suppressed because $|M_{R_1R_2}| \ll |\Delta_{ex} - \Delta_\nu|$ (Sect.6.5.2).

If $\Delta_{ex} > 0$, the most interesting situations arise. Table 6.3 lists the energy parameters of the vibronic states $B_{2u} \cdot e_{2g}$ of some IDC. It is seen that $|\Delta'| \gg \mathcal{M}$ for traces of bd_6 in bd_0 and of bd_1 in bd_0. The absorption spectra of the local vibron approach the molecular limit and the M' bands must dominate. For bd_3 in bd_0, $\Delta' < \Delta_{cr}^+$. Therefore, the level of the local vibron cannot split off. Hence, the M' band will not appear and impurity absorption will only contribute to a broad D_M^* band. For bd_2 in bd_0, $|\Delta'| \approx |\Delta_{cr}^-|$ and a quantum level of the local vibron must exist at the edge of the dissociated-states spectrum.

Table 6.3. Energy parameters of vibronic states $B_{2u} \cdot e_{2g}$ of isotopically doped crystals of d-benzenes [cm⁻¹] [6.88][a]

Guest	Host	E_ρ' [b]	v'	$(v')^*$	Δ_v'	Δ_{ex}	Δ_{ph}	Δ_{ex-ph}	Δ'	D_M'	$E_\rho' + (v')^{*}$ [b]	$E_\rho' + v'$ [b]
d_6	d_0	38035	577	499	-78	196	-29	7	89	38380-38440	38534	38611
d_3	d_0	37933	593	513	-80	94	-13	5	1	38396-38456	38446	38526
d_2	d_0	37900	602	514	-88	61	-4	-3	-31	38405-38465	38414	38502
d_1	d_0	37870	603	515	-88	31	-3	-3	-60	38406-38466	38385	38473
benzene-d_0 [c]		37839	606	520	-85					38409-38469	38359	38445

[a] Notation for substances is the same as in Table 3.2. Isotopic shifts have been determined with respect to benzene-d_0. $2\mathcal{M} = 60$ cm⁻¹.

[b] E_ρ', $E_\rho' + (v')^*$ and $E_\rho' + v'$ are the electronic terms of the guest in the host matrix and the terms of the joint and separated vibronic configurations, respectively; the phonon is located on the guest molecule.

[c] Parameters of the pure bd_0-crystal.

The pattern of the absorption spectra of these crystals [6.88] shown in Fig.6.21 is in complete agreement with the analysis carried out. The absorption due to the excitation of a local vibron is denoted by solid lines. It is obtained by subtracting the absorption of the host crystal from the absorption of the IDC. The spectra shown in Fig.6.21 illustrate the gradual transition from Case c in Fig.6.19 to Case d, where the M' band "passes through" the D_M continuum.

b) IDC of d-Naphthalenes

The absorption spectra of local vibrons in IDC of d-naphthalenes and d-benzenes are very similar. The d-naphthalene spectra in the region of the vibronic transition $B_{3u} \cdot b_{1g} \leftarrow A_g$ are also divided into two groups according to the sign of the isotopic shift Δ_{ex}.

If $\Delta_{ex} < 0$, bands of the bound state of the vibron always exist in the vibronic region, even if the level of the local exciton in the electronic region does not split off. Indeed, the second term in the sum $\Delta' = \Delta_{ex} + \Delta_v'$ for the phonon v_{17} is already large enough ($|\Delta_v'| > |\Delta_{cr}^-|$) to ensure the splitting off of the level. The absorption spectrum of a crystal of nd_0 in nd_8, which is typical for the entire group, is shown in Fig.6.22. The energy parameters and the positions of the absorption bands of several crystals are listed in Tables 6.4 and 6.5. In contrast to the spectra of d-benzenes with $\Delta_{ex} < 0$ one

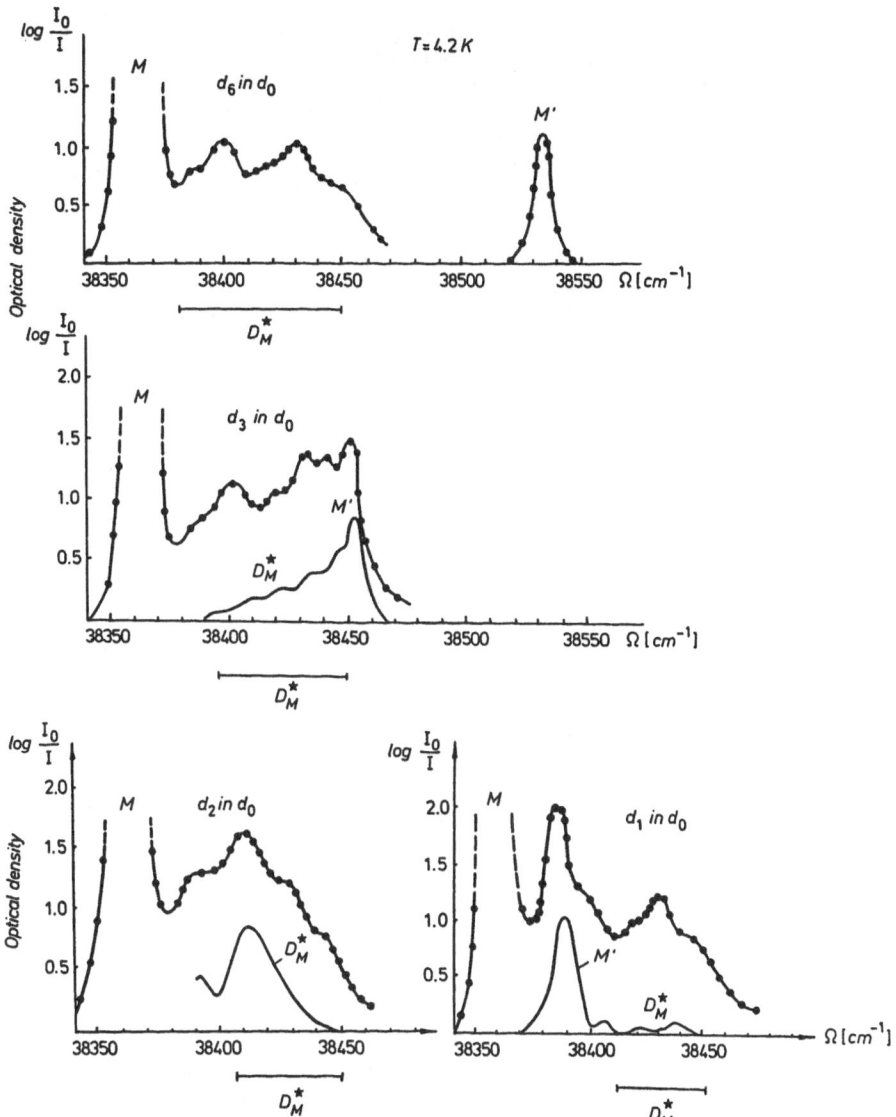

<u>Fig.6.21.</u> Absorption spectra of IDC of d-benzenes in the range of trans-
ition $B_{2u} \cdot e_{2g} \leftarrow A_{1g}$ (the direction of light polarization is close to the
b-axis of the crystal) [6.88]. The horizontal segments denote the regions
D_M^* of the dissociated states. Impurity concentration 5%

observes an M" band in addition to an M' band. The former corresponds to the
excitation of a state in which the NTS phonon is located on a host molecule
(Fig.6.19b). The M" bands can be observed due to the higher degree of exciton
delocalization in the d-naphthalenes as compared to d-benzenes (Sect.3.3.6).

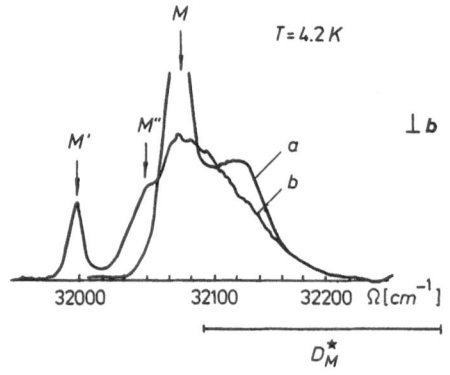

$T = 4.2\,K$

M

M' M''

$\perp b$

a
b

32000   32100   32200 $\Omega\,[cm^{-1}]$

D_M^*

Fig.6.22. Absorption spectra of crystals of nd$_8$ (a) and of IDC of nd$_0$ in
nd$_8$ (b) in the range of the vibronic
transition $B_{3u} \cdot b_{1g} \leftarrow A_g$. Impurity concentration $\approx 5\%$. The light is polar-
ized perpendicular to **b** [6.89]. The
horizontal segments under the frequency scale denote the dissociated-state
regions

An analogy may be seen between the vibronic spectra of the IDC of d-
naphthalenes with $\Delta_{ex} < 0$ and the vibrational two-phonon spectra of doped
crystals. Two internal phonons may be bound to an impurity even when they
cannot be bound to it separately. Such a case has recently been observed for
the first time [6.90]. Figure 6.23 exhibits Raman-scattering spectra of
$^{15}N_x^{14}N_{1-x}H_4Br$ in the region of the first overtone of the phonon ν_4. The
region of two-particle transitions is underlined on the abscissa. The low-
frequency edge of this region coincides with the doubled frequency of the
TO phonon ν_4 ($k = 0$) and is indicated with an arrow. Band M resembles the
M band in Fig.6.22 and corresponds to transitions to the one-particle (bi-
phonon) state. The magnitude of the anharmonic shift Δ_{ah}, analogous to Δ_ν
for vibrons, obeys $|\Delta_{ah}| > |\Delta_{cr}^-|$. Band M' is analogous to band M' in Fig.6.22
and corresponds to the bound state of two phonons on the impurity (local bi-

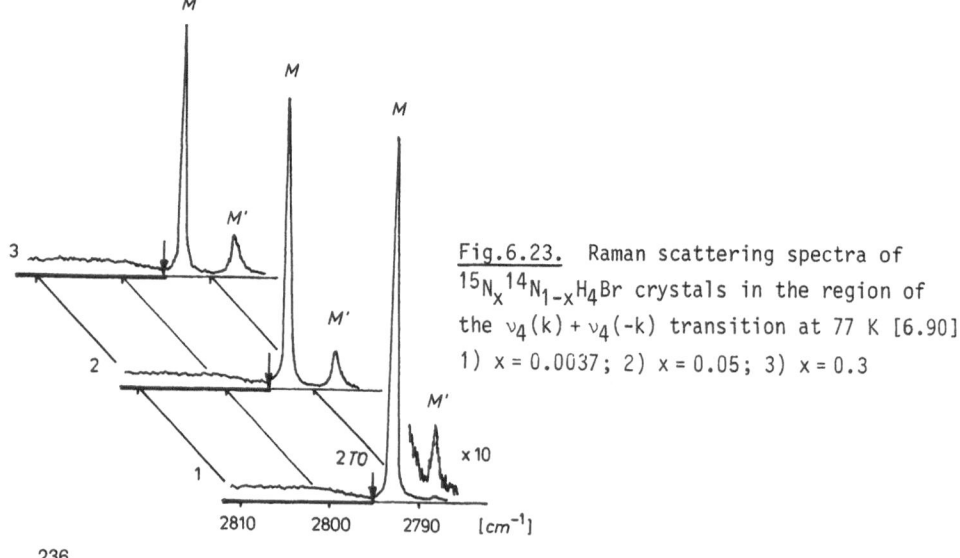

M

M

M

M'

3

M'

2

M'

$2\,TO$ $\times 10$

1

2810   2800   2790 $[cm^{-1}]$

Fig.6.23. Raman scattering spectra of
$^{15}N_x^{14}N_{1-x}H_4Br$ crystals in the region of
the $\nu_4(k) + \nu_4(-k)$ transition at 77 K [6.90]
1) $x = 0.0037$; 2) $x = 0.05$; 3) $x = 0.3$

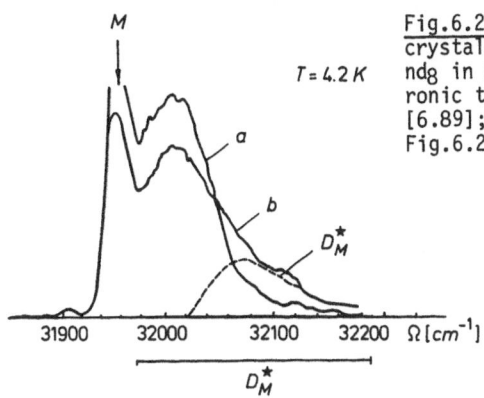

Fig.6.24. Absorption spectra of crystals of nd_0 (a) and of IDC of nd_8 in nd_0 in the range of the vibronic transition $B_{3u} \cdot b_{1g} \leftarrow A_g$ [6.89]; see also the caption to Fig.6.22

phonon). No local phonon exists in the one-phonon spectrum since $|\Delta_{ph}| < |\Delta_{cr}^-|$, where Δ_{ph} is the isotopic shift of the phonon frequency.

If $\Delta_{ex} > 0$, then for a large number of crystals $|\Delta'| < |\Delta_{cr}|$. The data for three such crystals are given in Tables 6.4,5. In this case the spectra of the local vibron only contain the D_M^* band. The spectrum of nd_8 in nd_0 is shown in Fig.6.24. The introduction of an impurity causes a deformation in the high-frequency wing of the composite side band that follows the M band of the host crystal (Sect.6.4.2b). The difference spectrum shows the broad D_M^* band of impurity absorption corresponding to the excitation of dissociated states.

c) Mixed Crystals of Naphthalenes nd_0 and nd_8

Figure 6.25 shows the vibronic absorption spectra of mixed crystals of nd_0 and nd_8 in the region of the vibronic transition $B_{3u} \cdot b_{1g} \leftarrow A_g$. The strong asymmetry of the spectrum with respect to the two isotopic components is striking. Indeed, the spectrum of nd_0 deforms slowly. The relevant M band remains narrow and is clearly visible over the entire concentration range. It is gradually displaced towards the high-frequency side with decreasing concentrations C_1 of nd_0. The fraction of the absorption that may be assigned to the dissociated states is reduced. As a result, the spectrum of the component nd_0, which includes a considerable contribution from the dissociated states at $C_1 = 1$, may be completely attributed to the bound states at $C_1 \rightarrow 0$ (Fig.6.22).

On the other hand, the spectrum of the component nd_8 rapidly rearranges with decreasing concentrations. Thus, the M band is already smeared at an impurity concentration $C_1 = 0.1$. At $C_1 \approx 1$ the spectrum may be completely ascribed to the dissociated states (Fig.6.24). The detailed shape of the mixed-crystal spectrum can only be found using numerical methods which have not yet been made.

Table 6.4. Energy parameters of electronic and vibronic $B_{3u} \cdot b_{1g}$ states of isotopically doped crystals of d-naphthalenes [cm^{-1}] [6.89]a

Guest	Host	E'_ρ	ν'	$(\nu')^*$	Δ_ν	Δ_{ex}	Δ_{ph}	Δ_{ex-ph}	Δ'	D_M	$D_{M'}$	$E'_\rho + (\nu')^*$	$E'_\rho + \nu'$
d_0	$\alpha-d_1$	31557	509	420	-89	-115	15	7	-204	32085-32265	32100-32280	31977	32051
$\alpha-d_1$	d_8	31569	509	418	-91	-103	15	9	-194	32085-32265	32100-32280	31987	32063
$\beta-d_1$	d_8	31576	507	416	-91	-96	13	9	-187	32085-32265	32090-32278	31992	32070
d_0	$\beta-d_4$	31557	509	420	-89	-74	8	7	-163	32051-32231	32059-32239	31977	32058
$\alpha-d_4$	d_8	31601	504	413	-91	-71	10	9	-162	32085-32265	32095-32275	32014	32095
d_8	d_0	31672	494	412	-82	115	-15	-7	33	31988-32168	31973-32153	32084	32181
d_8	$\alpha-d_1$	31672	494	412	-82	103	-15	-9	21	31997-32177	31982-32162	32084	32181
d_8	$\beta-d_1$	31672	494	412	-82	96	-13	-9	14	32002-32182	31989-32169	32084	32181

a) See notation to Tables 6.3 and 3.3

Table 6.5. Band positions of the electronic and vibronic $B_{3u} \cdot b_{1g}$ transitions in the IDC spectra of d-naphthalenes [cm^{-1}] [6.89]

Guest	Host	K'_0	M	M'	M''	D^*_M
d_0	$\alpha-d_1$	31542	32067	31977	32031	not isolated
$\alpha-d_1$	d_8	31551	32067	31985	32039	not isolated
$\beta-d_1$	d_8	31558	32067	31987	32047	not isolated
d_0	$\beta-d_4$	31533	32029	31971	31999	not isolated
$\alpha-d_4$	d_8	31572	32067	32009	32024	not isolated
d_8	d_0	31685	31960	not observed		32020-32200
d_8	$\alpha-d_1$	31695	31970	not observed		32010-32190
d_8	$\beta-d_1$	not observed	31978	not observed		32000-32180

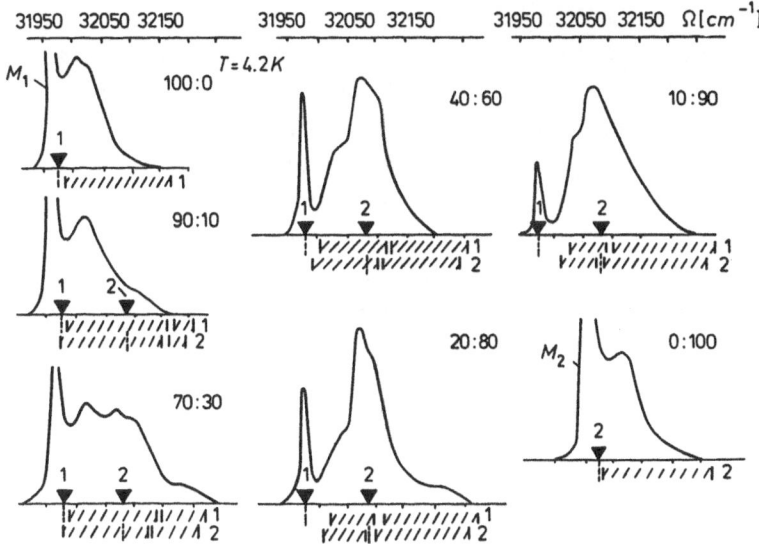

Fig.6.25. Microphotometer tracings of absorption spectra of $nd_0 + nd_8$ in the range of the vibronic transition $B_{3u} \cdot b_{1g} \leftarrow A_g$. The light is polarized perpendicular to **b** [6.91]. Spectra of the components nd_0 and nd_8 are marked 1 and 2, respectively. Composition of the crystals $C_1 : C_2$ is indicated at the upper right. M_1 and M_2 are the M bands of one-component crystals

The general qualitative trends can be understood by combining the dynamical theory with a calculation of the exciton spectrum using the CPA (Sect.4.4).

In a mixed crystal the NTS phonon is localized on one of the isotopic components. Accordingly, two spectra of two-particle states are obtained which are shifted Δ_{ph} from one another. Their edges can readily be found using the CPA for the energy spectrum of the excitons.

In the case of the two-band exciton spectrum under molecular-limit conditions (Sect.4.4.1), the vibronic absorptions of the individual components do not overlap and are symmetrical (unless, of course, Δ_ν cancels Δ_{ex}). Such a pattern is observed, for example, in mixed crystals of bd_0 and bd_6.

The situation is more complicated in a mixed crystal of nd_0 and nd_8. The exciton spectrum of this crystal is a pseudo-two-band one with a narrow pseudo-gap [6.92] (Fig.4.16). The dissociated-state regions are indicated under the frequency axis of Fig.6.25. When the phonon is localized on molecules of nd_0 and nd_8, the energy scales are shifted relative to one another. Therefore, they are shown separately for the two possible phonon localizations and are labelled 1 and 2, respectively, in Fig.6.25. Since the pseudo-gap is narrow, the region of two-particle states is practically continuous. The term of the joint vibronic configuration of nd_0 (triangle 1 in Fig.6.25) finds itself outside of the spectrum of dissociated states for the entire range of C_1 values (Fig.6.25). Hence, the corresponding M band is also observed outside of the region of two-particle states. On the other hand, the vibronic term of nd_8 (triangle 2 in Fig.6.25) falls within the region of the two-particle spectrum due to the cancellation of $\Delta_{ex} > 0$ and $\Delta_v < 0$. This results in a dissociation of the vibronic states and a rapid broadening of the corresponding M band. It is precisely for this reason that the M' band is absent in Fig.6.24. The M band of the nd_0 component becomes narrow when its term finds itself in the pseudo-gap of the two-particle spectrum (with phonon localization on nd_8). This occurs at a composition of 20:80 (Fig.6.25).

Recently, TOKURA et al. [6.93] measured the reflection spectra of the mixed crystals of nd_0 and nd_8. The observed behaviour of both M bands is quite analogous to that shown in Fig.6.25. The authors ascribe the smearing of the high-frequency M band, caused by the decay of a corresponding one-particle state into two-particle states, in terms of the Anderson localization of the excitons (Sect.4.6). They suppose that the decay of the vibron to an exciton and a phonon is strongly enhanced when the mobility edge is passed and the exciton is created in an extended (non-localized) state.

For the NTS phonon the absorption spectrum of a mixed crystal is also described by (6.30) if the isotopic shift of the phonon frequencies is neglected. This spectrum may be utilized to determine the density of exciton states of a mixed crystal. The large contribution of external phonons to the absorption in the region of dissociated states of both pure and mixed d-naphthalene crystals prevents one using the spectra in Fig.6.25 to determine the density of exciton states of a mixed crystal $nd_0 + nd_8$. However, the density of phonon states of a mixed crystal has recently been measured in this way. This was done for $K^{15}N_x^{14}N_{1-x}O_3$ crystals in the region of the composite two-phonon transition $\nu_1 + \nu_2$ where the dispersion of the phonon ν_1 had to be small [6.94]. Figure 6.26 illustrates the spectra of the phonon state density obtained in these isotopically mixed crystals for various concentrations of the isotope ^{15}N. The spectra reveal a quasi-one-dimensional behaviour, which is responsible for the fact that the bands of monomers and dimers persist in the region of high concentrations. There is a similarity between these spectra and the exciton spectra of the $nd_0 + nd_8$ mixed crystals shown in Fig.4.22.

d) Local Vibron on a Defect Centre

The properties of local excitons near defect centres formed by distorted host molecules were discussed in detail in Sect.3.5. It was shown that they are, in many respects, similar to those of a local exciton near an isotopic im-

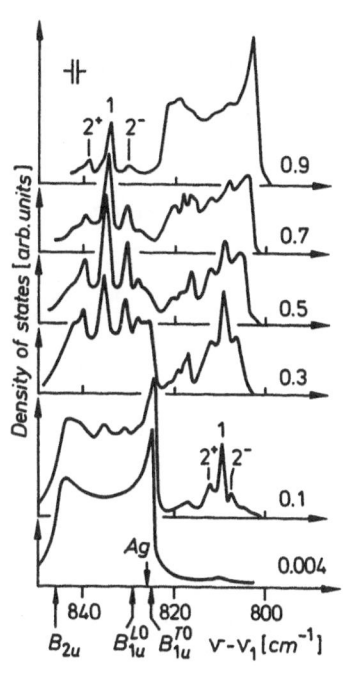

Fig.6.26. Density of states of isotopically mixed crystals $K^{15}N_x{}^{14}N_{1-x}O_3$ in the frequency range of the ν_2-phonon [6.94]. The relative concentration of ^{15}N is indicated in the figure. Arrows denote the frequencies of the components of the Davydov multiplet in the ν_2 ($k=0$) state in a pure crystal. 1, 2^+ and 2^- correspond to local phonons for a monomer and for symmetrical and antisymmetrical dimers, respectively

purity. The analogy can be extended to vibronic spectra. If an NTS phonon is found on a "defect" molecule, the equation for determining the position of the exciton level is obtained from (3.50) by replacing Δ by $\Delta' = \Delta_{ex} + \Delta_\nu$ (quite similar to the calculation of the positions of the M' bands, Sect.6.5.2). Since Δ_ν is the same in a "defect molecule" as in a "normal" molecule, an independent equation for Δ_{ex} arises. This circumstance was used to determine the value of Δ_{ex} in crystals of naphthalene for traces of indol, benzofurane, and thionaphthene [6.95]. The values of Δ_{ex} used in Sect.3.5 to describe the spectrum of a local exciton on a defect centre were obtained in this way.

6.5.5 Vibronic Absorption in the IDC of Naphthalene with a TS Phonon

The absorption of nd_0 in nd_8 in the region of the K_1 transition is illustrated in Fig.6.27 for different concentrations of nd_0. The same figure shows the theoretical positions of the impurity bands [6.76,96]. For the K_1 transition, $\Delta_{ex} = -115$ cm^{-1}, $\Delta_{ph} = 69$ cm^{-1}, $\Delta_\nu = -57$ cm^{-1}, and $\Delta_{ex-ph} = -5$ cm^{-1}. Due to the large value of $|\Delta_{ex}|$ and the mutual cancellation of Δ_ν and Δ_{ph}, the lower states are dominated by the configurations 1 and 2 (Fig.6.20) in which an exciton is found on an impurity molecule. The separated configuration 2 in

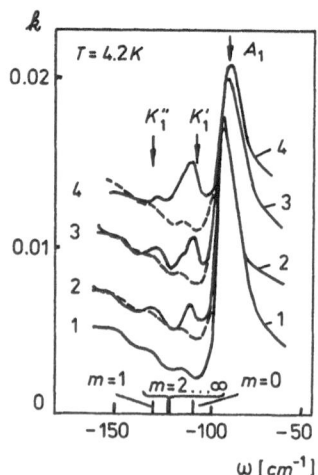

Fig.6.27. Absorption spectra of IDC of nd_0 in nd_8 in the range of the K_1 transition $B_{3u} \cdot a_g \leftarrow A_g$ at different concentrations of nd_0: 1) 0%, 2) 2%, 3) 5%, and 4) 10%, light being polarized perpendicular to **b** [6.76,96]. Curves 2-4 are given with displaced scales. For comparison, the dashed line indicates absorption of the nd_8 crystal. The energy is calculated from the term of the separated vibronic configuration $E_\rho + \nu$ (32167 cm^{-1})

which a phonon is found on a host molecule corresponds to the absolute minimum of energy. The results of numerical calculation are given in Table 6.6.

The sequence of the levels $m = 1, 2, \ldots, \infty$ in which the contribution from the configuration 2 prevails converges towards the higher frequencies. The

Table 6.6. Calculated values of the positions and intensities of the IDC bands of nd_0 in nd_8 in the K_1-transition region [6.76]

Band notation	Energy[a] of dominating configuration [cm^{-1}]	Energy[a] [cm^{-1}] and quantum number m	Intensity[b]	
			$\perp$**b**	||**b**
K_1'	-103	-111.4	1.381	0.66
		m = 0		
K_1'', K_1'''	-115	-134.2	0.53	0.1
		m = 1		
		-132.88	0.015	0.04
		m = 2		
		-132.83	0.005	0.017
		m = 3		

[a] The energy is counted from the term of separated vibronic configuration $E_\rho + \nu_8 = 32167$ cm^{-1}.

[b] The intensity of the vibronic transition in the molecule is taken as unity; the difference of the intensity from unity is due to the fact that the exciton-phonon wave function is partly delocalized from the guest site.

levels are very closely spaced. The total width of the sequence is only about 2 cm^{-1}, thus the individual levels coalesce. Since the exciton and the phonon are separated in the configuration 2, the intensities of the bands are low. Only the K_1'' band with m = 1 exhibits an appreciable intensity. A well-isolated level m = 0 lies above this sequence. It is dominated by the configuration 1, in which a phonon resides on a guest molecule. As a result, the intensity of the relevant transition is high and the K_1 band must dominate the impurity absorption spectrum. The arrows labelled A_1, K_1', and K_1'' indicate the calculated positions of the corresponding bands.

It is seen in Fig.6.27 that the K_1' band is already clearly defined at a concentration $C_{d_0} \approx 2\%$. With a further increase in concentration, up to 10%, the increase in the intensity of the band is almost linear. By contrast, the intensity of the K_1'' band at first increases and the band becomes more and more visible against the background. But then the band is rapidly broadened and becomes hardly visible at concentrations $C_{d_0} \gtrsim 5\%$. This last fact is not surprising, since a configuration with a phonon localized predominantly outside the guest molecule must be very unstable. It should be emphasized that the calculation of the positions of the K_1' and K_1'' bands does not involve any unknown parameters (Sect.6.5.3). The good agreement of theoretical and experimental values for them serves as an independent checkup on the analysis of intrinsic absorption in the region of the K_1 transition.

6.5.6 Absorption Spectra of Amorphous Films

Amorphous films are a particular type of disordered system. They may show regularities which significantly differ from those described above. The experimental data on such films are scarce. MARUYAMA et al. [6.97] and LEE and GAN [6.98] measured the absorption spectra of tetracene and pentacene. The spectra of crystalline and amorphous pentacene are compared in Fig.6.28. Both components of the exciton doublet are seen to totally merge in the spectrum of the amorphous film and the absorption in the exciton region becomes structureless. In contrast, the vibronic absorption that has a broad band shape in a crystal changes only slightly. One may estimate the width of the exciton energy band from the magnitude of the Davydov splitting; it exceeds 2000 cm^{-1}. As a result, the entire absorption in the vibronic regions should be two-particle. When analysing this spectrum in more detail, KLAFTER and JORTNER arrived at the same conclusion [6.99]. They also proposed using the weak effect of disorder on the absorption in the vibronic region as a diagnostic criterion for the identification of two-particle excitations.

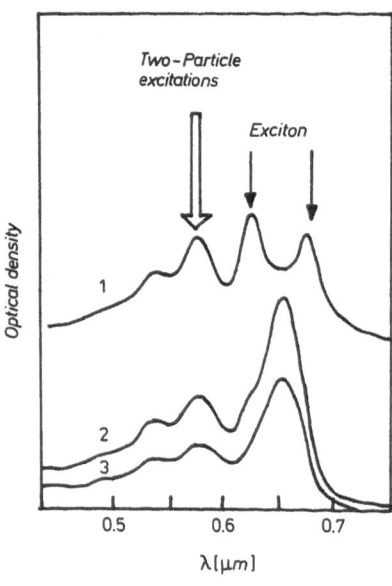

Fig.6.28. Optical spectra of solid pentacene films, obtained by deposition at various temperatures [6.98,99]. 1) Crystalline solid, T = 155 C; 2) Amorphous film, T = -90 C; 3) Amorphous film, T = -155 C; T-deposition temperature

6.5.7 Positions of the Electronic and Vibronic Bands

The summarized data on the positions of the bands of the local excitons and vibrons show that the concepts developed for the description of the electronic and vibronic states of perfect and doped crystals are self consistent. The results for the spectra of d-naphthalenes are given in Fig.6.29.

One set of points indicates the positions of the K_0' bands of isotopic guests in the exciton spectrum. The potential energy of the defect, i.e., the isotopic shift Δ_{ex} that determines the positions of the levels (3.8) is plotted on the abscissa. The second set of points corresponds to the vibronic M and M' bands. For them, the local potential of the exciton is equal to the phonon-frequency shift Δ_ν or the sum $\Delta_{ex} + \Delta_\nu$ for the intrinsic M band and the impurity M' bands, respectively. It is also plotted on the abscissa. Tables 3.3 and 6.4 list a selection of the numerical data. Within the framework of the dynamical theory these two sets of points are determined by the same "isotopic" equation (3.8), and, therefore, must fit the same curve. Figure 6.29 shows that this is actually the case. Similar data for d-benzenes are presented in Fig.6.30 (Tables 3.2 and 6.3).

The situation is more complicated for the group of bands involving a TS phonon. Their positions are determined by more complicated equations due to the first term in (6.11) and H_{imp}^γ in (6.39). Nevertheless, it can be seen that the positions of the bands 2, 3, and 4, which correspond to large values of Δ, deviate slightly from those calculated using the "isotopic" equation.

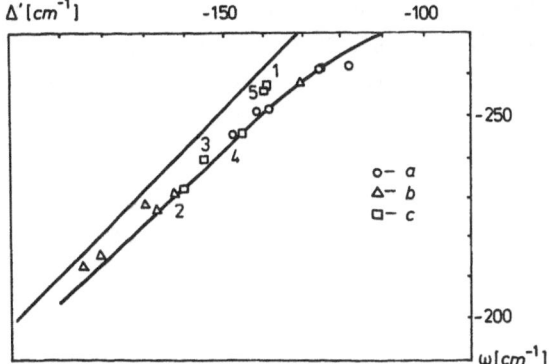

Fig.6.29. Band positions of the local exciton, local vibron, and one-particle absorption of pure and ICD d-naphthalenes versus Δ' - the local potential for the exciton [6.76]. a) the K_0' bands of IDC, the position is counted from the electronic term E_ρ. b) The M' bands of local vibrons with the NTS phonon ν_{17} and the M bands of the crystals of nd_0 and nd_8, the position is counted from the terms of the separated vibronic configurations $E_\rho' + \nu_{17}$ and $E_\rho + \nu_{17}$, respectively. c) Transitions involving the TS phonon ν_8. 1) The K_1' band of the IDC nd_0 in nd_8, the position is counted from the term $E_\rho + \nu_8$. 2) and 3) The MK bands of composite transitions in crystals of nd_0 and nd_8. 4) and 5) The K_2 bands of the overtone transition in crystals of nd_0 and nd_8, respectively. The positions of bands 2 - 5 are counted from the terms of separated vibronic configurations $E_\rho + \nu_8 + \nu_{17}$ and $E_\rho + 2\nu_8$. The straight line is the linear dependence $\omega = \Delta'$ according to the oriented gas model. The curve shows a calculation by the "isotopic" equation (3.8) with the density of states $\rho(\omega)$ given in Fig.3.5

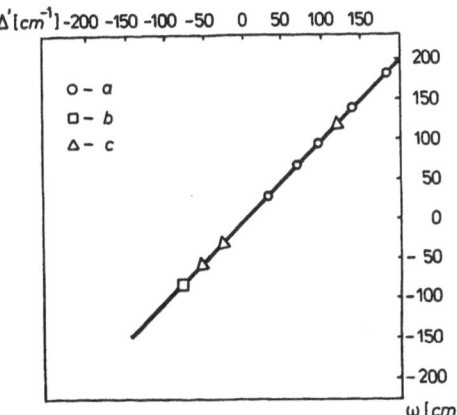

Fig.6.30. Band positions of a local exciton, a local vibron, and one-particle vibronic absorption of pure and IDC d-benzenes versus Δ' - the local potential for the exciton [6.88,100]. a) K_0' bands, b) M band of bd_0, c) M' bands of IDC. Calculation of energies is indicated in the caption to Fig.6.29

In particular, the positions of the MK bands of the composite transitions 2 and 3 (calculated in Sect.6.4.4) differ from the ones found from (3.8) by only 1 and 2 cm^{-1}, respectively.

The strong deviation of point 1 is quite natural, since the K_1' band with $m = 0$ is "repelled" upward from the group of bands with $m \geq 1$ (the K_1'' band in

Fig.6.27) disposed under it. At the same time the position of this band co-incides with the calculated position (Table 6.6). Therefore, the strong de-viation of this band from an "isotopic" dependence is an excellent demonstra-tion of the role of the mobility of TS phonons. An exact calculation for band 5 has not yet been made.

The success in the interpretation of the intrinsic and impurity vibronic bands is a quantitative confirmation of the dynamical theory of vibronic spectra.

Conclusion

This book, which is dedicated to the spectroscopy of molecular excitons, considers optical spectra which are determined, either predominantly or com-pletely, by the energy spectrum of excitons in the perfect crystal. The spec-tra belonging to this category proved to be quite extensive. It includes the spectra of doped and mixed molecular crystals in the electronic region and also the vibronic spectra of perfect, doped, and mixed crystals. It covers both absorption and fluorescence spectra. An analysis of all this diversity of spectra can be done quantitatively, from the unified point of view, with-in the framework of the same scope of ideas and concepts.

Two concepts are used in an analysis of the spectra. The first one is the energy spectrum of an exciton in a perfect crystal, i.e., the dispersion law $\varepsilon(\mathbf{k})$ and the density of states $\rho(\varepsilon)$ in the exciton band. The second concept is the possibility to describe in the same terms the exciton spectra of doped crystals and the vibronic spectra of crystals (pure and doped). Its field of application is restricted to crystals whose exciton bands are not too broad and whose vibronic spectra can be described by dynamical theory.

The basic initial information on a crystal which must be built into the theory is contained in the dispersion law $\varepsilon(\mathbf{k})$. The other parameters can usually be determined from the spectra of isolated molecules. The peculiarity of the situation lies in the fact that all of the presently available informa-tion in the dispersion law $\varepsilon(\mathbf{k})$ has been exclusively obtained from an analysis of the same optical spectra. Therefore, the problem is posed and solved self-consistently from the very start. A definite part of the experimental data yields information on the energy spectrum, so that $\rho(\varepsilon)$ and $\varepsilon(\mathbf{k})$ can be re-constructed from it. On this basis one can describe the other experimental data quantitatively. The successful interpretation of the benzene, hexamethyl-benzene, and naphthalene spectra demonstrates the efficiency of this procedure.

It should be noted that the accuracy of the dispersion laws found by these methods is relatively low. The reason is that so far all of the methods used to determine $\varepsilon(\mathbf{k})$ from the experimental data are based on a measurement of various integral characteristics, which are only slightly sensitive to the details of the behaviour of $\varepsilon(\mathbf{k})$. In the anthracene crystal, the unfavourable parameter relation has until now prevented a reconstruction of $\varepsilon(\mathbf{k})$. Therefore, it is necessary to develop methods that are preciser and more versatile. It is obvious that, in principle, the problem can be solved by investigating the creation of excitons by the scattering of electrons, X-rays, neutrons, etc. The transfer of momentum must be of the order of the Brillouin momentum, while the transfer of energy must be equal to the energy of the electronic excitation (of the order of several eV). The future will show which of these methods will be the most effective and to what extent. It is clear, however, that such experiments are very difficult because a high-energy resolution ($\sim 10^{-3}$ eV) is required. Their preparation will take much time. Hence, it is expedient that, for the time being, the accuracy of the existing methods is improved, for example, by using modulation spectroscopy.

We believe that the success in the interpretation and quantitative description of the exciton and vibronic spectra is quite convincing. It indicates that the basic assumptions and models that were used are correct and that the theory is efficient and heuristic. This allows to enlarge the scope of the experiments which may be treated at the quantitative level. They have to cover the interaction of the excitons with the external phonons, the shapes of the absorption and fluorescence spectra of pure crystals and determine the kinetics of the exciton. Rather little is known about this interaction at the moment. Systematic quantitative investigations are in progress. Much can be done using the methods of stationary spectroscopy. However, with respect to the kinetics of excitons it is difficult to overestimate the possibilities offered by modern time-resolved spectroscopy with times of the order of picoseconds.

Appendix A Intramolecular Vibronic Interaction in Terms of the Secondary-Quantization Representation

The vibrational Hamiltonian of a molecule in the ground electronic state has the following form in the harmonic approximation:

$$H_0 = \frac{v}{2}(q^2 + p^2) = \frac{v}{2}\left(q^2 - \frac{\partial^2}{\partial q^2}\right) \quad , \tag{A.1}$$

where q and p are dimensionless operators of the coordinate and the momentum. The creation and annihilation operators can be introduced in the usual way:

$$b^+ = \frac{1}{\sqrt{2}}\left(q - \frac{\partial}{\partial q}\right) \quad , \quad b = \frac{1}{\sqrt{2}}\left(q + \frac{\partial}{\partial q}\right) \quad . \tag{A.2}$$

They satisfy the Bose commutation relations

$$[b, b^+] \equiv bb^+ - b^+b = 1 \quad . \tag{A.3}$$

The operator H_0, expressed in terms of b and b^+, is equal to

$$H_0 = \frac{v}{2}(bb^+ + b^+b) = v\left(b^+b + \frac{1}{2}\right) \quad . \tag{A.4}$$

The action of the operators b and b^+ on the normalized eigenfunctions of the oscillator $|n\rangle$ is defined as follows:

$$b|n\rangle = \sqrt{n}|n-1\rangle \quad , \quad b^+|n\rangle = \sqrt{n+1}|n+1\rangle \quad , \quad n = 0, 1, \ldots, \infty \quad . \tag{A.5}$$

It follows that

$$|n\rangle = \frac{1}{\sqrt{n!}}(b^+)^n|0\rangle \quad , \tag{A.6}$$

and that

$$b|0\rangle = 0 \quad , \tag{A.7}$$

which is actually a definition of the ground state $|0\rangle$.

If we restrict ourselves to the linear vibronic interaction, $H_v = \sqrt{2}\,\gamma v q$, the vibrational Hamiltonian of the excited electronic state is equal to

$$\tilde{H} = \frac{\nu}{2}\left[\left(q + \sqrt{2}\,\gamma\right)^2 - \frac{\partial^2}{\partial q^2}\right] - \gamma^2 \nu \quad . \tag{A.8}$$

The last term $E_{FC} = \gamma^2 \nu$ represents the Franck-Condon energy. In this case, the creation and annihilation operators must evidently be introduced by the relations

$$\tilde{b}^+ = b^+ + \gamma \quad , \quad \tilde{b} = b + \gamma \quad . \tag{A.9}$$

They satisfy the same commutation relations as b^+ and b. In terms of the new operators the Hamiltonian H has the following form:

$$\tilde{H} = \nu(\tilde{b}^+\tilde{b} + \frac{1}{2}) - \gamma^2 \nu \quad . \tag{A.10}$$

The transformation (A.9) corresponds to a shift in the equilibrium position of the oscillator. The wave functions of all of the vibrationally excited states $|\tilde{n}\rangle$ can be expressed in terms of the wave function of the ground vibrational level $|\tilde{0}\rangle$ using a relation similar to (A.6).

Let us establish a relationship between the eigenfunctions of the operators H_0 and $\tilde{H}$. To do this, we first of all determine the operator

$$b(\tau) = e^{\tau b^+} b \, e^{-\tau b^+} \quad . \tag{A.11}$$

With due consideration for (A.3)

$$\frac{db(\tau)}{d\tau} = e^{\tau b^+} [b^+, b] e^{-\tau b^+} = -1 \quad , \tag{A.12}$$

and, since $b(0) = b$,

$$b(\tau) = b - \tau \quad , \quad \tilde{b} = b(-\gamma) \quad . \tag{A.13}$$

Hence,

$$\tilde{b} \cdot e^{-\gamma b^+} = e^{-\gamma b^+} b \quad .$$

Acting on both sides of the equation on the vector $|0\rangle$ and taking into account that $b|0\rangle = 0$ and $\tilde{b}|\tilde{0}\rangle = 0$, we obtain

$$|\tilde{0}\rangle = e^{-\gamma^2/2} e^{-\gamma b^+} |0\rangle \quad . \tag{A.14}$$

Using (A.6) it can easily be shown that $|\tilde{0}\rangle$ is normalized.

One can calculate the non-orthogonality integrals using the above equations. In the Condon approximation they determine the matrix elements of the optical transitions. For example, the matrix element of the transition from the ground state of a crystal to an excited electronic state, which is accompanied by the creation of n vibrational quanta, contains the non-orthogonality integral

$$\langle \tilde{n}|0\rangle = \langle \tilde{0}|\frac{1}{\sqrt{n!}}\,\tilde{b}^n\, e^{-\gamma^2/2}\, e^{\gamma\tilde{b}^+}|\tilde{0}\rangle = e^{-\gamma^2/2}\,\frac{\gamma^n}{\sqrt{n!}} \quad ; \tag{A.15}$$

To obtain this expression, (A.5,6,14) were used. Therefore, the ratio of the intensity of the n-quantum transition $I(n)$ to that of the pure electronic transition I_0 (in the absence of electron-vibrational interaction) is equal to

$$\frac{I(n)}{I_0} = \frac{\gamma^{2n}}{n!}\, e^{-\gamma^2} \quad . \tag{A.16}$$

The factor $e^{-\gamma^2}$ is an analogue of the *Debye-Waller factor*.

Let us now consider a more general electronic-vibrational interaction containing the quadratic term $\delta q^2/2$, which represents a change in the vibrational frequency of the excited electronic state.

In order to use the formalism that conveniently describes the vibronic spectra (Chap.6), let us consider the electron subsystem of a molecule as a two-level system and introduce the creation and annihilation operators a^+ and a. The corresponding occupancy numbers $(N = a^+a)$ can take one of two values, $N = 0, 1$. In terms of this notation the Hamiltonian of the molecule is equal to

$$H = \frac{\nu}{2}\left(q^2 - \frac{\partial^2}{\partial q^2}\right) + \left\{E_e + \sqrt{2}\,\gamma\nu q + \delta q^2\right\}.a^+a \quad , \tag{A.17}$$

or, in terms of the operators b^+ and b,

$$H = \nu\left(b^+b + \frac{1}{2}\right) + \left\{E_e + \gamma\nu(b^+ + b) + \frac{\delta}{2}(b^+ + b)^2\right\}a^+a \quad . \tag{A.18}$$

To diagonalize this Hamiltonian, one must perform a canonical transformation:

$$H(\tau) = e^{\tau S}H e^{-\tau S} \quad , \tag{A.19}$$

where [A.1]

$$S = \{(b - b^+) + \xi(b^2 - b^{+2})\}\,a^+a \quad . \tag{A.20}$$

S is a generalization of the well-known canonical transformation used in the small-polaron theory [A.2]. Transforming the operators b^+ and b in a similar way, we obtain by analogy with (A.12)

$$\frac{db(\tau)}{d\tau} = (1 + 2\xi b^+(\tau))\,\hat{N} \quad , \qquad \frac{db^+(\tau)}{d\tau} = (1 + 2\xi b(\tau))\,\hat{N} \quad . \tag{A.21}$$

The solution that satisfies the conditions $b(0) = b$, $b^+(0) = b^+$ has the following form:

$$b(\tau) = b \cosh (2\xi\hat{N}\tau) + b^+ \sinh (2\xi\hat{N}\tau) + \frac{1}{2\xi}(e^{2\xi\hat{N}\tau} - 1) \quad . \tag{A.22}$$

$b^+(\tau)$ has a similar form. The meaning of (A.22) becomes apparent if we consider $\hat{N}$ to be a projection operator, i.e., $\hat{N}^p = \hat{N}$ for any integral p. Therefore, for an arbitrary function f

$$f(\hat{N}\tau) = f(0) + \hat{N}(f(\tau) - f(0)) \quad . \tag{A.23}$$

If $b(\tau)$ is substituted into (A.18) and the parameters τ and ξ are chosen so that the operators b^+ and b only appear together as b^+b, the following expressions for τ and ξ appear:

$$e^{-8\xi\tau} = 1 + \frac{2\delta}{\nu} \quad ; \quad \frac{1}{\xi} = \frac{2\gamma}{\left(1 + \frac{2\delta}{\nu}\right)\left(1 - e^{2\xi\tau}\right)} \quad . \tag{A.24}$$

Hence, the transformed Hamiltonian has the following form:

$$H = \nu(b^+b + \frac{1}{2}) + \left[E_{mol} + \left(\sqrt{1 + \frac{2\delta}{\nu}} - 1\right)b^+b\right]a^+a \quad , \quad \text{where} \tag{A.25}$$

$$E_{mol} = E_e - \frac{\gamma^2\nu}{1 + \frac{2\delta}{\nu}} + \frac{\nu}{2}\left(\sqrt{1 + \frac{2\delta}{\nu}} - 1\right) \quad . \tag{A.26}$$

E_{mol} is the energy of a pure electronic transition. It differs from E_e (which is the difference in electronic terms at $q = 0$) by the Franck-Condon energy (second term) and the change in the zero-point energy (third term).

The vibrational frequency of the excited state is equal to

$$\nu^* = \nu\sqrt{1 + \frac{2\delta}{\nu}} \quad . \tag{A.27}$$

If $2\delta \ll \nu$, the change in vibrational frequency is

$$\Delta_\nu \approx \delta \quad . \tag{A.28}$$

To calculate the matrix element of the optical transition, the operator a^+ must be similarly transformed:

$$a^+(\tau) = e^{\tau S}a^+e^{-\tau S} = a^+\exp[\tau(b - b^+) + \xi\tau(b^2 - b^{+2})] \quad . \tag{A.29}$$

Using (A.24) and setting $\delta = 0$, we obtain $\xi = 0$, $\tau = -\gamma$. In this case it follows from the Weyl identity [A.3] that

$$a^*(\tau) = a^+e^{-\tau^2/2}e^{-\tau b^+}e^{\tau b} \quad , \tag{A.30}$$

and we readily obtain (A.15). In the opposite limiting case, i.e., as $\gamma \to 0$, it follows from (A.24) that $\tau \to 0$ and $\xi\tau \to$ constant. Rewriting $a^{+}(\tau)$ as

$$a^{+}(\tau) = a^{+} \exp[-\frac{1}{2} \tanh(2\xi\tau) b^{+2}] \tag{A.31}$$

$$\cdot \exp[-\ln \cosh(2\xi\tau)(b^{+}b + \frac{1}{2})] \exp[-\frac{1}{2}\tanh(2\xi\tau)b^{2}] \quad , \tag{A.31}$$

we see that the vibrational operators only appear in pairs, i.e., as b^2, b^{+2} and $b^{+}b$ [A.3]. It follows that transitions are only possible if an even number of quanta are absorbed or emitted. The matrix element of the 2n-quantum transition from the ground state is equal to

$$\langle 2n | a^{+} | 0 \rangle = (-1)^{n} \sqrt{\frac{(2n-1)!!}{2^{n}n!}} \, [\tanh(2\xi\tau)]^{n} \exp[-\frac{1}{2}\ln \cosh(2\xi\tau)] \quad . \tag{A.32}$$

The parameter $\xi\tau$ is determined from (A.24). A comparison of the Debye-Waller factors in (A.16,32) shows that for a quadratic interaction the role of the coupling constant γ^2 is played by $2(\xi\tau)^2 \approx \frac{1}{8}\left(\frac{\Delta\nu}{\nu}\right)^2$. This value is usually small, and, therefore, the effect of the quadratic coupling on the intensities of the bands can usually be neglected.

Appendix B Theory of Degenerate Perturbations

Let us consider the operator $H = H_0 + U$. The perturbation operator matrix U possesses the following structure:

$$U_{nm} = \sum_{s=1}^{S} f_n^{(s)} h_m^{(s)} \quad , \tag{B.1}$$

where $f_n^{(s)}$ and $h_m^{(s)}$ depend arbitrarily on the subscripts n and m. It is easy to see that the operator U makes it possible to change arbitr̖ily the matrix elements in S columns and S rows. Taking advantage of the diadic symbols, one can rewrite U as follows:

$$U = \sum_{s=1}^{S} \widehat{f^{(s)} h}^{(s)} \quad . \tag{B.2}$$

In this case, the caret indicates that the diadic product of the vectors $f^{(s)}$ and $h^{(s)}$ is to be taken. Such a perturbation with a finite S is said to be degenerate. S is called the rank of the perturbation.

The following statement is true for such operators: if the Green's function $G^0(E) = (E - H_0)^{-1}$ for the operator H_0 is known, then the determination of the Green's function $G(E)$ for the operator H is reduced to solving a set of S linear equations. To prove this we note that G obeys the operator equation

$$G = G^0 + G^0 H G \quad . \tag{B.3}$$

Its validity can readily be proven if we consider that $G = (E - H)^{-1}$ and multiply both sides of (B.3) from the right by $(E - H)$. Substituting U into the right-hand side of (B.3), we obtain

$$G = G^0 + \sum_{s=1}^{S} (G^0 \widehat{f^{(s)}})(h^{(s)} G) \quad . \tag{B.4}$$

The function G appears on the right-hand side of (B.4) only in the vectors $(h^{(s)} G)$:

$$(h^{(s)} G)_m = \sum_{n} h_n^{(s)} G_{nm} \quad . \tag{B.5}$$

This fact can be used to simplify (B.4). Indeed, let us multiply (B.4) from the left by $h^{(s)}$:

$$(h^{(s)}G) = (h^{(s)}G^0) + \sum_{s'=1}^{S} (h^{(s)}G^0f(s'))(h^{(s')}G) \quad . \tag{B.6}$$

Thus, we have obtained a set of linear equations for S vectors $(h^{(s)}G)$. Solving this set of equations by way of the determinants and substituting the result into (B.4), we obtain the explicit expression for the Green's function G:

$$G_{nm} = \frac{1}{g} \begin{vmatrix} G^0_{nm} & (G^0f^{(1)})_n & \cdots & (G^0f^{(S)})_n \\ -(h^{(1)}G^0)_m & g_{11} & \cdots & g_{1S} \\ \vdots & \vdots & & \\ -(h^{(S)}G^0)_m & g_{S1} & & g_{SS} \end{vmatrix} \quad , \tag{B.7}$$

$$g_{ss'} = \delta_{ss'} - (h^{(s)}G^0f(s')) \quad , \quad g = \mathrm{Det}\| g_{ss'} \| \quad . \tag{B.8}$$

A similar equation in the integral equation theory is called the Bateman formula. Thus, we obtain a closed general expression for the Green's function of a perturbed system. It is valid at an arbitrary magnitude of the perturbation.

Let us assume that the spectrum of the Hamiltonian H_0 is continuous and is concentrated on limited sections of the E-axis (as in perfect crystals). The Hamiltonian H may, along with the continuous spectrum on the same sections, have isolated eigenvalues E_i which lie outside of the spectrum of the operator H_0. These isolated eigenvalues are the poles of G, i.e., are the real roots of the equation $g(E_i) = 0$. It can be shown that the number of eigenvalues E_i split off from each section of the continuous spectrum does not exceed S.

The Green's function G can also be used to determine the eigenfunctions $\psi_i(n)$ corresponding to the eigenvalues E_i. To this end the general bilinear representation for the Green's function must be applied:

$$G_{nm}(E) = (E - H)^{-1}_{nm} = \sum_j \frac{\psi_j(n)\psi_j^*(m)}{E - E_j} \quad . \tag{B.9}$$

The summation over j includes all of the eigenvalues of H that belong to the discrete and continuous spectrum. The validity of (B.9) can be proven by applying the left- and right-hand sides of (B.9) on an arbitrary eigenfunction ψ_j. By definition of the operator $(E - H)^{-1}$,

$$(E - H)^{-1} \psi_j = (E - E_j)^{-1} \psi_j \quad .$$
(B.10)

At the same time, if we take advantage of the fact that the system of functions ψ_j is orthonormalized, then

$$\sum_{j'm} \frac{\psi_{j'}(n)\psi_{j'}^*(m)}{E - E_{j'}} \psi_j(m) = \frac{\psi_j(n)}{E - E_j} \quad ,$$
(B.11)

which is in complete agreement with (B.10). Therefore, by expanding the G defined in (B.7) in the vicinity of its pole $E = E_i$, we obtain the eigenfunction ψ_i.

The fact that the positions of the points of the spectrum are all determined by the zeros of the function g(E) suggests that the density of states $\rho(E)$ must be closely related to the function g(E). The density of states per unit volume and per energy interval is related to the trace of the matrix G(E) by the following simple expression

$$\rho(E) = -\frac{1}{\pi V} \, \text{Im} \, \text{Tr}\{G(E)\} \quad ,$$
(B.12)

where V is the volume of the system. This can be immediately seen from (B.9) if we set $E \to E + i0$. If the determinant of the right-hand side of (B.7) is expanded in the elements of the first row and the first column, the trace is calculated, and the identity,

$$G^2 = -\frac{dG}{dE}$$
(B.13)

[which is easy to check, taking advantage of (B.9)] is used, we find that

$$\rho(E) = \rho_0(E) - \frac{1}{\pi V} \, \text{Im} \left\{ \frac{1}{g} \sum_{ss'} (-)^{s+s'} (h^{(s)} G^0 G^0 f^{(s')}) \tilde{g}_{ss'} \right\}$$

$$= \rho_0(E) - \frac{1}{\pi V} \, \text{Im} \left\{ \frac{1}{g} \sum_{ss'} (-)^{s+s'} \frac{dg_{ss'}}{dE} \tilde{g}_{ss'} \right\}$$

$$= \rho_0(E) - \frac{1}{\pi V} \, \text{Im} \left\{ \frac{d \ln g}{dE} \right\} \quad ,$$
(B.14)

where $\tilde{g}_{ss'}$ is the minor of the element $g_{ss'}$. From (B.14) it can be seen that the contribution to the density of states from the impurities is completely determined by g(E). The last term in (B.14) is of the order of V^{-1}, i.e., it disappears at large volumes. This is to be expected, since only one impurity centre was taken into consideration. If we assume that the number of centres is proportional to the volume, then the correction to the density will be proportional to the concentration C of the impurity.

The degenerate-perturbation theory can be translated from the language of the G functions to that of the T matrix. The latter is related to the Green's function by the expression

$$G = G^0 + G^0 T G^0 \quad .$$
(B.15)

After transforming the determinant in the numerator of (B.7) it is seen that the G that is determined using this equation has the structure of (B.15) and that the elements of the T matrix are as follows

$$T_{nm} = \frac{1}{g} \begin{vmatrix} 0 & f_n^{(1)} & \cdots & f_n^{(S)} \\ -h_m^{(1)} & g_{11} & \cdots & g_{1S} \\ -h_m^{(S)} & g_{S1} & \cdots & g_{SS} \end{vmatrix} \quad .$$
(B.16)

As an example, we consider a first-rank perturbation ($S = 1$) whose Hamiltonian H differs from H_0 by one diagonal matrix element:

$$f_n h_m = \Delta \delta_{n1} \delta_{1m} \quad .$$
(B.17)

In this case, (B.6) is reduced to a single equation and the Green's function is equal to

$$G_{nm} = G_{nm}^0 + \frac{(G^0 f)_n (h G^0)_m}{1 - (h G^0 f)} = G_{nm}^0 + \frac{\Delta G_{n1}^0 G_{1m}^0}{1 - \Delta G_{11}^0} \quad .$$
(B.18)

The isolated eigenvalues E_i are determined from the equation

$$\Delta G_{11}^0 (E_i) = 1 \quad .$$
(B.19)

To find the corresponding eigenfunctions ψ_i, we must expand the denominator in the vicinity of E_i:

$$1 - \Delta G_{11}^0 (E) \approx -\Delta \left(\frac{d G_{11}^0}{dE} \right)_{E=E_i} (E - E_i) \quad .$$
(B.20)

It is easy to show that dG_{11}/dE is always negative for values of E that lie outside of the spectrum of the operator H_0. Therefore, from (B.9,18,20) it follows that

$$\psi_i(n) = G_{n1}^0(E_i) \left/ \left| \frac{d G_{11}^0(E_i)}{dE_i} \right|^{1/2} \right. \quad .$$
(B.21)

References

Chapter 1

1.1 F. Seitz: *The Modern Theory of Solids* (McGraw-Hill, New York 1940)
1.2 I.V. Obreimov, W.J. de Haas: Proc. Roy. Soc. Amsterdam **31**, 53 (1929), and **32**, 1 (1929)
1.3 P. Pringsheim, A. Kronenberger: Z. Phys. **40**, 75 (1927)
1.4 A. Kronenberger: Z. Phys. **63**, 494 (1930)
1.5 M.A. Elyashevich: *Atomnaya i molekulyarnaya spektroskopiya* [*Atomic and Molecular Spectroscopy*] (Fizmatgiz, Moscow 1962)
1.6 G. Herzberg: *Molecular Spectra and Molecular Structure. II. Infrared and Raman Spectra of Polyatomic Molecules* (Van Nostrand, Princeton 1945)
1.7 V.L. Broude, G.V. Klimusheva, A.L. Liberman, M.I. Onoprienko, A.F. Prikhotjko, A.I. Shatenshtein: *Spektry pogloshcheniya molekulyarnykh kristallov. Benzol i nekotorye ego gomologi* [*Absorption Spectra of Molecular Crystals. Benzene and Some of Its Homologues*] (Naukova Dumka, Kiev 1965)
1.8 V.L. Broude, G.V. Klimusheva, A.F. Prikhotjko, E.F. Sheka, L.P. Yatsenko: *Spektry pogloshcheniya molekulyarnykh kristallov (polizameshchennyie benzola)* [*Absorption Spectra of Molecular Crystals; Polysubstituted Benzene Compounds*] (Naukova Dumka, Kiev 1972)
1.9 I.N. Krupennikova, M.M. Raikhshtadt, L.A. Gribov: Izv. Timiryazev Skh. Akad. Vyp. **1**, 196 (1977)
1.10 T.E. Peacock: *Electronic Properties of Aromatic and Heterocyclic Molecules* (Academic, London 1965)
1.11 A.F. Prikhotjko: Zh. Eksp. Teor. Fiz. **19**, 487 (1949)
1.12 O.P. Kharitonova: Opt. Spektrosk. **5**, 29 (1958)
1.13 L.D. Landau, E.M. Lifshitz: *Kvantovaya mekhanika* (Fizmatgiz, Moscow 1963) [English transl.: *Quantum Mechanics* (Addison-Wesley, London 1965)]
1.14 M. Born, R. Oppenheimer: Ann. Phys. Leipzig **84**, 457 (1927)
1.15 M. Born, K. Huang: *Dynamical Theory of Crystal Lattice* (Clarendon, Oxford 1954)
1.16 G. Herzberg, E. Teller: Z. Phys. Chem. Leipzig **21**, 410 (1933)
1.17 G. Orlandi, W. Siebrand: J. Chem. Phys. **58**, 4513 (1973)
1.18 L.M. Sverdlov, M.A. Kovner, E.P. Krainov: *Kolebatel'nye spektry mnogo-atomnykh molekul* [*Vibration Spectra of Multiatomic Molecules*] (Nauka, Moscow 1970)
1.19 N.O. Lipari, C.B. Duke, L. Pietronero: J. Chem. Phys. **65**, 1165 (1976)
1.20 T.N. Bolotnikova: *Kvazilineichatyie spektry i nekotoryie voprosy molekulyarnoi spektroskopii* [*Quasi-line Spectra and Some Problems of Molecular Spectroscopy*] Dokt. Dissert., Mosk. Gosud. Pedagogich. Inst. Im. Lenina (1972)
1.21 E.F. McCoy, I.G. Ross: Aust. J. Chem. **15**, 573 (1962)
1.22 H. Sponner, G. Nordheim, A.L. Sklar, B. Geller: J. Chem. Phys. **7**, 207 (1939)
1.23 D.P. Craig: J. Chem. Soc. London 2309 (1955)

1.24 D.S. McClure: J. Chem. Phys. **22**, 1668 (1954)
1.25 I.V. Obreimov, A.F. Prikhotjko: Phys. Z. Sowjetunion **9**, 34, 48 (1936)
1.26 A.F. Prikhotjko: J. Phys. **8**, 257 (1944)
1.27 A.F. Prikhotjko: Izv. Akad. Nauk SSSR, Ser. Fiz. **12**, 499 (1948)
1.28 I.V. Obreimov, A.F. Prikhotjko, I.V. Rodnikova: Zh. Eksp. Teor. Fiz. **18**, 409 (1948)
1.29 A.F. Prikhotjko: Zh. Eksp. Teor. Fiz. **19**, 383 (1949)
1.30 D.S. McClure, O. Schnepp: J. Chem. Phys. **23**, 1575 (1955)
1.31 A.F. Prikhotjko: Opt. Spektrosk. **3**, 434 (1957)
1.32 V.L. Broude: Opt. Spektrosk. **30**, 89 (1971)
1.33 S.V. Marisova: *Opticheskiye svoistva kristalla antratsena i energetiches-kaya struktura ego singletnykh eksitonov [Optical Properties of the Anthracene Crystal and the Energy Structure of Its Singlet Excitons]* Kand. Dissert. Inst. Fiz., Acad. Sci. Ukr. SSR, Kiev (1969)
1.34 A.S. Davydov: Zh. Eksp. Teor. Fiz. **18**, 210 (1948)
1.35 A.S. Davydov: *Teoriya pogloshcheniya sveta v molekulyarnykh kristallakh* (Izd. AN Ukr. SSR, Kiev 1951) [English transl.: *Theory of Molecular Excitons* (McGraw-Hill, New York 1962)]
1.36 A.S. Davydov: *Teoriya molekulyarnykh eksitonov* (Nauka, Moscow 1968) [English transl.: *Theory of Molecular Excitons* (Plenum, New York 1971)]
1.37 E.F. Sheka: Usp. Fiz. Nauk **104**, 593 (1971) [English transl.: Sov. Phys. Usp. **14**, 484 (1972)]
1.38 D.P. Craig, S.H. Walmsley: *Excitons in Molecular Crystals* (Benjamin, New York 1968)
1.39 F. Gutman, L.E. Lyons: *Organic Semiconductors* (Wiley, New York 1967)
1.40 R.M. Hochstrasser: Rev. Mod. Phys. **34**, 531 (1962); Ann. Rev. Phys. Chem. **17**, 457 (1966)
1.41 D.S. McClure: In *Solid State Physics*, Vol.8, ed. by F. Seitz, D. Turnbull (Academic, New York 1958) p.1
1.42 G.W. Robinson: Ann. Rev. Phys. Chem. **21**, 429 (1970)
1.43 O. Schnepp: Ann. Rev. Phys. Chem. **14**, 35 (1963)
1.44 H.C. Wolf: In *Solid State Physics*, Vol.9, ed. by F. Seitz, D. Turnbull (Academic, New York 1959) p.1
1.45 D.P. Craig, J.M. Hollas: Philos. Trans. Roy. Soc. London **253**, 569 (1961)
1.46 D.S. McClure: J. Chem. Phys. **24**, 1 (1956)
1.47 E.F. Sheka, I.P. Terenetskaya: Chem. Phys. **8**, 99 (1975)
1.48 T.L. Muchnik, R.F. Turner, S.D. Colson: Chem. Phys. Lett. **42**, 570 (1976)
1.49 S.M. Kochubei: Zh. Prikl. Spektrosk. **7**, 74 (1967)
1.50 V.L. Broude, A.F. Prikhotjko, E.I. Rashba: Usp. Fiz. Nauk **67**, 99 (1959) [English transl.: Sov. Phys. Usp. **67(2)**, 38 (1959)]
1.51 E.R. Bernstein, S.D. Colson, D.S. Tinti, G.W. Robinson: J. Chem. Phys. **48**, 4632 (1968)
1.52 E.V. Shpolsky: Usp. Fiz. Nauk **71**, 215 (1960); **77**, 321 (1962); **80**, 255 (1963) [English transl.: Sov. Phys. Usp. **3**, 372 (1961); **5**, 522 (1962); **6**, 411 (1963)]
1.53 L.I. Zagainova, G.V. Klimusheva, G.M. Soroka, R.V. Yaremko: Izv. Akad. Nauk SSSR, Ser. Fiz. **39**, 1913 (1975)
1.54 A.I. Kitaigorodsky: *Organicheskaya kristallokhimiya [Organic Crystallo-chemistry]* (Izd. AN SSSR, Moscow 1955)
1.55 K.K. Rebane: *Elementarnaya teoriya kolebatel'noi struktury spektrov primesnykh tsentrov kristallov* (Nauka, Moscow 1968) [English transl.: *Impurity Spectra of Solids* (Plenum, New York 1970]
1.56 V.L. Broude, E.I. Rashba, E.F. Sheka: Dokl. Acad. Nauk SSSR **139**, 1085 (1961) [English transl.: Sov. Phys. Dokl. **6**, 718 (1962)]
1.57 E.F. Sheka: Fiz. Tverd. Tela **5**, 2316 (1963) [English transl.: Sov. Phys. Solid State **5**, 1718 (1963)]
1.58 E.I. Rashba: Opt. Spektrosk. **2**, 568 (1957)
1.59 N.I. Ostapenko, V.I. Sugakov, M.T. Shpak: In *Eksitony v molekulyarnykh*

kristallakh [*Excitons in Molecular Crystals*] (Naukova Dumka, Kiev 1973)
p.92
1.60 D.P. Craig, J.M. Hollas, M.F. Redies, S.C. Wait: Philos. Trans. Roy. Soc.
London A**253**, 543 (1961)

Chapter 2

2.1 J. Frenkel: Phys. Rev. **37**, 17, 1276 (1931)
2.2 A.S. Davydov: *Teoriya molekulyarnykh eksitonov* (Nauka, Moscow 1968)
     [English transl.: *Theory of Molecular Excitons* (Plenum, New York 1971)]
2.3 V.M. Agranovich: *Teoriya eksitonov* [*Exciton Theory*] (Nauka, Moscow 1968)
2.4 M.R. Philpott: Adv. Chem. Phys. **23**, 227 (1973)
2.5 A.S. Davydov: Zh. Eksp. Teor. Fiz. **18**, 210 (1948)
2.6 M.H. Cohen, F. Keffer: Phys. Rev. **99**, 1128 (1955)
2.7 M. Born, K. Huang: *Dynamical Theory of Crystal Lattices* (Clarendon, Oxford
     1954)
2.8 S.I. Pekar: Zh. Eksp. Teor. Fiz. **35**, 522 (1958) [English transl.: Sov.
     Phys. JETP **8**, 360 (1959)]
2.9a V.M. Agranovich, V.L. Ginzburg: *Kristallooptika s uchetom prostranstvennoi
     dispersii i teoriya eksitonov*, 2nd ed. (Nauka, Moscow 1979)
2.9b V.M. Agranovich, V.L. Ginzburg: *Crystal Optics with Spatial Dispersion,
     and Excitons*, Springer Ser. Solid-State Sci., Vol.42 (Springer, Berlin,
     Heidelberg, New York, Tokyo 1984)
2.9c S.I. Pekar: *Kristallooptika i dobavochnye svetovye volny* (Naukova Dumka,
     Kiev 1982) [English transl.: *Crystal Optics and Additional Light Waves*
     (Benjamin, Reading, MA 1983)]
2.10 A.S. Davydov: *Teoriya pogloshcheniya sveta v molekulyarnykh kristallakh*
     (Izdatel'stvo Akademii Nauk Ukrainskoi SSR, Kiev 1951) [English transl.:
     Theory of Molecular Excitons (McGraw-Hill, New York 1962)]
2.11 H. Winston: J. Chem. Phys. **19**, 156 (1951)
2.12 R.M. Hochstrasser: *Molecular Aspects of Symmetry* (Benjamin, New York 1966)
2.13 S.I. Pekar: Zh. Eksp. Teor. Fiz. **38**, 1786 (1960) [English transl.: Sov.
     Phys. JETP **11**, 1286 (1960)]
2.14 A.A. Abrikosov, L.P. Gor'kov, I.E. Dzyaloshinksii: *Metody kvantovoi teorii
     polya v statisticheskoi fizike* (Fizmatgiz, Moscow 1962) [English transl.:
     Methods of Quantum Field Theory in Statistical Physics (Prentice Hall,
     Englewood Cliffs, NY 1963)]
2.15 K.B. Tolpygo: Zh. Eksp. Teor. Fiz. **20**, 497 (1950)
2.16 K. Huang: Proc. Roy. Soc. London Ser. A**208**, 352 (1951)
2.17 J.J. Hopfield: Phys. Rev. **112**, 1555 (1958)
2.18 V.M. Agranovich: Zh. Eksp. Teor. Fiz. **37**, 430 (1959) [English transl.:
     Sov. Phys. JETP **10**, 307 (1960)]
2.19 S.I. Pekar: Zh. Eksp. Teor. Fiz. **33**, 1022 (1957) [English transl.:
     Sov. Phys. JETP **6**, 785 (1958)]
2.20 A.S. Davydov: *Teoriya tverdogo tela* (Nauka, Moscow 1976) [French
     transl.: *Théorie des Solides* (Mir, Moscou 1980)]
2.21 D.P. Craig, S.H. Walmsley: Mol. Phys. **4**, 113 (1961)
2.22 D. Fox, O. Schnepp: J. Chem. Phys. **23**, 767 (1955)
2.23 E.G. Cox, J.A.S. Smith: Nature **173**, 75 (1954)
2.24 J. Zak, W.A. Benjamin (eds.): *The Irreducible Representations of Space
     Groups* (Benjamin, New York 1969)
2.25 G.L. Bir, G.E. Pikus: *Simmetriya i deformatsionnyie effekty v poluprovod-
     nikakh* (Nauka, Moscow 1972) [English transl.:*Symmetry and Strain-Induced
     Effects in Semiconductors* (Halsted, New York 1974)]
2.26 J.L. Birman: *Theory of Crystal Space Groups and Infrared and Raman Lattice
     Processes in Insulating Crystals* (Springer, Berlin, Heidelberg, New York
     1974)

2.27 O.V. Kovalev: *Neprivodimyie predstavleniya prostranstevennykh grupp* (Izdatel'stvo Akademii Nauk Ukr. SSSR, Kiev 1961) [English transl.: *Irreducible Representations of the Space Groups* (Gordon and Breach, New York 1965)]

2.28 S.D. Colson, D.M. Hanson, R. Kopelman, G.W. Robinson: J. Chem. Phys. **48**, 2215 (1968)

2.29 A.I. Kitaigorodsky: *Organicheskaya kristallokhimiya [Organic Crystallochemistry]* (Izdatel'stvo Akademii Nauk SSSR, Moscow 1955)

2.30 A.F. Prikhotjko, M.S. Soskin: Opt. Spektrosk. **13**, 522 (1962) [English transl.: Sov. Opt. Spectrosc. **13**, 291 (1962)]

2.31 M.S. Soskin: Ukr. Fiz. Zh. (Ukr. Ed.) **7**, 180, 635 (1962)

2.32 A.S. Davydov, E.N. Myasnikov: Dokl. Akad. Nauk SSSR **173**, 1040 (1967)

2.33 A.S. Davydov, E.F. Sheka: Phys. Status Solidi **11**, 877 (1965)

2.34 G.D. Mahan: In *Electronic Structure of Polymers and Molecular Crystals*, ed. by J.-M. André, L. Ladik, J. Delhalle (Plenum, New York 1975)

2.35 M.R. Philpott: J. Chem. Phys. **54**, 111 (1971)

2.36 J. Schroeder, R. Silbey: J. Chem. Phys. **55**, 5418 (1971)

2.37 V.L. Broude, G.V. Klimusheva, A.F. Prikhotjko, E.F. Sheka, L.P. Yatsenko: *Spektry pogloshcheniya molekulyarnykh kristallov (polizameshchennyie benzola) [Absorption Spectra of Molecular Crystals; Polysubstituted Benzene Compounds* (Naukova Dumka, Kiev 1972)]

2.38 S.D. Woodruff, R. Kopelman: Chem. Phys. **22**, 1 (1977)

2.39 S.D. Woodruff, R. Kopelman: J. Cryst. Mol. Struct. **7**, 29 (1977)

2.40 D. Fox, S. Yatsiv: Phys. Rev. **108**, 938 (1957)

2.41 D.V. Schlosser, M.R. Philpott: Chem. Phys. **49**, 181 (1980)

Chapter 3

3.1 B.S. Sommer, J. Jortner: J. Chem. Phys. **50**, 187 (1969)

3.2 E.I. Rashba: In *Physics of Impurity Centres in Crystals*, ed. by G.S. Zavt (Acad. Sci. Estonia SSR, Tallin 1972) p.415

3.3 E.F. Sheka: In *Physics of Impurity Centres in Crystals*, ed. by G.S. Zavt (Acad. Sci. Estonia SSR, Tallin 1972) p.431

3.4 J. Hoshen, J. Jortner: J. Chem. Phys. **56**, 933 (1972)

3.5 N.I. Ostapenko, V.I. Sugakov, M.T. Shpak: In *Eksitony v molekulyarnykh kristallakh [Excitons in Molecular Crystals]*, ed. by M.S. Brodin (Naukova Dumka, Kiev 1973) p.92

3.6 V.L. Broude, E.I. Rashba: Pure Appl. Chem. **37**, 21 (1974)

3.7 R. Kopelman: Rec. Chem. Progr. **31**, 211 (1970)

3.8 R. Kopelman: In *Radiationless Processes in Molecules and Condensed Phases*, ed. by K.F. Fong, Topics Appl. Phys., Vol.15 (Springer, Berlin, Heidelberg, New York 1976) p.296

3.9 I.M. Lifshitz: Zh. Eksp. Teor. Fiz. **17**, 1017, 1076 (1947)

3.10 G.F. Koster, J.C. Slater: Phys. Rev. **95**, 1167 (1954)

3.11 Yu.A. Izyumov: Adv. Phys. **14**, 569 (1965)

3.12 E.I. Rashba: Opt. Spektrosk. **2**, 568 (1957)

3.13 V.L. Broude, E.I. Rashba, E.F. Sheka: Dokl. Akad. Nauk SSSR **139**, 1085 (1961) [English transl.: Sov. Phys. Dokl. **6**, 718 (1962)]

3.14 L.D. Landau, E.M. Lifshitz: *Kvantovaya mekhanika* (Fizmatgiz, Moscow 1963) [English transl.: *Quantum Mechanics* (Addison-Wesley, London 1965)] Sect.45

3.15 D.G. Thomas, J.J. Hopfield: Phys. Rev. **175**, 1021 (1968)

3.16 E.I. Rashba: Fiz. Tekh. Poluprovodn. **8**, 1241 (1974) [English transl.: Sov. Phys. Semicond. **8**, 807 (1975)]

3.17 M.V. Belousov, D.E. Pogarev, A.A. Shultin: Fiz. Tverd. Tela **20**, 1415 (1978) [English transl.: Sov. Phys. Solid State **20**, 814 (1978)]

3.18 M.V. Belousov: In *Excitons*, ed. by E.I. Rashba, M.D. Sturge (North-Holland, Amsterdam 1982) p.771

3.19 V.V. Yeremenko, V.M. Naumenko, S.V. Petrov, V.V. Pishko: Zh. Eksp. Teor. Fiz. **82**, 813 (1982) [English transl.: Sov. Phys. JETP **55**, 481 (1982)]
3.20 E.I. Rashba: Fiz. Tverd. Tela **4**, 3301 (1962) [English transl.: Sov. Phys. Solid State **4**, 2417 (1962)]
3.21 A.I. Larkin, D.E. Khmel'nitsky: Zh. Eksp. Teor. Fiz. **56**, 2087 (1969) [English transl.: Sov. Phys. JETP **29**, 1123 (1969)]
3.22 V.I. Sugakov: Fiz. Tverd. Tela **10**, 2995 (1968) [English transl.: Sov. Phys. Solid State **10**, 2363 (1968)]
3.23 V.I. Sugakov: Fiz. Tverd. Tela **12**, 216 (1970) [English transl.: Sov. Phys. Solid State **12**, 172 (1970)]
3.24 B.S. Sommer, J. Jortner: J. Chem. Phys. **50**, 822 (1969)
3.25 V.V. Ryazanov, V.I. Sugakov: Opt. Spektrosk. **32**, 932 (1972) [English transl.: Opt. Spectrosc. **32**, 499 (1972)]
3.26 R. Peierls: Ann. Phys. **13**, 905 (1932)
3.27 V.M. Agranovich: Usp. Fiz. Nauk **71**, 141 (1960) [English transl.: Sov. Phys. Usp. **3**, 427 (1961)]
3.28 V.I. Sugakov: Opt. Spektrosk. **24**, 477 (1968)
3.29 J.J. Hopfield: Phys. Rev. **182**, 945 (1969)
3.30 R.E. Merrifield: J. Chem. Phys. **38**, 920 (1963)
3.31 R.G. Body, I. Ross: Aust. J. Chem. **19**, 1 (1966)
3.32 S. Takeno: J. Chem. Phys. **44**, 853 (1966)
3.33 D.P. Craig, M.R. Philpott: Proc. Roy. Soc. London A**290**, 583, 602 (1966); **293**, 213 (1966)
3.34 E.R. Bernstein, S.D. Colson, D.S. Tinti, G.W. Robinson: J. Chem. Phys. **48**, 4632 (1968)
3.35 D.M. Hanson: J. Chem. Phys. **51**, 653 (1969)
3.36 I. Natkaniec, A.V. Belushkin, W. Dyck, H. Fuess, C.M.E. Zeyen: Z. Kristallogr. **163**, 285 (1983)
3.37 S.C. Abrahams, J.M. Robertson, J.G. White: Acta Crystallogr. **2**, 233 (1949)
3.38 V.I. Ponomaryov, O.S. Filippenko, L.O. Atovmjan: Kristallografiya **21**, 392 (1976)
3.39 V.K. Dolganov: *Eksiton-fononnyie spektry kristalla benzola* [*Exciton phonon Spectra of the Benzene Crystal*] Kand. Diss. Inst. Solid State Phys., Acad. Sci. USSR, Chernogolovka (1973)
3.40 I.P. Terenetskaya: *Spektral'nyie issledovania izotopicheskogo effecta v naftaline* [*Spectral Investigation into Isotopic Effect in Naphthalene*] Kand. Diss. Inst. Phys. Acad. Sci. UkrSSR, Kiev (1973)
3.41 E.F. Sheka, I.P. Terenetskaya: Chem. Phys. **8**, 99 (1975)
3.42 S.D. Colson, T.L. Netzel: Chem. Phys. Lett. **16**, 555 (1972)
3.43 G.F. Bacon, N.A. Curry, S.A. Wilson: Proc. Roy. Soc. London A**279**, 98 (1964)
3.44 A.I. Kolesnikov: Dipl. Work, Inst. Solid State Phys., Acad. Sci. USSR, Chernogolovka (1977)
3.45a M.S. Lehman, G.S. Pawley: Acta Chem. Scand. **26**, 1996 (1972)
3.45b S.L. Chaplot, N. Lehner, G.S. Pawley: Acta Cryst. B**38**, 483 (1982)
3.46a R. Mason: Acta Crystallogr. **17**, 547 (1964)
3.46b V.I. Ponomaryov, G.V. Shilov: Kristallografiya **28**, 674 (1983)
3.47 V.K. Dolganov, E.F. Sheka: Zh. Eksp. Teor. Fiz. **60**, 2230 (1971) [English transl.: Sov. Phys. JETP **33**, 1198 (1971)]
3.48 S.D. Colson, D.M. Hanson, R. Kopelman, G.W. Robinson: J. Chem. Phys. **48**, 2215 (1968)
3.49 N.V. Rabin'kina, E.I. Rashba, E.F. Sheka: Fiz. Tverd. Tela **12**, 3569 (1970) [English transl.: Sov. Phys. Solid State **12**, 2898 (1970)]
3.50 E.F. Sheka: Fiz. Tverd. Tela **12**, 1167 (1970) [English transl.: Sov. Phys. Solid State **12**, 911 (1970)]
3.51 F.W. Ochs, P.N. Prasad, R. Kopelman: Chem. Phys. **6**, 253 (1974)
3.52 V.L. Broude: Fiz. Tverd. Tela **11**, 1159 (1969) [English transl.: Sov. Phys. Solid State **11**, 930 (1969)]

3.53   E.F. Sheka: Fiz. Tverd. Tela 5, 2361 (1963) [English transl.: Sov. Phys.
       Solid State 5, 1718 (1963)]
3.54   K.P. Meletov, E.F. Sheka: Fiz. Tverd. Tela 21, 1291 (1979) [English
       transl.: Sov. Phys. Solid State 21, 749 (1979)]
3.55   V.L. Broude, A.I. Vlasenko, E.I. Rashba, E.F. Sheka: Fiz. Tverd. Tela 7,
       2094 (1965) [English transl.: Sov. Phys. Solid State 7, 1686 (1965)]
3.56   M.R. Philpott: J. Chem. Phys. 54, 111 (1971)
3.57   M.V. Belousov, D.E. Pogarev, A.A. Shultin: Phys. Status Solidi B80,
       417 (1977)
3.58   M.V. Belousov, D.E. Pogarev, A.A. Shultin: Fiz. Tverd. Tela 20, 1415
       (1978) [English transl.: Sov. Phys. Solid State 20, 814 (1978)]
3.59   M.V. Belousov, D.E. Pogarev: Fiz. Tverd. Tela 20, 3461 (1978) [English
       transl.: Sov. Phys. Solid State 20, 1999 (1978)]
3.60   A.V. Krol', N.V. Levichev, A.L. Natadze, A.I. Ryskin: Solid State Commun.
       24, 151 (1977)
3.61   D.M. Hanson: J. Chem. Phys. 52, 3409 (1970)
3.62   H.K. Hong, R. Kopelman: Phys. Rev. Lett. 25, 1030 (1970)
3.63   H.K. Hong, R. Kopelman: J. Chem. Phys. 55, 724 (1971); 57, 3888 (1972)
3.64   V.L. Broude, A.V. Leiderman, T.G. Tratas: Fiz. Tverd. Tela 13, 3624 (1971)
       [English transl.: Sov. Phys. Solid State 13, 3058 (1971)]
3.65   F.W. Ochs, R. Kopelman: J. Chem. Phys. 66, 1599 (1977)
3.66   H. Port, D. Vogel, H.C. Wolf: Chem. Phys. Lett. 34, 23 (1975)
3.67   V.L. Broude, A.V. Leiderman: Zh. Eksp. Teor. Fiz. Pis'ma 13, 426 (1971)
       [English transl.: Sov. Phys. JETP Lett. 13, 302 (1971)]
3.68   V.L. Broude, A.F. Prikhotjko, E.I. Rashba: Usp. Fiz. Nauk 67, 99 (1959)
       [English transl.: Sov. Phys. Usp. 67(2), 38 (1959)]
3.69   M.T. Shpak, N.I. Sheremet: Opt. Spektrosk. Sbornik statei I Lyuminest-
       sentsiya 110, (1963)
3.70   A.F. Prikhotjko, I.Ya. Fugol': Opt. Spektrosk. 4, 335 (1958)
3.71   V.I. Sugakov: Opt. Spektrosk. 21, 574 (1966) [English transl.: Opt.
       Spectrosc. 21, 319 (1966)]
3.72   A. Propstl, H.C. Wolf: Z. Naturforsch. A18, 724 (1963)
3.73   N.N. Malykhina, M.T. Shpak: Ukr. Fiz. Zh. 9, 991 (1964)
3.74   N.I. Ostapenko, M.T. Shpak: Phys. Status Solidi 36, 515 (1969)
3.75   N.I. Ostapenko, M.T. Shpak: Izv. Akad. Nauk SSSR, Ser. Fiz. 34, 553
       (1970)
3.76   M.T. Shpak, N.I. Sheremet: Ukr. Fiz. Zh. 8, 6, 667 (1963)
3.77   I.S. Osad'ko: Fiz. Tverd. Tela 11, 441 (1969) [English transl.: Sov.
       Phys. Solid State 11, 347 (1969)]
3.78   N.I. Ostapenko, V.I. Sugakov, M.T. Shpak: Phys. Status Solidi B45,
       729 (1971)
3.79   N.I. Ostapenko, V.I. Sugakov, M.T. Shpak: J. Lumin. 4, 261 (1971)
3.80   N.I. Ostapenko, V.I. Sugakov: Izv. Akad. Nauk SSSR, Ser. Fiz. 36, 1042
       (1972)
3.81   H. Port: In *Organic Molecular Aggregates*, ed. by P. Reineker, H. Haken,
       H.C. Wolf, Springer Ser. Solid-State Sci., Vol.49 (Springer, Berlin,
       Heidelberg, New York, Tokyo 1983) p.22

Chapter 4

4.1   I.M. Lifshitz: Usp. Fiz. Nauk 83, 617 (1964) [English transl.: Sov.
       Phys. Usp. 7, 549 (1965)]
4.2   A.A. Maradudin: In *Solid State Physics*, Vol.18, ed. by F. Seitz,
       D. Turnbull (Academic, New York 1966) p.274; Vol.19, p.2
4.3   N.F. Mott: Adv. Phys. 16, 49 (1967)
4.4   R. Bell: Rep. Prog. Phys. 35, 1315 (1972)
4.5   P. Dean: Rev. Mod. Phys. 44, 127 (1972)
4.6   R.J. Elliott, J.A. Krumhansl, P.L. Leath: Rev. Mod. Phys. 46, 465 (1974)

4.7 D.S. Saxon, R.A. Hunter: Philips Res. Rep. **4**, 81 (1949)
4.8 Y. Onodera, Y. Toyozawa: J. Phys. Soc. Jpn. **24**, 341 (1968)
4.9 P. Soven: Phys. Rev. **156**, 809 (1967)
4.10 P. Dean: Proc. Phys. Soc. London **73**, 413 (1959)
4.11 N.-K. Hong, R. Kopelman: J. Chem. Phys. **55**, 5380 (1971)
4.12 V.L. Broude, M.I. Onoprienko: Opt. Spektrosk. **10**, 634 (1961) [English
     transl.: Opt. Spectrosc. **10**, 334 (1961)]
4.13 A.A. Abrikosov, L.P. Gor'kov, I.E. Dzyaloshinskii: *Metody kvantovoi teorii
     polya v statisticheskoi fizike* (Fizmatgiz, Moscow 1962) [English transl.:
     Methods of Quantum Field Theory in Statistical Physics (Prentice Hall,
     Englewood Cliffs, NY 1963)]
4.14 J. Hoshen, J. Jortner: Chem. Phys. Lett. **5**, 351 (1970)
4.15 V.M. Agranovich: Usp. Fiz. Nauk **112**, 143 (1974) [English transl.: Sov.
     Phys. Usp. **17**, 103 (1974)]
4.16 T.A. Krivenko, A.V. Leiderman, E.I. Rashba: Fiz. Tverd. Tela **17**, 137
     (1975) [English transl.: Sov. Phys. Solid State **17**, 78 (1975)]
4.17 L. Nordheim: Ann. Phys. **9**, 607 (1931)
4.18 V.L. Broude, S.M. Kochubei: Fiz. Tverd. Tela **6**, 354 (1964) [English
     transl.: Sov. Phys. Solid State **6**, 285 (1964)]
4.19 V.L. Broude, E.I. Rashba: Fiz. Tverd. Tela **3**, 1941 (1961) [English
     transl.: Sov. Phys. Solid State **3**, 1415 (1961)]
4.20 E.F. Sheka: Opt. Spektrosk. **10**, 684 (1961) [English transl.: Opt.
     Spectrosc. **10**, 360 (1961)]
4.21 E.F. Sheka: Izv. Akad. Nauk SSSR, Ser. Fiz. **27**, 503 (1963) [English
     transl.: Bull. Acad. Sci. USSR, Phys. Ser. **27**, 501 (1963)]
4.22 I. Hubbard: Proc. Roy Soc. London A**276**, 238 (1963)
4.23 Y.S. Chen, W. Shockley, G.L. Pearson: Phys. Rev. **151**, 648 (1966)
4.24 I.F. Chang, S.S. Mitra: Adv. Phys. **20**, 359 (1971)
4.25 A.S. Barker, A.J. Sievers: Rev. Mod. Phys. **47**, Suppl.2, 4941 (1975)
4.26 V.L. Broude: Fiz. Tverd. Tela **11**, 1159 (1969) [English transl.: Sov.
     Phys. Solid State **11**, 930 (1969)]
4.27 T. Netzel, S. Colson, D. Fox: J. Chem. Phys. **59**, 475 (1973)
4.28 V.L. Broude, E.I. Rashba: Pure Appl. Chem. **37**, 21 (1974)
4.29 V.L. Broude, A.V. Leiderman, T.G. Tratas: Fiz. Tverd. Tela **13**, 3624
     (1971) [English transl.: Sov. Phys. Solid State **13**, 3058 (1971)]
4.30 F. Yonezawa, K. Morigaki: Prog. Theor. Phys. Suppl. **53**, 1 (1973)
4.31 B. Velicky, S. Kirkpatrick, H. Ehrenreich: Phys. Rev. **175**, 747 (1968)
4.32 L.D. Landau, E.M. Lifshitz: *Kvantovaya mekhanika* (Fizmatgiz, Moscow 1963)
     [English transl.: *Quantum Mechanics* (Addison-Wesley, London 1965)]
4.33 H.-K. Hong, R. Kopelman: J. Chem. Phys. **55**, 3491 (1971)
4.34 R. Kopelman: In *Excited States*, Vol.2, ed. by E.C. Lim (Academic, New
     York 1975) p.33
4.35 B. Velicky: Phys. Rev. **184**, 614 (1969)
4.36 J. Hoshen, J. Jortner: J. Chem. Phys. **56**, 933 (1972)
4.37 J. Hoshen, J. Jortner: J. Chem. Phys. **56**, 5550 (1972)
4.38 O.A. Dubovsky, Yu.V. Konobeyev: Fiz. Tverd. Tela **12**, 405 (1970) [English
     transl.: Sov. Phys. Solid State **12**, 321 (1970)]
4.39 H.-K. Hong, G.W. Robinson: J. Chem. Phys. **52**, 825 (1970)
4.40 H.-K. Hong, G.W. Robinson: J. Chem. Phys. **54**, 1369 (1971)
4.41 S.D. Colson, D.M. Hanson, R. Kopelman, G.W. Robinson: J. Chem. Phys.
     48, 2215 (1968)
4.42 V.L. Broude, A.V. Leiderman: Zh. Eksp. Teor. Fiz. Pis'ma **13**, 426 (1971)
     [English transl.: Sov. Phys. JETP Lett. **13**, 302 (1971)]
4.43 N.V. Rabin'kina, E.I. Rashba, E.F. Sheka: Fiz. Tverd. Tela **12**, 3569 (1970)
     [English transl.: Sov. Phys. Solid State **12**, 2898 (1970)]
4.44 P. Dean, J.L. Martin: Proc. Roy. Soc. London A**259**, 409 (1960)
4.45 S.D. Woodruff, R. Kopelman: Chem. Phys. **22**, 1 (1977)
4.46 Y. Tokura, T. Koda, I. Nakada: J. Lumin. **18/19**, 467 (1979)

4.47 P. Argyrakis, E.M. Monberg, R. Kopelman: Chem. Phys. Lett. **36**, 349 (1975)
4.48 H.C. Wolf, H. Port: In *Molecular Spectroscopy of Dense Phases*, Proc.
12th European Congress on Molecular Spectroscopy, ed. by M. Grosmann,
S.G. Elkomoss, J. Ringeissen (Elsevier, Amsterdam 1976) p.31
4.49 M.S. Brodin, S.V. Marisova, S.A. Shturkhetskaya: Ukr. Fiz. Zh. **13**, 353
(1968)
4.50 M.S. Brodin, S.V. Marisova: Opt. Spektrosk. **10**, 473 (1961) [English
transl.: Opt. Spectrosc. **10**, 271 (1961)]
4.51 E.F. Sheka: Mol. Cryst. Liq. Cryst. **29**, 323 (1975)
4.52 J. Schroeder, R. Silbey: J. Chem. Phys. **55**, 5418 (1971)
4.53 R. Kopelman, E.M. Monberg, F.W. Ochs, P.N. Prasad: Phys. Rev. Lett. **34**,
1506 (1975)
4.54 R. Kopelman, E.M. Monberg, F.W. Ochs, P.N. Prasad: J. Chem. Phys. **62**,
292 (1975)
4.55 R. Kopelman, E.M. Monberg, F.W. Ochs: Chem. Phys. **21**, 373 (1977)
4.56 R. Kopelman: In *Radiationless Processes in Molecules and Condensed Phases*,
Topics Appl. Phys., Vol.15, ed. by K.F. Fong (Springer, Berlin, Heidelberg, New York 1976) p.296
4.57 T.B. El-Kareh, H.C. Wolf: Z. Naturforsch. A**22**, 1242 (1967)
4.58 M.A. El-Sayed, H.C. Wauk, G.W. Robinson: Mol. Phys. **5**, 205 (1962)
4.59 M. Schwörer, H.C. Wolf: Mol. Cryst. **3**, 177 (1967)
4.60 S.D. Colson, S. George, T. Keyes, V. Vaida: J. Chem. Phys. **67**, 4941 (1977)
4.61 V.L. Bonch-Bruevich, I.P. Zvyagin, R. Keiper, A.G. Mironov, R. Enderlein,
B. Esser: *Elektronnaya teoria neuporyadochennych poluprovodnikov* (Nauka,
Moscow 1981) p.384 [*Electron Theory of the Disordered Semiconductors*]
[German transl.: *Elektronentheorie der ungeordneten Halbleiter* (Deutscher
Verlag der Wissenschaften, Berlin 1984)
4.62 B. Shklovskii, A.L. Efros: *Electronic Properties of Doped Semiconductors*, Springer Ser. Solid State Sci., Vol.45 (Springer, Berlin,
Heidelberg, New York 1983)
4.63 I.M. Lifshitz, S.A. Gredeskul, L.A. Pastur: *Vvedenie v teoriyu neuporyadochennych sistem* (Nauka, Moscow 1982) [English transl.: *Introduction to the Theory of the Disordered Systems* (Wiley, New York, in press)]
4.64 N.F. Mott: Adv. Phys. **16**, 49 (1967)
4.65 N.F. Mott, E.A. Davis: *Electronic Processes in Noncrystalline Materials*
(Clarendon, Oxford 1971)
4.66 V.L. Bonch-Bruevich, A.G. Mironov, I.P. Zviagin: Riv. Nuovo Cimento **3**,
321 (1973)
4.67 A.L. Efros: Usp. Fiz. Nauk **126**, 41 (1978) [English transl.: Sov. Phys.
Usp. **21**, 746 (1978)]
4.68 P.W. Anderson: Phys. Rev. **109**, 1492 (1958)
4.69 V.K.S. Shante, S. Kirkpatrick: Adv. Phys. **20**, 325 (1971)
4.70 S. Kirkpatrick: Rev. Mod. Phys. **45**, 574 (1973)
4.71 B.I. Shklovskii, A.L. Efros: Usp. Fiz. Nauk **117**, 401 (1975) [English
transl.: Sov. Phys. Usp. **117**, 401 (1975)]
4.72 R. Kopelman: In *Spectroscopy and Excitation Dynamics of Condensed Molecular Systems*, ed. by V.M. Agranovich, R.M. Hochstrasser (North-Holland,
Amsterdam 1983) p.139
4.73 D.M. Hanson: J. Chem. Phys. **52**, 3409 (1970)
4.74 J. Klafter, J. Jortner: Chem. Phys. Lett. **49**, 410 (1977)
4.75 D.D. Smith, R.D. Mead, A.H. Zewail: Chem. Phys. Lett. **50**, 358 (1977)

Chapter 5

5.1 N.I. Ostapenko, V.I. Sugakov, M.P. Chernomorets, M.T. Shpak: Phys. Status
Solidi B**93**, 493 (1979)
5.2 V.I. Sugakov: In *Excitons*, ed. by E.I. Rashba, M.D. Sturge (North Holland,
Amsterdam 1982) p.709

5.3 G. Vankataraman, V.C. Sahni: Rev. Mod. Phys. **42**, 409 (1970)
5.4 G. Taddei, H. Bonadeo, M.P. Marzocchi, S. Califano: J. Chem. Phys. **58**, 966 (1973)
5.5 V.L. Broude, A.F. Prikhotjko, E.I. Rashba: Usp. Fiz. Nauk **67**, 99 (1959) [English transl.: Sov. Phys. Usp. **67**(2), 38 (1959)]
5.6 G.L. Bir, G.E. Pikus: *Simmetriya i deformatsionnyie effekty v polu-provodnikakh* (Nauka, Moscow 1972) [English transl.: *Symmetry and Strain-Induced Effects in Semiconductors* (Halsted, New York 1974)]
5.7 E.I. Rashba: Fiz Tverd. Tela **5**, 1040 (1963) [English transl.: Sov. Phys. Solid State **5**, 757 (1963)]
5.8 V.L. Broude, E.F. Sheka, M.T. Shpak: Izv. Akad. Nauk SSSR Ser. Fiz. **27**, 596 (1963) [English transl.: Bull. Acad. Sci. USSR Phys. Ser. **27**, 597 (1963)]
5.9 V.L. Broude, E.F. Sheka, M.T. Shpak, L.G. Shpakovskaya: Opt. Spektrosk. Sb. Statei **1**, 98 (1963)
5.10 L. Van Hove: Phys. Rev. **89**, 1189 (1953)
5.11 M. Cardona: *Modulation Spectroscopy* (Academic, New York 1969)
5.12 B.H. Loo, A.H. Francis, K.W. Hipps: J. Chem. Phys. **65**, 5068 (1976)
5.13 M. Tomioka: J. Phys. Soc. Jpn. **44**, 489 (1978)
5.14 A. Twarowsky, A. Abrecht: Phys. Status Solidi A**52**, K35 (1979)
5.15 K.P. Meletov, E.F. Sheka: Fiz. Tverd. Tela **25**, 1612 (1983) [English transl.: Sov. Phys. Solid State **25**, 930 (1983)]
5.16 E.F. Sheka: Opt. Spektrosk. **29**, 78 (1970)
5.17 A.I. Kolesnikov, V.A. Dementjev, E.L. Bokhenkov, T.A. Krivenko, E.F. Sheka: Fiz. Tverd. Tela **25**, 2881 (1983) [English transl.: Sov. Phys. Solid State **25**, 1663 (1983)]
5.18 S.D. Colson, D.M. Hanson, R. Kopelman, G.W. Robinson: J. Chem. Phys. **48**, 2215 (1968)
5.19 A.B. Zimin, E.I. Rashba: Fiz. Tverd. Tela **16**, 856 (1974) [English transl.: Sov. Phys. Solid State **16**, 550 (1974)]
5.20 S.D. Woodruff, R. Kopelman: Chem. Phys. **22**, 1 (1977)
5.21 V.L. Broude, E.I. Rahsba: Pure Appl. Chem. **37**, 21 (1974)
5.22 J. Schroeder, R. Silbey: J. Chem. Phys. **55**, 5418 (1971)
5.23 B. Leong, D.M. Hanson: Chem. Phys. **81**, 21 (1983)
5.24 J. Schmidt: In *Organic Molecular Aggregates*, ed. by P. Reineker, H. Haken, H.C. Wolf, Springer Ser. Solid-State Sci., Vol.49 (Springer, Berlin, Heidelberg, New York, Tokyo 1983) p.56
5.25 A.H. Francis, C.B. Harris: Chem. Phys. Lett. **9**, 181, 188 (1971)
5.26 B. Velicky, S. Kirkpatrick, H. Ehrenreich: Phys. Rev. **175**, 747 (1968)
5.27 T. Netzel, S. Colson, D. Fox: J. Chem. Phys. **59**, 475 (1973)
5.28 S.D. Colson, T.L. Netzel: Mol. Phys. **26**, 119 (1973)
5.29 E.F. Sheka: Opt. Spektrosk. **10**, 684 (1961) [English transl.: Opt. Spectrosc. **10**, 360 (1961)]
5.30 E.F. Sheka: Opt. Spektrosk. **12**, 137 (1962) [English transl.: Opt. Spectrosc. **12**, 72 (1962)]
5.31 H.-K. Hong, G.W. Robinson: J. Chem. Phys. **52**, 825 (1970)
5.32 H.C. Wolf, H. Port: In *Molecular Spectroscopy of Dense Phase*, Proc. 12th Europ. Congr. on Molecular Spectroscopy, ed. by M. Grosmann, S.G. Elkomoss, J. Ringeissen (Elsevier, Amsterdam 1976) p.31

Chapter 6

6.1 E.G. McRay: Austral. J. Chem. **14**, 329, 344, 354 (1961)
6.2 M.V. Belousov: In *Excitons*, ed. by E.I. Rashba, M.D. Sturge (North-Holland, Amsterdam 1982) p.771
6.3 A. Suna: Phys. Rev. A**135**, 111 (1964)
6.4 V. Fedosejev: "The Anomalous Regions in the Spectral Density of Excitons Interacting with Phonons," Preprint of the Institute of Physics and Astronomy, Acad. Sci. Est. SSR, F.A.I-5 (Tartu 1970)

6.5 Y.B. Levinson, E.I. Rashba: Rep. Progr. Phys. **36**, 1449 (1973)
6.6 A.S. Davydov: *Teoriya pogloshcheniya sveta v molekulyarnykh kristallakh*
    (Izdatel'stvo Akademii Nauk Ukrainskoi SSR, Kiev 1951) [English transl.:
    Theory of Molecular Excitons (McGraw-Hill, New York 1962)]
6.7 W.A. Bingel: Can. J. Phys. **37**, 680 (1959)
6.8 D.P. Craig. P.C. Hobbins: J. Chem. Soc. 539 (1955)
6.9 D.P. Craig, S.H. Walmsley: Mol. Phys. **4**, 113 (1961)
6.10 R. Peierls: Ann. Physik **13**, 905 (1932)
6.11 Y.I. Frenkel: Zh. Eksp. Teor. Fiz. **6**, 647 (1936); Phys. Zs. Sowjet. **9**,
    158 (1936)
6.12 L.D. Landau: Phys. Zs. Sowjet. **3**, 664 (1933)
6.13 S.I. Pekar: *Issledovaniya po elektronnoi teorii kristallov [Investi-
    gations on the Electronic Theory of Crystals]* Gostekhizdat, Moscow 1951)
    [German transl.: *Untersuchungen über die Elektronentheorie der
    Kristalle* (Akademie Verlag, Berlin 1954)]
6.14 T. Holstein: Ann. Phys. **8**, 343 (1959)
6.15 S.I. Pekar: Usp. Fiz. Nauk **50**, 197 (1953)
6.16 Yu.E. Perlin: Usp. Fiz. Nauk **80**, 553 (1963) [English transl.: Sov. Phys.
    Usp. **6**, 542 (1964)]
6.17 K.K. Rebane: *Elementarnaya teoriya kolebatel'noi struktury spektrov
    primesnykh tsentrov kristallov* (Nauka, Moscow 1968) [English transl.:
    Impurity Spectra of Solids (Plenum, New York 1970)]
6.18 J. Frank, E. Teller: J. Chem. Phys. **6**, 861 (1938)
6.19 W.T. Simpson, D.L. Peterson: J. Chem. Phys. **26**, 588 (1957)
6.20 E.I. Rashba: Opt. Spektrosk. **2**, 75, 88 (1957); **3**, 568 (1957)
6.21 E.I. Rashba: Izv. Akad. Nauk SSSR, Ser. Fiz. **21**, 37 (1957)
6.22 R.E. Merrifield: J. Chem. Phys. **40**, 445 (1964)
6.23 A.B. Zimin, E.I. Rashba: Fiz. Tverd. Tela **16**, 856 (1974) [English transl.:
    Sov. Phys. Solid State **16**, 550 (1974)]
6.24 D. Emin: Adv. Phys. **22**, 57 (1973)
6.25 T. Toyozawa: In *Vacuum Ultraviolet Radiation Physics*, ed. by E.E. Koch,
    R. Haensel, G. Kunz (Pergamon-Vieweg, Braunschweig 1974) p.317
6.26 E.I. Rashba: Izv. Akad. Nauk SSSR, Ser. Fiz. **40**, 1793 (1976) [English
    transl.: Bull. Acad. Sci. USSR, Phys. Ser. **40**, N9, 20 (1976)]
6.27 E.I. Rashba: In *Excitons*, ed. by E.I. Rashba, M.D. Sturge (North-Holland,
    Amsterdam 1982) p.543
6.28 I.Ya. Fugol': Adv. Phys. **27**, 1 (1978)
6.29 Th. Förster, K. Kasper: Z. Phys. Chem. N. F. **1**, 275 (1954)
6.30 J. Ferguson: J. Chem. Phys. **28**, 765 (1958)
6.31 M.D. Cohen: Mol. Cryst. Liq. Cryst. **50**, 1 (1979)
6.32 H. Sumi: J. Phys. Soc. Jpn. **36**, 770 (1974); **38**, 825 (1975)
6.33 H. Sumi: In *Defects in Insulating Crystals*, ed. by V.M. Tuchkevich,
    K.K. Shvarts (Zinatne, Riga and Springer, Berlin, Heidelberg, New York.
    1981) p.267
6.34 E.I. Rashba: Zh. Eksp. Teor. Fiz. **50**, 1064 (1966) [English transl.:
    Sov. Phys. JETP **23**, 708 (1966)]
6.35 E.I. Rashba: Zh. Eksp. Teor. Fiz. **54**, 542 (1968) [English transl.: Sov.
    Phys. JETP **27**, 292 (1968)]
6.36 D.S. McClure: Can. J. Chem. **36**, 59 (1958)
6.37 A. Witkowski, W. Moffit: J. Chem. Phys. **33**, 872 (1960)
6.38 R.L. Fulton, M. Gouterman: J. Chem. Phys. **35**, 1059 (1961)
6.39 R.E. Merrifield: Radiat. Res. **20**, 154 (1963)
6.40 A. Witkowski, M.Z. Zgierski: Phys. Status Solidi **46**, 429 (1971)
6.41 M.R. Philpott: J. Chem. Phys. **47**, 2534, 4437 (1967)
6.42 M.R. Philpott: J. Chem. Phys. **51**, 2616 (1969)
6.43 M.R. Philpott: J. Chem. Phys. **52**, 5842 (1970)
6.44 M.R. Philpott: J. Chem. Phys. **53**, 136 (1970)
6.45 A.S. Davydov, A.A. Serikov: Phys. Status Solidi **42**, 603 (1970); **44**,
    127 (1971)

6.46 L. Valkunas, V.I. Sugakov: Ukr. Fiz. Zh. **17**, 1561 (1972)
6.47 J. Klafter, J. Jortner: Chem. Phys. **47**, 25 (1980)
6.48 G.C. Nieman, G.W. Robinson: J. Chem. Phys. **39**, 1298 (1963)
6.49 J. Van Kranendonk: Physica **25**, 1080 (1959); Can. J. Phys. **38**, 240 (1960)
6.50 J. Jortner, S.A. Rice: J. Chem. Phys. **44**, 3364 (1966)
6.51 V.M. Agranovich: Fiz. Tverd. Tela **12**, 562 (1970) [English transl.:
     Sov. Phys. Solid State **12**, 430 (1970)]
6.52 R.W. Munn, W. Siebrand: J. Chem. Phys. **52**, 47, 6391 (1970)
6.53 A.F. Prikhotjko: Zh. Eksp. Teor. Fiz. **19**, 383 (1949)
6.54 A.F. Prikhotjko: Opt. Spektrosk. **3**, 434 (1957)
6.55 N.V. Rabin'kina, E.I. Rashba, E.F. Sheka: Fiz. Tverd. Tela **12**, 3569 (1970)
     [English transl.: Sov. Phys. Solid State **12**, 2898 (1970)]
6.56 E.I. Rashba: Fiz. Tverd. Tela **12**, 1801 (1970) [English transl.: Sov.
     Phys. Solid State **12**, 1426 (1970)]
6.57 V.L. Broude, E.I. Rashba, E.F. Sheka: Phys. Status Solidi **19**, 395
     (1967); Zh. Eksp. Teor. Fiz. Pis'ma **3**, 429 (1966) [English transl.:
     Sov. Phys. JETP Lett. **3**, 281 (1966)]
6.58 E.F. Sheka: Opt. Spektrosk. **29**, 275 (1970)
6.59 S.M. Kochubei: *Issledovania detal'noi struktury spektra kristallicheskogo*
     benzola Investigation into Detailed Structure of the Crystalline
     Benzene Spectrum Kand. Dissert. Inst. Fiz. AN Ukr. SSSR, Kiev (1964)
6.60 V.L. Broude: Usp. Fiz. Nauk **74**, 577 (1961) [English transl.: Sov. Phys.
     Usp. **4**, 584 (1962)]
6.61 R.M. Hochstrasser, C.M. Klimcak, G.R. Meredith: J. Chem. Phys. **70**,
     870 (1979)
6.62 M.S. Soskin: Ukr. Fiz. Zh. **7**, 180, 635 (1962)
6.63 A.F. Prikhotjko, M.S. Soskin, A.K. Tomashchik: Opt. Spektrosk. **16**, 615
     (1964) [English transl.: Opt. Spectrosc. **16**, 337 (1964)]
6.64 I.P. Terenetskaya: Opt. Spektrosk. **36**, 518 (1974)
6.65 N. Mikami, M. Ito: Chem. Phys. **23**, 141 (1977)
6.66 M.S. Brodin, S.V. Marisova, S.A. Shturkhetskaya: Ukr. Fiz. Zh. **13**, 353
     (1968)
6.67 M.R. Philpott, J.-M. Tourlet: J. Chem. Phys. **64**, 3852 (1976)
6.68 Y. Tokura, T. Mitani, T. Koda: J. Phys. Soc. Jpn. **51**, 1551 (1982)
6.69 K. Sumi: J. Phys. Soc. Jpn. **36**, 770 (1974); **38**, 825 (1975)
6.70 Y. Tokura, T. Koda, I. Nakada: J. Phys. Soc. Jpn. **47**, 1936 (1979)
6.71 P.M. Saari, T.B. Tamm: Izv. Akad. Nauk SSSR, Ser. Fiz. **42**, 562 (1978)
6.72 Ya. Yu. Aaviksoo, P.M. Saari, T.B. Tamm: Zh. Eksp. Teor. Fiz. Pis'ma
     29, 388 (1979) [English transl.: Sov. Phys. JETP Lett. **29**, 351 (1979)]
6.73 Ya.Yu. Aaviksoo, P.M. Saari, T.B. Tamm: Izv. Akad. Nauk SSSR, Ser.
     Fiz. **44**, 848 (1980)
6.74 V.L. Broude, G.V. Klimusheva, A.F. Prikhotjko, E.F. Sheka, L.P. Yatsenko:
     Spektry pogloshcheniya molekulyarnykh kristallov (polizameshchennyie
     benzola) [Absorption Spectra of Molecular Crystals: Polysubstituted
     Benzene Compounds] (Naukova Dumka, Kiev 1972)
6.75 G.V. Klimusheva: In *Eksitony v molekulyarnykh kristallakh [Excitons*
     in Molecular Crystals] (Naukova Dumka, Kiev 1973) p.141
6.76 T.A. Krivenko, E.I. Rashba, E.F. Sheka: Mol. Cryst. Liq. Cryst. **47**,
     119 (1978)
6.77 V.K. Dolganov, E.F. Sheka: Fiz. Tverd. Tela **12**, 1450 (1970) [English
     transl.: Sov. Phys. Solid State **12**, 1138 (1970)]
6.78 K.P. Meletov, E.F. Sheka: Mol. Cryst. Liq. Cryst. **43**, 203 (1977)
6.79 I. Natkaniec, E.L. Bokhenkov, B. Dorner, J. Kalus, G.A. Mackenzie,
     G.S. Pawley, U. Schmelzer, E.F. Sheka: J. Phys. C: Solid State Phys.
     13, 4265 (1980)
6.80 D.A. Dows, V.J. Schettino: J. Chem. Phys. **58**, 5009 (1973)
6.81 M.V. Belousov, D.E. Pogarev, A.A. Shultin: Fiz. Tverd. Tela **18**, 521
     (1976) [English transl.: Sov. Phys. Solid State **18**, 300 (1976)]

6.82   J.C. Decius, R.M. Hexter: *Molecular Vibrations in Crystals* (McGraw-Hill, New York 1977)
6.83   E.R. Bernstein, G.R. Meredith: Chem. Phys. **24**, 301 (1977)
6.84   S.D. Colson, K.N. Wong: Chem. Phys. **69**, 223 (1982)
6.85   T. Holstein, R. Orbach, S. Alexander: Phys. Rev. B**26**, 4271 (1982)
6.86   E.F. Sheka, J.P. Terenetskaya: Chem. Phys. **8**, 99 (1975)
6.87   T.A. Krivenko, A.V. Leiderman, I.P. Terenetskaya, E.F. Sheka: Opt. Spektrosk. **45**, 30 (1978)
6.88   V.K. Dolganov, E.F. Sheka: Fiz. Tverd. Tela **11**, 2427 (1969) [English transl.: Sov. Phys. Solid State **11**, (1969)]
6.89   E.F. Sheka, I.P. Terenetskaya: Fiz. Tverd. Tela **12**, 720 (1970) [English transl.: Sov. Phys. Solid State **12**, 558 (1970)]
6.90   M.V. Belousov, B.E. Vol'f, E.A. Ivanova, D.E. Pogarev: Zh. Eksp. Teor. Fiz. Pis'ma **35**, 457 (1982) [English transl.: Sov. Phys. JETP Lett. **35**, 565 (1982)]
6.91   E.F. Sheka, I.P. Terenetskaya: Fiz. Tverd. Tela **13**, 1071 (1971) [English transl.: Sov. Phys. Solid State **13**, 889 (1971)]
6.92   H.K. Hong, G.W. Robinson: J. Chem. Phys. **52**, 825 (1970)
6.93   Y. Tokura, T. Koda, I. Nakada: J. Phys. Soc. Jpn. **49**, Suppl.A, 417 (1980)
6.94   M.V. Belousov, D.E. Pogarev: Zh. Eksp. Teor. Fiz. Pis'ma **36**, 152 (1982) [English transl.: JETP Lett. **36**, 189 (1982)]
6.95   N.I. Ostapenko, V.I. Sugakov, M.T. Shpak: In *Eksitony v molekulyarnykh kristallakh* [*Excitons in Molecular Crystals*] (Naukova Dumka, Kiev 1973) p.92
6.96   T.A. Krivenko, A.V. Leiderman, E.I. Rashba, E.F. Sheka: Zh. Eksp. Teor. Fiz. Pis'ma **25**, 538 (1977) [English transl.: Sov. Phys. JETP Lett. **25**, 503 (1977)]
6.97   Y. Maruyama, N. Iwasaki: Chem. Phys. Lett. **24**, 26 (1974)
6.98   K.O. Lee, T.T. Gan: Chem. Phys. Lett. **51**, 120 (1977)
6.99   J. Klafter, J. Jortner: J. Chem. Phys. **77**, 2812 (1982)
6.100 V.K. Dolganov, E.F. Sheka: Zh. Eksp. Teor. Fiz. **60**, 2230 (1971) [English transl.: Sov. Phys. JETP **33**, 1198 (1971)]

Appendix A

A.1   E.I. Rashba: Zh. Eksp. Teor. Fiz. **54**, 542 (1968) [English transl.: Sov. Phys. JETP **27**, 292 (1968)]
A.2   Yu.A. Firsov: In *Polyarony* [*Polarons*] (Nauka, Moscow 1975) p.207
A.3   D.A. Kirzhnits: *Polevyie metody teorii mnogikh chastits* [*Field Methods of the Many-Particles Theory*] (Gosatomizdat, Moscow 1963)

Subject Index

271

V.M.Agranovich, V.Ginzburg

Crystal Optics with Spatial Dispersion, and Excitons

2nd corrected and updated edition. 1984. 46 figures.
XI, 441 pages. (Springer Series in Solid-State Sciences,
Volume 42). ISBN 3-540-11520-X
(First English edition published in 1966 by Interscience
London, NewYork, Sidney

Contents: Introduction. - The Complex Dielectric-Constant Tensor $\varepsilon_{ij}(\omega,k)$ and Normal Waves in a Medium. - The Tensor $\varepsilon_{ij}(\omega,k)$ in Crystals. - Spatial Dispersion in Crystal Optics. - Surface Excitons and Polaritons. - Microscopic Theory. - Calculation of the Tensor $\varepsilon_{ij}(\omega,k)$. - Conclusion. - Appendix. - Notation. - References. - Subject Index.

Organic Molecular Aggregates

Electronic Excitation and Interaction Processes

Proceedings of the International Symposium on Organic Materials at Schloss Elmau, Bavaria, June 5-10, 1983

Editors: **P.Reineker, H.Haken, H.C.Wolf**

1983. 113 figures. IX, 285 pages. (Springer Series in Solid-State Sciences, Volume 49). ISBN 3-540-12843-3

Contents: Basic Concepts, Methods and Results. - Interaction of Electronic Excitations with Electromagnetic Radiation. - Electronic Excitations and Spin Dynamics. - Interaction of Electronic Excitations with Lattice Vibrations. - Excimers, Charge Transfer Excitons and Exciton Fission. - Electronic Excitations in Disordered Systems. - Electronic Excitation of Impurities in Glasses and Polymers. - Conductivity and Superconductivity in Organic Materials. - Electronic Excitations in Photosynthetic Systems. - Index of Contributors.

Springer-Verlag
Berlin
Heidelberg
New York
Tokyo

Secondary Ion Mass Spectrometry SIMS IV

Proceedings of the Fourth International Conference, Osaka, Japan, November 13–19, 1983

Editors: **A. Benninghoven, J. Okano, R. Shimizu, H. W. Werner**

1984. 415 figures. XV, 503 pages. (Springer Series in Chemical Physics, Volume 36). ISBN 3-540-13316-X

Contents: Fundamentals. – Ouantification. – Instrumentation. – Combined and Static SIMS. – Application to Semiconductor and Depth Profiling. – Organic SIMS. – Application: Metallic and Inorganic Materials. Geology. Biology. – Index of Contributors.

Excitons

Editor: **K. Cho**

1979. 118 figures, 8 tables. XI, 274 pages. (Topics in Current Physics, Volume 14). ISBN 3-540-09567-5

Contents: *K. Cho:* Introduction. – *K. Cho:* Internal Structure of Excitons. – *P. J. Dean, D. C. Herbert:* Bound Excitons in Semiconductors. – *B. Fischer, J. Lagois:* Surface Exciton Polaritons. – *P. Y. Yu:* Study of Excitons and Exciton-Phonon Interactions by Resonant Raman and Brillouin Spectroscopies.

Springer-Verlag
Berlin
Heidelberg
New York
Tokyo

Exciton Dynamics in Molecular Crystals and Aggregates

1982. 37 figures. IX, 226 pages. (Springer Tracts in Modern Physics, Volume 94). ISBN 3-540-11318-5

Contents: *Y. M. Kenkre:* The Master Equation Approach: Coherence, Energy Transfer, Annihilation, and Relaxation. – *P. Reineker:* Stochastic Liouville Equation Approach: Coupled Coherent and Incoherent Motion, Optical Line Shapes, Magnetic Resonance Phenomena.